Synthesis, Chracterization and Applications of Coated Composite Materials for Energy Applications

Synthesis, Chracterization and Applications of Coated Composite Materials for Energy Applications

Editors

Awais Ahmad
Deepak Dubal

MDPI • Basel • Beijing • Wuhan • Barcelona • Belgrade • Manchester • Tokyo • Cluj • Tianjin

Editors
Awais Ahmad
University of Lahore
Lahore
Pakistan

Deepak Dubal
Queensland University of
Technology
Brisbane
Australia

Editorial Office
MDPI
St. Alban-Anlage 66
4052 Basel, Switzerland

This is a reprint of articles from the Special Issue published online in the open access journal *Coatings* (ISSN 2079-6412) (available at: www.mdpi.com/journal/coatings/special_issues/energyappl).

For citation purposes, cite each article independently as indicated on the article page online and as indicated below:

LastName, A.A.; LastName, B.B.; LastName, C.C. Article Title. *Journal Name* **Year**, *Volume Number*, Page Range.

ISBN 978-3-0365-1660-8 (Hbk)
ISBN 978-3-0365-1659-2 (PDF)

Contents

About the Editors

Awais Ahmad is an early career researcher currently working in the Department of Chemistry at the University of Lahore, Pakistan. He graduated from the Government College University Faisalabad, Pakistan, applied chemistry, then earned a Master's in inorganic chemistry from the University of Lahore. His main research interests are nanomaterials, nanocomposites, wastewater treatment, energy applications, water splitting, supercapacitors, solar cells, metal organic framework (MOF), and single atom catalysis. He is also researching electrochemical sensors, electrocatalysis, and biosensors. He has eight books as an editor at Elsevier in the pipeline and one in Material Research LLC.

Deepak Dubal is Associate Professor (Australian Future Fellow) at QUT and has a strong research background in the field of nanomaterials for clean energy conversion and storage systems. Briefly, Deepak's research interest is focused on designing and engineering of functional materials such as New Oxides/Nitrides, Polyoxometalates (POMs), Conducting polymers and their hybrids for energy storage applications with special emphasis on Supercapacitors, Li-ion batteries, beyond Li-ion batteries, etc. At QUT, his research is to develop an integrated system as a self-charging power source for wearable electronics and implantable medical devices. Before joining QUT, Deepak worked as Vice Chancellor Fellow at the School of Chemical Engineering, University of Adelaide. He is also a former recipient of two most prestigious fellowships: Marie-Curie at ICN2, Spain (2014) and Alexander von Humboldt at TUC, Germany (2012). So far, Deepak has co-authored more than 140 research articles, including Chem. Soc.

coatings

MDPI

Article

Lanthanum-Zinc Binary Oxide Nanocomposite with Promising Heterogeneous Catalysis Performance for the Active Conversion of 4-Nitrophenol into 4-Aminophenol

Ikram Ahmad [1], Muhammad Asghar Jamal [2,*], Miara Iftikhar [2], Awais Ahmad [3], Shahid Hussain [4], Humaira Asghar [2], Muhammad Saeed [2], Ammar Bin Yousaf [5], Rama Rao Karri [6,*], Nada Sulaymaniyah Al-kadhi [7], Mohamed Ouladsmane [8], Ayman Ghfar [8] and Safia Khan [9]

1 Department of Chemistry, University of Sahiwal, Sahiwal 54000, Pakistan; drikramahmad@uosahiwal.edu.pk
2 Department of Chemistry, Government College University, Faisalabad 38000, Pakistan; timsheet123@gmail.com (M.I.); humairaasghar@live.com (H.A.); msaeed@gcuf.edu.pk (M.S.)
3 Department of Chemistry, The University of Lahore, Lahore 54590, Pakistan; awaisahmed@gcuf.edu.pk
4 School of Materials Science and Engineering, Jiangsu University, Zhenjiang 212013, China; shahid@ujs.edu.cn
5 Centre for Advanced Materials, Qatar University, Doha 2713, Qatar; ammar.chemist18@gmail.com
6 Petroleum and Chemical Engineering, Faculty of Engineering, Universiti Teknologi Brunei, Mukim Gadong 1410, Brunei
7 Department of Chemistry, College of Science, Princess Nourah Bint Abdulrahman University, Riyadh 11671, Saudi Arabia; nsalkadhi@pnu.edu.sa
8 Advanced Materials Research Chair, Chemistry Department College of Science, King Saud University, P.O. Box 2455, Riyadh 11451, Saudi Arabia; mouladsmane@ksu.edu.sa (M.O.); Aghafr@ksu.edu.sa (A.G.)
9 Department of Chemistry, Quaid-i-Azam University Islamabad, Islamabad 45320, Pakistan; safiakhan@chem.qau.edu.pk
* Correspondence: asgharjamal@gcuf.edu.pk (M.A.J.); kramarao.iitd@gmail.com (R.R.K.)

Citation: Ahmad, I.; Jamal, M.A.; Iftikhar, M.; Ahmad, A.; Hussain, S.; Asghar, H.; Saeed, M.; Yousaf, A.B.; Karri, R.R.; Al-kadhi, N.S.; et al. Lanthanum-Zinc Binary Oxide Nanocomposite with Promising Heterogeneous Catalysis Performance for the Active Conversion of 4-Nitrophenol into 4-Aminophenol. *Coatings* **2021**, *11*, 537. https://doi.org/10.3390/coatings11050537

Academic Editors: Luca Valentini and Ioannis V. Yentekakis

Received: 25 March 2021
Accepted: 23 April 2021
Published: 30 April 2021

Abstract: This work intended to enhance the unique and outstanding properties of lanthanum by synthesizing its nanocomposite. A lanthanum-based nanocomposite was prepared by a simple and cost-effective "co-precipitation" method. Lanthanum nitrate ($La(NO_3)_3$) and zinc nitrate ($Zn(NO_3)_2$) were used as precursors. The lanthanum/zinc oxide nano composite formed was then calcined at 450 °C for 4 h in order to obtain a fine powder with size in the nano range of 1–100 nm. Characterization of the prepared catalyst was done by ultraviolet/visible spectroscopy, Fourier transform infrared spectroscopy, and photoluminescence. Crystallinity and morphology were found by X-ray diffraction and scanning electron microscopy. The synthesized nanocomposite material was also tested for heterogeneous catalytic applications of 4-nitrophenol (4-NP) reduction into 4-aminophenol (4-AP). It was found to be successful in complete reduction of 4-NP with enhanced catalytic performance.

Keywords: nanocomposite structure; XRD; photoluminescence; rare earth element REE; heterogeneous catalysis

1. Introduction

4-aminophenol (*p*-aminophenol), an imperative intermediate in the manufacturing of multiple pharmaceutical products, mainly including phenacetin, acetanilide, and paracetamol, also has numerous applications in anticorrosion, photography, and lubrication as a dyeing agent [1,2]. Due to its increased demand by pharmacists, the direct catalytic conversion of easily available 4-nitrophenol is the center of huge research [3,4]. Several reduction methods have been reported to date for conversion of 4-nitrophenol into 4-aminophenol [5–7]. The use of precious metals for this catalytic reduction is frequently seen, e.g., Pt, Au, etc. [8,9], involving hydrazine as a reducing agent. However, the discovery and development of more compatible, ecofriendly, and efficient reduction catalysts for this specific reaction is still under research.

Consequently, rare earth metal compounds are being considered by scientists across the globe for investigation of their catalytic performance.

Over the last decade, rare earth metals with nanostructures of size 1–100 nm have been studied by researchers around the world. Owing to the presence of 4f orbital electrons, they have unique properties and offer promising applications in many fields. Their nanostructures are defenseless against compositional or structural effects [10–12]. The electrons in 4f shell orbitals of rare earth metals are continuously shielded by the electrons of 5p and 4 d; this feature plays an important role in the catalytic applications of these oxides [13]. The particle size and electronic configuration of metal oxides are interlinked, and the band gaps and energy excitation levels change when the particle size of the metal oxide is reduced [14].

Among nanostructured rare earth metal compounds, Lanthanum metal is an important p-type semiconductor metal, used in numerous fields, such as solar cells [15,16], photocatalysis [17,18], photo detectors [19,20], sensors, light-emitting diodes [21,22], and laser communications [23,24]. Lanthanum oxides with metals is a perovskite type oxide i.e., its ABO_3 type structure makes it special for catalytic applications [25,26]. In addition, Lanthanum metal oxide in powder form, with particle size in nano range of 1–100 nm, exhibits a lot of industrial properties as well. Lanthanum oxides are used in the synthesis of organic catalysts in the field of electrochemistry; it is used as an electrode, and the burning rate of propellants can also be enhanced by lanthanum oxide [27,28].

Moreover, zinc is an important transition metal, due to its feasibility and low cost, and researchers have extensively studied its oxides. Zinc oxide is a non-toxic, highly stable n-type semiconductor having a wide band gap (3.37 eV), which is an attractive feature because it can absorb sufficient amounts of UV light at room temperature [29–31]. Zinc oxides are used in several fields: in solar cells [32,33], super capacitors [34], field effect transistors [35], light-emitting materials [36,37], gas sensors [38], drug delivery [39]; as anticancer [40,41], antibacterial [42], diabetes treatment [43], bio- imaging [44], photocatalysts [45,46], base material for magnetic semiconductors [47], food additive [48], transparent UV- protection films [49,50], luminescence [51,52], solar energy conversion [53], etc. Zinc oxides are also used in care products like sunscreen and cosmetics, as they can strongly absorb UV light [54,55].

Additionally, the properties of metals can be enhanced in combination with another metal i.e., transition metal. The combination of metal oxides as nanocomposite, hence, increase interactions and effectiveness [56]. Various methods have been reported for the synthesis of nanocomposites, such as the sol-gel method [57,58], chemical vapor deposition (CVD) [59], hydrothermal methods [60], and the solid state method [61]. Among all preparation methods, co-precipitation is the easiest and most cost-effective method, and no special or complicated material and equipment is required [62].

In this work, the reduction of 4-nitrophenol using aqueous NaOH as a reducing agent in the presence of an La_2ZnO_4 catalyst could be a substitute for conservative reduction processes. Up till now, no work has done on the direct combination of Lanthanum and Zinc metals in the form of a combined oxide. Thus, in the present work, we report the synthesis of Lanthanum Zinc oxide nanocomposite La_2ZnO_4 through a simple co-precipitation method, keeping in mind the properties of Lanthanum and Zinc. The prepared nanocomposite was characterized by many useful techniques, i.e., FTIR Fourier transform infrared spectroscopy, ultraviolet spectroscopy, XRD X-ray diffraction, SEM scanning electron microscopy, and PL photoluminescence. Herein, 4-nitrophenol was reduced to 4-aminophenol in the presence of the La_2ZnO_4 composite applying aq. NaOH as a reducing agent. This scheme did not involve a specific solvent and was established at room temperature. Hence, the current method offers an eco-friendly route for the catalytic hydrogenation of 4-nitrophenol to 4-aminophenol.

2. Materials and Methods

2.1. Materials

Lanthanum nitrate hexahydrate ($La(NO_3)_3 \cdot 6H_2O$) (99.99% pure), zinc nitrate hexahydrate ($Zn(NO_3)_2 \cdot 6H_2O$) (99% pure), sodium hydroxide anhydrous pellets (NaOH) (98% pure), and methanol (CH_3OH) (99.8%) were purchased from Sigma Aldrich and used as received without any further purification treatment. All chemicals, including deionized water (DI) used during the experiment, were of analytical reagent grade.

2.2. Synthesis of La_2ZnO_4

The process of synthesizing the La_2ZnO_4 nanocomposite, was followed according to the work reported by Tinwala et al. [63]. The La_2ZnO_4 nanocomposite was synthesized via the co-precipitation method, using the precursors lanthanum nitrate La $(NO_3)_3$, zinc nitrate Zn $(NO_3)_2$, deionized water, and methanol as starting materials. Sodium hydroxide NaOH was used as a precipitating agent. The solutions of $La(NO_3)_3$ and $Zn(NO_3)_2$ 150 mL each were prepared separately by dissolving them in the solvent. The concentration of mixtures prepared was 0.1 M for the Lanthanum nitrate solution and 0.003 M for the Zinc nitrate solution. 0.2 M concentrated solution of the precipitating agent was also prepared.

Co-precipitation was done by combining both the mixtures and titrating by adding the precipitating agent i.e., NaOH dropwise, while stirring at 200 rpm with a magnetic stirrer and heating at 70 °C temperature for 6 hrs. Initially the pH was ~4, and the precipitating agent was added until the pH reached ~11. On completion of the precipitation, the solution was filtered using Whattman filter paper and washed with acetone several times in order to remove all the byproducts. The as-prepared material was dried in an ordinary oven at 80 °C for 24 h, and then ground using a mortar and pestle. The dried powder was calcined at 400–450 °C for 4 h.

2.3. Characterization

The Lanthanum-based Zinc oxide nanocomposite was characterized with different analytical techniques. Crystalline size and structure were determined by X-ray diffraction on (JDX-3532, JEOL, Tokyo, Japan), X-Ray Diffractometer, using Cu-Kα (λ = 1.5046 Å) radiation at a tube voltage of 40 KV and 20 mA current. The morphology was found by using scanning electron microscopy SEM (JSM-5910, JEOL, Tokyo, Japan); a band gap was found via ultraviolet spectroscopy UV. Photoluminescence spectroscopy PL was also used to check the optical properties of the prepared sample as a semiconductor. Fourier transform infrared spectroscopy FTIR (IR Prestige 21, Shimadzu, Kyoto, Japan) was used to identify the presence of any unwanted substance in the sample after calcination.

2.4. Catalysis for P-Nitrophenol Conversion

In a typical test, to 20 mL of a 0.1 mmol·L^{-1} aqueous solution of p-nitrophenol taken in a beaker, a freshly prepared aqueous solution of NaOH in distilled water was introduced. To the mixture, 0.1 g of the La_2ZnO_4 oxide was added and stirred at room temperature. The disappearance of the yellow color of p-nitrophenol was monitored by UV-VIS spectroscopy (Tensor II BRUKER, Billerica, MA, USA).

3. Results and Discussion

3.1. X-ray Diffraction

The morphological and structural analysis of the as-developed catalyst material was carried out by various physicochemical characterizations. The crystallographic and structural confirmation of the La_2ZnO_4 nanocomposite was done by means of powder X-ray diffraction analysis. The XRD pattern of the La_2ZnO_4 nanocomposite treated at 400–450 °C is shown in Figure 1 with 2θ range from 20° to 60° at room temperature. A mixed phase of La_2O_3 and ZnO is formed, which is evident from the graph. All the diffraction peaks refer to hexagonal phase La_2ZnO_4 and are in good agreement with the standards (ICCD card No. 00-002-0688) for lanthanum oxide phase and (ICCD card No. 00-001-1136) for zinc oxide phase, respectively. In Figure 1, the peaks at 22° (100), 26° (002), 44° (101), and 46° (103) correspond to La and peaks at 36° (002) and 57° (110) match up with the multiple similar reported XRD patterns. Besides, the peaks at 31° (101) and 36° (100) appeared due to the synergistic impact of La/Zn oxides, as both of these oxides give peaks around such 2 theta positions, singly. Moreover, few additional peaks are observed in the XRD pattern i.e., 40° (101), 42.3° (100), 51° (101), and 55° (111), which is indication of the successful formation of the mixed metal oxide composite [64–66].

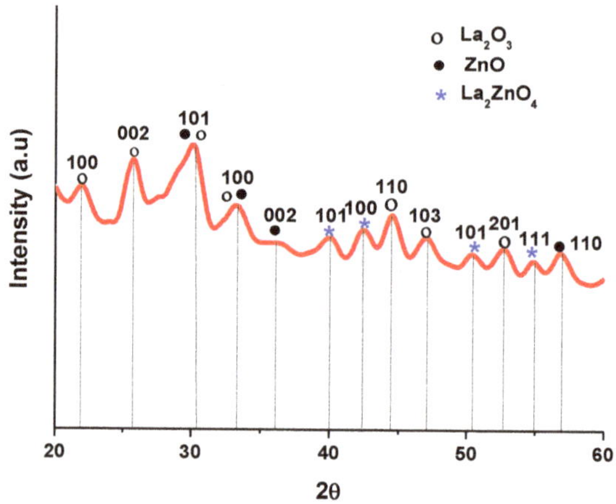

Figure 1. X-ray diffraction spectrum of La_2ZnO_4.

The crystallite size of the sample calcined at 400 °C was calculated by using Debye-Scherer's equation [67], which is given below:

$$D_{\mathrm{hkl}} = \frac{k\lambda}{\beta \cos\theta} \quad (1)$$

where D_{hkl} is the average crystalline size perpendicular to the crystal phase (hkl) K is constant, λ is 1.5406 Å, and β is full width half maxima of the peak at (100) plane (Figure 1) [67]. The average crystallite size of the nanocomposite calculated using the above equation is 8.62 nm for the sample calcined at 400–450 °C.

3.2. Scanning Electron Microscopy

The morphological analysis was done with scanning electron microscopy (SEM, Hitachi High-Tech, Seoul, Korea). Figure 2 represents the hexagonal morphology of the nanocomposite prepared La_2ZnO_4 via scanning electron microscopy (SEM). At some places, aggregates of the nanocomposite were observed due to the presence of binary oxides of lanthanum and zinc. It is clearly seen that the composite is in crystalline form and four pictures shows the presence of Zn particles on the surface of Lanthanum crystals, forming a unique nanocomposite.

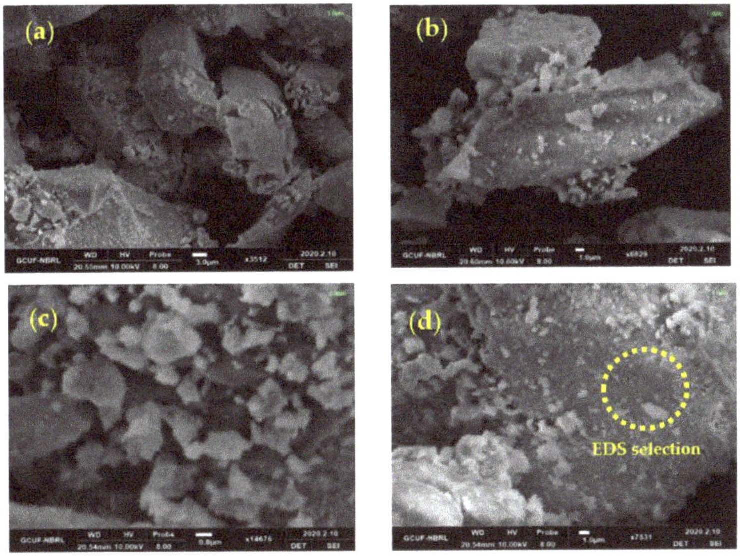

Figure 2. Scanning electron microscopy images of the La_2ZnO_4 nanocomposite at different resolutions (**a**) at 3.0 µm (**b**) 1.0 µm (**c**) 0.8 µm (**d**) Selection of EDX on SEM Pattern.

3.3. *Energy Dispersive Spectroscopy*

Elemental composition and purity of the as-synthesized La_2ZnO_4 composite was determined by EDS analysis (MIRA3 TESCAN, Brno, Czech Republic), as presented in Figure 3. Uniform distribution of Zn, La, and O throughout the whole matrix is evident from the EDS spectrum and mapping. No other peak corresponding to any impurity was observed. It is witnessed that the composite contained a smooth surface, which could be associated with the homogenous mixing of Zn and La oxides in the composite, resulting in a single phase surface morphology.

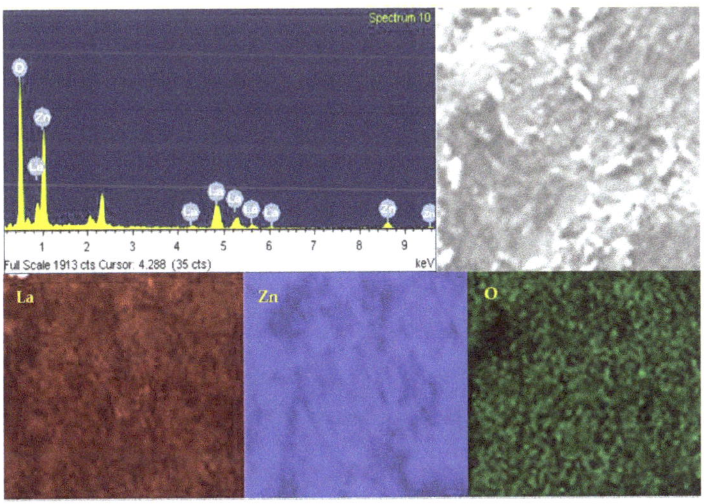

Figure 3. EDS spectrum and mapping for the La_2ZnO_4 nanocomposite.

3.4. FTIR Analysis

Figure 4 shows the FTIR spectrum of the La_2ZnO_4 nanocomposite. The characteristic peaks at 529.37 and 831 cm^{-1} clearly indicate the presence of Zn–O stretching vibration and peaks at 717.98–591.70 cm^{-1} confirms the presence of ZnO and La [68]. Slight shifting of peaks is noticed due to the formation of the nanocomposite. The peak at 3434.85 cm^{-1} represents the La_2O_3 stretching vibration. A peak with very weak intensity at 854.24 cm^{-1} can be assigned to the residual nitrate ion. Other characteristic absorption bands from 2359.95 to 1458.21 cm^{-1} may be due to the presence of water molecule, C=C, stretching of C–H and C–C, respectively. The difference in peak positions of the starting lanthanum nitrate and the as-synthesized composite can be witnessed in the reported FTIR spectrum [69]. The appearance of new peaks at 649, 1393, and 2361 cm^{-1} and the disappearance of specific nitrate peaks indicate the formation of the La oxide composite [70].

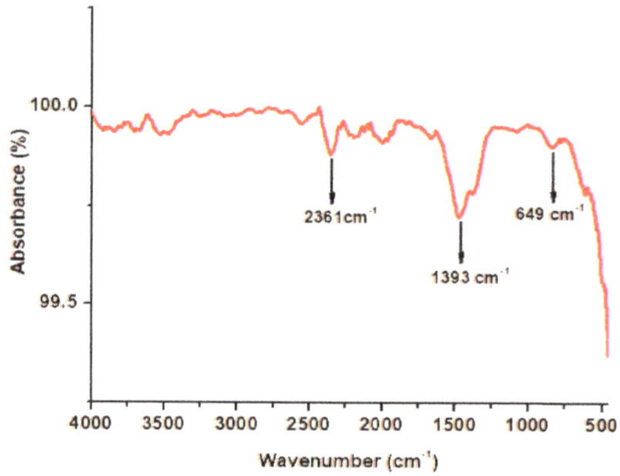

Figure 4. FTIR spectra of the La_2ZnO_4 nanocomposite.

3.5. UV-Visible Spectroscopy

Optical properties of nanocomposite La_2ZnO_4 were investigated by UV-Vis absorption spectra using a UV-Visible spectrophotometer (PharmaSpec UV-1700, Shimadzu, Kyoto, Japan), as shown in Figure 5a. It was noted that band-edge absorption of the synthesized La_2ZnO_4 is located in the near UV region. The optical band gap was calculated by using the following equation [71], as shown in Figure 5b:

$$\alpha = A(h\nu - E_g)^n / h\nu \quad (2)$$

where A and n is a constant, equal to $\frac{1}{2}$ for the direct band gap semiconductor. The spectra identify UV active optical properties of the nanocomposite. It was noted that due to very low concentrations of Zinc, as compared to Lanthanum, the band gap had minute changes, although both metals are good semiconductors; however, for better efficiency, the concentration of Zn metal could be increased. The prepared composite could be applied as a UV light photocatalyst.

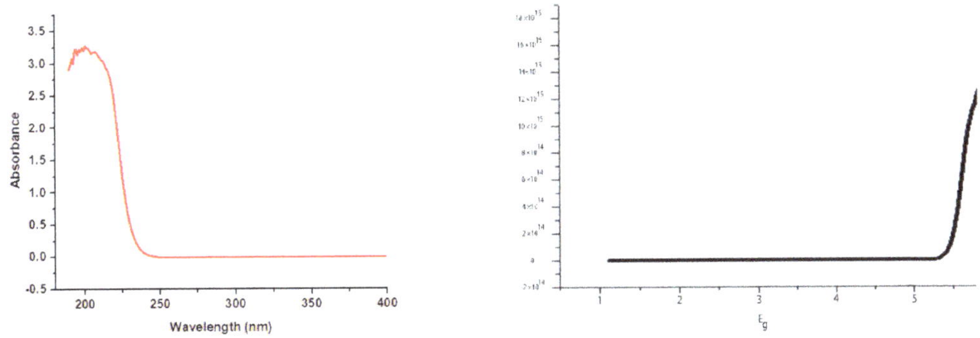

(a) (b)

Figure 5. (**a**) Absorption spectra (UV-Vis) of the La_2ZnO_4 nanocomposite (**b**) Tauc plots of $(\alpha h\nu)^2$ versus $h\nu$ of the La_2ZnO_4 nanocomposite.

3.6. Photoluminescence Observation

The PL spectra of the La_2ZnO_4 nanocomposite is shown in Figure 6. Small variations in the absorption peaks, due to the formation of nanocomposites, were noticed. The position of emission bands becoming less intense may be due to the strain in crystal lattice to accommodate larger Lanthanum atoms with Zinc metal as oxide. If the UV spectrum of the as-proposed composite is compared with pure lanthanum nitrate, a blue shift is observed, a result of the addition of zinc, attributed to wider band gap of ZnO [72]. The results suggest that the as-synthesized nanoparticles can absorb radiation in UV and in the visible region, as well from solar light, indicating that the La_2ZnO_4 composite could be useful as a visible light photocatalyst.

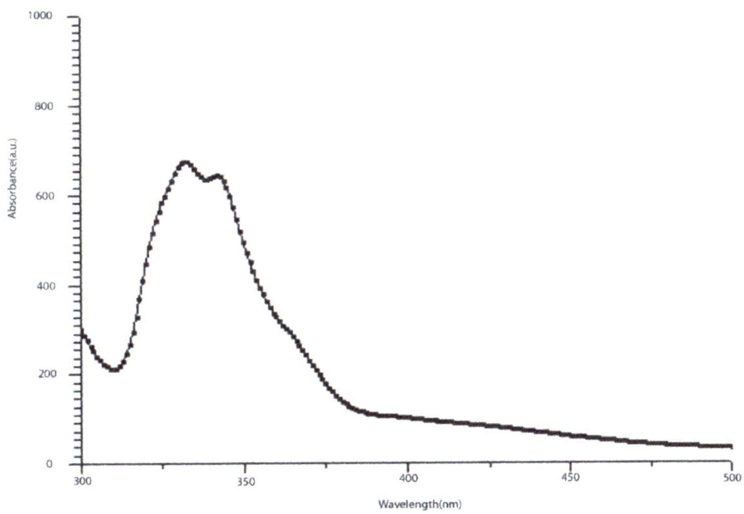

Figure 6. PL spectra of the La_2ZnO_4 nanocomposite.

3.7. Catalysis Process

The catalytic performance for hydrogenation reaction was examined by the catalytic reduction of 4-nitrophenol into 4-aminophenol, shown in Figure 7. Before the explanation of results of 4-NP reduction, the common mechanism and logical reasons for the enhanced hydrogenation reactions performance can be ascribed due to the presence of mixed-phase binary oxides nanoparticles in nanocomposite form. The reduction capability of the catalyst material depends on the mutual electron transfer process within the composite system. In the case of the present catalyst, La_2ZnO_4 exhibits a higher catalytic performance, resulting from both the constituent elements. The brief mechanism could be explained as follows; in La_2ZnO_4, the La_2O_3 and ZnO sites exhibited selectivity in initiating the reduction reactions at the mixed-phase binary oxide of the La_2ZnO_4 interfaces; further, this composite improved the electron transfer phenomenon as well boosted the reaction synergistically. These collective factors enhanced the overall catalytic performance of the nanocomposite material.

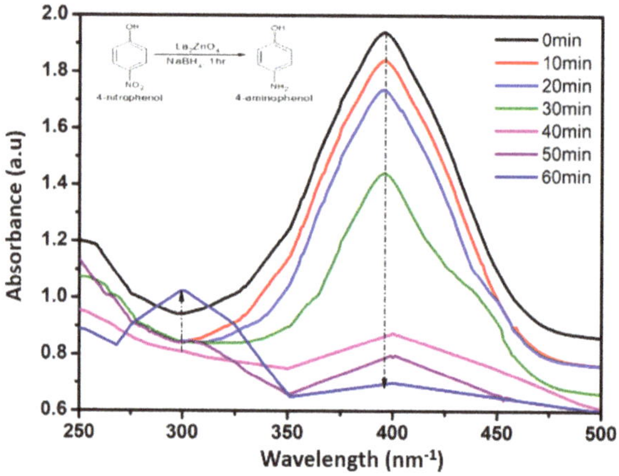

Figure 7. Photocatalytical reduction of 4-nitophenol into 4-aminophenol.

It was observed that the conversion of 4-nitrophenol to 4-aminophenol with NaOH as a reducing agent in the presence of La_2ZnO_4 reducing catalysts took 60 min (when the color of the mixture changed from yellow to colorless). Comparatively, this decolarization requires 5.5 h with NaOH, in the absence of a catalyst. The appearance of a new peak at 300 nm^{-1} corresponds to the absorbance by 4-aminophenol and is an indication of successful conversion [73,74]. The observed UV-visible absorbance peaks of the reactants and the product are significantly isolated at ~400 nm and ~300 nm, respectively. Furthermore, the concentrations of 4-nitrophenol were estimated from absorbance at ~400 nm using the respective calibration curve. Therefore, it can be deduced that the mild reducing ability of NaOH could be efficiently enhanced with the addition of the La_2ZnO_4 composite.

3.8. Kinetics of the Reduction Catalysis

The kinetic mechanism of 4-nitrophenol reduction with NaOH was studied with the La_2ZnO_4 catalyst. It typically follows the pseudo-first-order reaction owing to the concentration of 4-nitrophenol. The equation for the pseudo-first-order reaction is given in Equation (3) [75]:

$$\ln(C_t/C_0) = \ln(A_t/A_0) = -kt \quad (3)$$

Here C_t and C_0 represent the concentrations of 4-nitrophenol at time t and $t = 0$, respectively. A_t and A_0 corresponds to the absorbance of 4-nitrophenol at time t and $t = 0$ respectively, at peak position of 400 nm^{-1}. k is the rate constant of the reaction.

Moreover, when natural log of concentration/absorbance was plotted as a function of time, a significant decrease can be witnessed (Figure 8). The slope of the graph gave the rate constant of the reaction, which is found out to be $k = 0.012\ min^{-1}$.

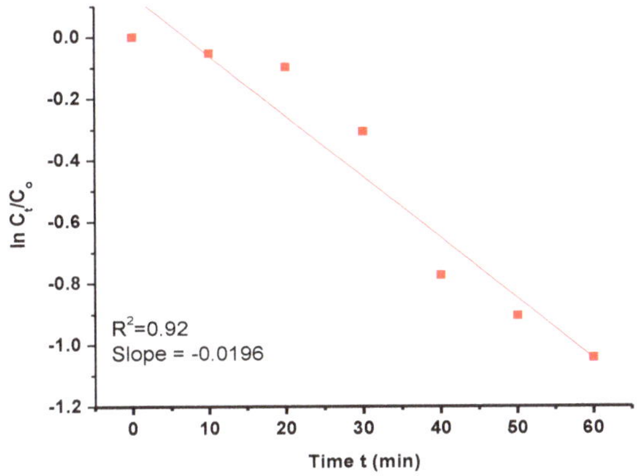

Figure 8. The kinetic plot for the adsorption-catalytic reduction of 4-nitrophenol.

4. Conclusions

Lanthanum-based Zinc oxide nanocomposites were synthesized by co-precipitation method using water as a solvent. The reaction between both the precursors in the presence of water produces hydroxyl ions for precipitation. The hexagonal crystalline structure with crystallite size of 8.62 nm was observed after calcination at 450 °C, detected by XRD. The morphology was confirmed by SEM; all the images show the presence of Zinc on the surface of Lanthanum, hence forming the La_2ZnO_4 nanocomposite. No evidence of the presence of separate La and Zn was found. The Pl spectra indicated that the band gap of La metal can be reduced by the addition of Zn metal, if an efficient amount is used. The nanocomposite prepared was found to have strong photocatalytic efficiency. Co-precipitation, the method that was adopted for synthesis, was cost-effective and simple, as compared to other reported methods. In addition, the as-developed nanocomposite material showed enhanced catalytic performance for complete reduction of 4-nitophenol into 4-aminophenol within 60 min. The enhanced 4-NP reduction may be ascribed to the synergistic effects of binary oxides sites of the La_2ZnO_4 nanocomposite. The findings herein could pave the way for the fruitful study and development of heterogeneous catalysts for eco-friendly environmental applications.

Author Contributions: Conceptualization, I.A.; methodology, H.A.; software, M.I.; validation, S.H.; formal analysis, A.B.Y.; investigation, R.R.K.; resources, N.S.A.-k.; data curation, M.O.; writing—original draft preparation, A.A.; writing—review and editing, S.K.; visualization, A.G.; supervision, M.A.J.; project administration, M.S.; funding acquisition, M.O. All authors have read and agreed to the published version of the manuscript.

Funding: This research was funded by the Deanship of Scientific Research at Princess Nourah bint Abdulrahman University through the Fast-track Research Funding Program.

Institutional Review Board Statement: Not applicable.

Informed Consent Statement: Not applicable.

Data Availability Statement: Data is contained within the article.

Conflicts of Interest: The authors declare no conflict of interest.

References

1. Škríba, A.; Janková, Š.; Váňa, J.; Barták, P.; Bednář, P.; Fryčák, P.; Kučera, L.; Kurka, O.; Lemr, K.; Macíková, P.; et al. Protonation sites and fragmentations of para-aminophenol. *Int. J. Mass Spectrom.* **2013**, *337*, 18–23. [CrossRef]
2. Schultz, S.; DeSilva, M.; Gu, T.T.; Qiang, M.; Whang, K. Effects of the analgesic acetaminophen (paracetamol) and its para-aminophenol metabolite on viability of mouse-cultured cortical. *Neurons. Basic Clin. Pharmacol. Toxicol.* **2012**, *110*, 141–144. [CrossRef]

3. Gao, L.; Li, R.; Sui, X.; Li, R.; Chen, C.; Chen, Q. Conversion of chicken feather waste to N-doped carbon nanotubes for the catalytic reduction of 4-nitrophenol. *Environ. Sci. Technol.* **2014**, *48*, 10191–10197. [CrossRef]
4. Ma, M.; Yang, Y.; Li, W.; Feng, R.; Li, Z.; Lyu, P.; Ma, Y. Gold nanoparticles supported by amino groups on the surface of magnetite microspheres for the catalytic reduction of 4-nitrophenol. *J. Mater. Sci.* **2019**, *54*, 323–334. [CrossRef]
5. Bano, M.; Naikoo, G.A.; Ahirwar, D.; Thomas, M.; Sheikh, M.U.-D.; Khan, F. Hierarchical synthesis of silver monoliths and their efficient catalytic activity for the reduction of 4-nitrophenol to 4-aminophenol. *N. J. Chem.* **2016**, *40*, 6787–6795. [CrossRef]
6. Li, J.; Liu, C.-Y.; Liu, Y. Au/graphene hydrogel: Synthesis, characterization and its use for catalytic reduction of 4-nitrophenol. *J. Mater. Chem.* **2012**, *22*, 8426–8430. [CrossRef]
7. Serrà, A.; Artal, R.; Pozo, M.; Garcia-Amorós, J.; Gómez, E. Simple environmentally-friendly reduction of 4-nitrophenol. *Catalysts* **2020**, *10*, 458. [CrossRef]
8. Lv, J.-J.; Wang, A.-J.; Ma, X.; Xiang, R.-Y.; Chen, J.-R.; Feng, J.-J. One-pot synthesis of porous P-Au nanodendrites supported on reduced graphene oxide nanosheets toward catalytic reduction of 4-nitrophenol. *J. Mater. Chem. A* **2015**, *3*, 290–296. [CrossRef]
9. Zhang, J.; Chen, G.; Guay, D.; Chaker, M.; Ma, D. Highly active PtAu alloy nanoparticle catalysts for the reduction of 4-nitrophenol. *Nanoscale* **2014**, *6*, 2125–2130. [CrossRef] [PubMed]
10. Yan, Z.-G.; Yan, C.-H. Controlled synthesis of rare earth nanostructures. *J. Mater. Chem.* **2008**, *18*, 5046–5059. [CrossRef]
11. Wang, X.; Li, Y. Rare-earth-compound nanowires, nanotubes, and fullerene-like nanoparticles: Synthesis, characterization, and properties. *Chem. A Eur. J.* **2003**, *9*, 5627–5635. [CrossRef] [PubMed]
12. Xin, Y.; Wang, Z.; Qi, Y.; Zhang, Z.; Zhang, S. Synthesis of rare earth (Pr, Nd, Sm, Eu and Gd) hydroxide and oxide nanorods (nanobundles) by a widely applicable precipitation route. *J. Alloys Compd.* **2010**, *507*, 105–111. [CrossRef]
13. Xiao, Y.; Feng, Z.; Huang, X.; Huang, L.; Long, Z.; Wang, Q.; Hou, Y. Synthesis of lanthanum oxide nanosheets by a green carbonation process. *Chin. Sci. Bull.* **2014**, *59*, 1864–1867. [CrossRef]
14. Iranmanesh, T.; Jahani, S.; Foroughi, M.M.; Zandi, M.S.; Nadiki, H.H. Synthesis of La_2O_3/MWCNT nanocomposite as the sensing element for electrochemical determination of theophylline. *Anal. Methods* **2020**, *12*, 4319–4326. [CrossRef]
15. Teo, S.; Guo, Z.; Xu, Z.; Zhang, C.; Kamata, Y.; Hayase, S.; Ma, T. The role of lanthanum in a nickel oxide-based inverted perovskite solar cell for efficiency and stability improvement. *ChemSusChem* **2019**, *12*, 518–526. [CrossRef]
16. Li, H.; Zheng, B.; Xue, Y.; Liu, S.; Gao, C.; Liu, X. Spray deposited lanthanum doped TiO_2 compact layers as electron selective contact for perovskite solar cells. *Sol. Energy Mater. Sol. Cells* **2017**, *168*, 85–90. [CrossRef]
17. Meng, W.; Hu, R.; Yang, J.; Du, Y.; Li, J.; Wang, H. Influence of lanthanum-doping on photocatalytic properties of $BiFeO_3$ for phenol degradation. *Chin. J. Catal.* **2016**, *37*, 1283–1292. [CrossRef]
18. Meksi, M.; Turki, A.; Kochkar, H.; Bousselmi, L.; Guillard, C.; Berhault, G. The role of lanthanum in the enhancement of photocatalytic properties of TiO_2 nanomaterials obtained by calcination of hydrogenotitanate nanotubes. *Appl. Catal. B Environ.* **2016**, *181*, 651–660. [CrossRef]
19. Ismail, R.A.; Fadhil, F.A.; Rashed, H.H. Novel route to prepare lanthanum oxide nanoparticles for optoelectronic devices. *Int. J. Mod. Phys. B* **2020**, *34*, 2050134. [CrossRef]
20. Surti, S.; Karp, J.; Muehllehner, G.; Raby, P. Investigation of lanthanum scintillators for 3D PET. In Proceedings of the 2002 IEEE Nuclear Science Symposium Conference Record, Norfolk, VA, USA, 10–16 November 2002.
21. Thomas, K.; Alexander, D.; Sisira, S.; Biju, P.R.; Unnikrishnan, N.V.; Ittyachen, M.A.; Joseph, C. NUV/blue LED excitable intense green emitting terbium doped lanthanum molybdate nanophosphors for white LED applications. *J. Mater. Sci. Mater. Electron.* **2017**, *28*, 17702–17709. [CrossRef]
22. Zhang, S.; Liu, H.; Li, X.; Wang, S. Enhancing quantum yield of CsPb (Br_xCl_{1-x}) 3 nanocrystals through lanthanum doping for efficient blue light-emitting diodes. *Nano Energy* **2020**, *77*, 105302. [CrossRef]
23. Luewarasirikul, N.; Kim, H.; Meejitpaisan, P.; Kaewkhao, J. White light emission of dysprosium doped lanthanum calcium phosphate oxide and oxyfluoride glasses. *Opt. Mater.* **2017**, *66*, 559–566. [CrossRef]
24. Madhu, A.; Eraiah, B.; Manasa, P.; Srinatha, N. Nd^{3+}-doped lanthanum lead boro-tellurite glass for lasing and amplification applications. *Opt. Mater.* **2018**, *75*, 357–366. [CrossRef]
25. Jiang, S.P. Development of lanthanum strontium cobalt ferrite perovskite electrodes of solid oxide fuel cells-A review. *Int. J. Hydrog. Energy* **2019**, *44*, 7448–7493. [CrossRef]
26. Talley, K.R.; Mangum, J.; Perkins, C.L.; Woods-Robinson, R.; Mehta, A.; Gorman, B.P.; Brennecka, G.L.; Zakutayev, A. Synthesis of lanthanum tungsten oxynitride perovskite thin films. *Adv. Electron. Mater.* **2019**, *5*, 1900214. [CrossRef]
27. Mogha, N.K.; Gosain, S.; Masram, D.T. Lanthanum oxide nanoparticles immobilized reduced graphene oxide polymer brush nanohybrid for environmental vitiation of organic dyes. *Arab. J. Chem.* **2020**, *13*, 1367–1376. [CrossRef]
28. Subhan, F.; Aslam, S.; Yan, Z.; Ahmad, A.; Etim, U.; Naeem, M.; Zhen, L.; Ikram, M.; Yaseen, M. Highly dispersive lanthanum oxide fabricated in confined space of SBA-15 for adsorptive desulfurization. *Chem. Eng. J.* **2020**, *384*, 123271. [CrossRef]
29. Sharma, G.; Kumar, A.; Sharma, S.; Al-Saeedi, S.I.; Al-Senani, G.M.; Nafady, A.; Ahamad, T.; Naushad, M.; Stadler, F.J. Fabrication of oxidized graphite supported La_2O_3/ZrO_2 nanocomposite for the photoremediation of toxic fast green dye. *J. Mol. Liq.* **2019**, *277*, 738–748. [CrossRef]
30. Ankamwar, B.G.; Kamble, V.B.; Annsi, J.I.; Sarma, L.S.; Mahajan, C.M. Photocatalytic Degradation *of* Methylene Blue *by* ZnO Nanoparticles. *J. Nanosci. Nanotechnol.* **2017**, *17*, 1185–1192. [CrossRef]

31. Ong, C.B.; Mohammad, A.W.; Ng, L.Y. Integrated adsorption-solar photocatalytic membrane reactor for degradation of hazardous Congo red using Fe-doped ZnO and Fe-doped ZnO/rGO nanocomposites. *Environ. Sci. Pollut. Res.* **2019**, *26*, 33856–33869. [CrossRef] [PubMed]
32. Zang, Z. Efficiency enhancement of ZnO/Cu_2O solar cells with well oriented and micrometer grain sized Cu_2O films. *Appl. Phys. Lett.* **2018**, *112*, 042106. [CrossRef]
33. Zhang, P.; Wu, J.; Zhang, T.; Wang, Y.; Liu, D.; Chen, H.; Jiang, W.; Liu, C.; Ahmad, W.; Chen, Z.D.; et al. Perovskite solar cells with ZnO electron-transporting materials. *Adv. Mater.* **2018**, *30*, 1703737. [CrossRef] [PubMed]
34. Jung, J.; Jeong, J.R.; Lee, J.; Lee, S.H.; Kim, S.Y.; Kim, M.J.; Nah, J.; Lee, M.H. In situ formation of graphene/metal oxide composites for high-energy microsupercapacitors. *Npg Asia Mater.* **2020**, *12*, 1–9. [CrossRef]
35. Ahmad, R.; Ahn, M.-S.; Hahn, Y.-B. ZnO nanorods array based field-effect transistor biosensor for phosphate detection. *J. Colloid Interface Sci.* **2017**, *498*, 292–297. [CrossRef]
36. Bao, R.; Wang, C.; Peng, Z.; Ma, C.; Dong, L.; Pan, C. Light-emission enhancement in a flexible and size-controllable ZnO nanowire/organic light-emitting diode array by the piezotronic effect. *ACS Photonics* **2017**, *4*, 1344–1349. [CrossRef]
37. Rahman, F. Zinc oxide light-emit. *Diodes A Rev. Opt. Eng.* **2019**, *58*, 010901.
38. Zhu, L.; Zeng, W. Room-temperature gas sensing of ZnO-based gas sensor: A review. *Sens. Actuators A Phys.* **2017**, *267*, 242–261. [CrossRef]
39. Huang, X.; Zheng, X.; Xu, Z.; Yi, C. ZnO-based nanocarriers for drug delivery application: From passive to smart strategies. *Int. J. Pharm.* **2017**, *534*, 190–194. [CrossRef] [PubMed]
40. Hussain, A.; Oves, M.; Alajmi, M.F.; Hussain, I.; Amir, S.; Ahmed, J.; Rehman, M.T.; El-Seedi, H.R.; Ali, I. Biogenesis of ZnO nanoparticles using Pandanus odorifer leaf extract: Anticancer and antimicrobial activities. *RSC Adv.* **2019**, *9*, 15357–15369. [CrossRef]
41. Alavi-Tabari, S.A.; Khalilzadeh, M.A.; Karimi-Maleh, H. Simultaneous determination of doxorubicin and dasatinib as two breast anticancer drugs uses an amplified sensor with ionic liquid and ZnO nanoparticle. *J. Electroanal. Chem.* **2018**, *811*, 84–88. [CrossRef]
42. Da Silva, B.L.; Caetano, B.L.; Chiari-Andréo, B.G.; Pietro, R.C.L.R.; Chiavacci, L.A. Increased antibacterial activity of ZnO nanoparticles: Influence of size and surface modification. *Colloids Surf. B Biointerfaces* **2019**, *177*, 440–447. [CrossRef]
43. Jaisutti, R.; Lee, M.; Kim, J.; Choi, S.; Ha, T.J.; Kim, J.; Kim, H.; Park, S.K.; Kim, Y.H. Ultrasensitive room-temperature operable gas sensors using p-type Na: ZnO nanoflowers for diabetes detection. *ACS Appl. Mater. Interfaces* **2017**, *9*, 8796–8804. [CrossRef] [PubMed]
44. Eixenberger, J.E.; Anders, C.B.; Wada, K.; Reddy, K.M.; Brown, R.J.; Moreno-Ramirez, J.; Weltner, A.E.; Karthik, C.; Tenne, D.A.; Fologea, D.; et al. Defect engineering of ZnO nanoparticles for bioimaging applications. *Acs Appl. Mater. Interfaces* **2019**, *11*, 24933–24944. [CrossRef]
45. Raja, A.; Rajasekaran, P.; Selvakumar, K.; Arunpandian, M.; Kaviyarasu, K.; Bahadur, S.A.; Swaminathan, M. Visible active reduced graphene oxide-$BiVO_4$-ZnO ternary photocatalyst for efficient removal of ciprofloxacin. *Sep. Purif. Technol.* **2020**, *233*, 115996. [CrossRef]
46. Zhang, Q.; Xu, M.; You, B.; Zhang, Q.; Yuan, H.; Ostrikov, K.K. Oxygen vacancy-mediated ZnO nanoparticle photocatalyst for degradation of methylene blue. *Appl. Sci.* **2018**, *8*, 353. [CrossRef]
47. Tan, C.F.; Su Su Zin, A.K.; Chen, Z.; Liow, C.H.; Phan, H.T.; Tan, H.R.; Xu, Q.H.; Ho, G.W. Inverse stellation of CuAu-ZnO multimetallic-semiconductor nanostartube for plasmon-enhanced photocatalysis. *ACS Nano* **2018**, *12*, 4512–4520. [CrossRef] [PubMed]
48. Jeon, Y.-R.; Yu, J.; Choi, S.-J. Fate determination of ZnO in commercial foods and human intestinal cells. *Int. J. Mol. Sci.* **2020**, *21*, 433. [CrossRef] [PubMed]
49. Zhao, Z.; Mao, A.; Gao, W.; Bai, H. A facile in situ method to fabricate transparent, flexible polyvinyl alcohol/ZnO film for UV-shielding. *Compos. Commun.* **2018**, *10*, 157–162. [CrossRef]
50. Babaei-Ghazvini, A.; Shahabi-Ghahfarrokhi, I.; Goudarzi, V. Preparation of UV-protective starch/kefiran/ZnO nanocomposite as a packaging film: Characterization. *Food Packag. Shelf Life* **2018**, *16*, 103–111. [CrossRef]
51. Lv, J.; Li, C.; Chai, Z. Defect luminescence and its mediated physical properties in ZnO. *J. Lumin.* **2019**, *208*, 225–237. [CrossRef]
52. Rodnyi, P.; Chernenko, K.; Venevtsev, I. Mechanisms of ZnO luminescence in the visible spectral region. *Reg. Opt. Spectrosc.* **2018**, *125*, 372–378. [CrossRef]
53. Majumder, T.; Dhar, S.; Debnath, K.; Mondal, S.P. Role of S, N co-doped graphene quantum dots as a green photosensitizer with Ag-doped ZnO nanorods for improved electrochemical solar energy conversion. *Mater. Res. Bull.* **2017**, *93*, 214–222. [CrossRef]
54. Subramaniam, V.D.; Prasad, S.V.; Banerjee, A.; Gopinath, M.; Murugesan, R.; Marotta, F.; Sun, X.-F.; Pathak, S. Health hazards of nanoparticles: Understanding the toxicity mechanism of nanosized ZnO in cosmetic products. *Drug Chem. Toxicol.* **2019**, *42*, 84–93. [CrossRef]
55. Bocca, B.; Caimi, S.; Senofonte, O.; Alimonti, A.; Petrucci, F. ICP-MS based methods to characterize nanoparticles of TiO_2 and ZnO in sunscreens with focus on regulatory and safety issues. *Sci. Total Environ.* **2018**, *630*, 922–930. [CrossRef]
56. Pachiyappan, J.; Gnanasundaram, N.; Rao, G.L. Preparation and characterization of ZnO, MgO and ZnO-MgO hybrid nanomaterials using green chemistry approach. *Results Mater.* **2020**, *7*, 100104. [CrossRef]

57. Deepthi, T.; Balamurugan, K. Effect of yttrium (20%) doping on mechanical properties of rare earth nano lanthanum phosphate ($LaPO_4$) synthesized by aqueous sol-gel process. *Ceram. Int.* **2019**, *45*, 18229–18235. [CrossRef]
58. Khan, M.; Janjua, N.K.; Khan, S.; Qazi, I.; Ali, S.; Saad Algarni, T. Electro-oxidation of ammonia at novel $Ag_2O{-}PrO_2/\gamma{-}Al_2O_3$ catalysts. *Coatings* **2021**, *11*, 257. [CrossRef]
59. Katsui, H.; Kondo, N. Preferred orientations and microstructures of lanthanum phosphate films prepared via laser chemical vapor deposition. *J. Cryst. Growth* **2019**, *519*, 46–53. [CrossRef]
60. Qian, Y.; Qiao, P.; Li, L.; Han, H.; Zhang, H.; Chang, G. Hydrothermal synthesis of lanthanum-doped MgAl-layered double hydroxide/graphene oxide hybrid and its application as flame retardant for thermoplastic polyurethane. *Adv. Polym. Technol.* **2020**, *2020*, 1018093. [CrossRef]
61. Mane, V.; Malavekar, D.; Ubale, S.; Bulakhe, R.; In, I.; Lokhande, C. Binder free lanthanum doped manganese oxide@ graphene oxide composite as high energy density electrode material for flexible symmetric solid state supercapacitor. *Electrochim. Acta* **2020**, *335*, 135613. [CrossRef]
62. Zakirov, M.I.; Semen'Ko, M.P.; Korotchenkov, O.A. A simple sonochemical synthesis of nanosized ZnO from zinc acetate and sodium hydroxide. *J. Nano- Electron. Phys.* **2018**, *10*, 05023-1. [CrossRef]
63. Tinwala, H.; Shah, D.V.; Menghani, J.; Pati, R. Synthesis of $La_2Ce_2O_7$ nanoparticles by co-precipitation method and its characterization. *J. Nanosci. Nanotechnol.* **2014**, *14*, 6072–6076. [CrossRef] [PubMed]
64. Sescu, A.M.; Harja, M.; Favier, L.; Berthou, L.O.; De Castro, C.G.; Pui, A.; Lutic, D. Zn/La mixed oxides prepared by coprecipitation: Synthesis, characterization and photocatalytic studies. *Materials* **2020**, *13*, 4916. [CrossRef] [PubMed]
65. Manikandan, A.; Meenatchi, B.; Vadivel, S.; Jaganathan, S.; Ladchumananandasivam, R.; Henini, M.; Maaza, M.; Aanand, J.S. Rare earth element (REE) lanthanum doped zinc oxide (La:ZnO) nanomaterials: Synthesis structural optical and antibacterial studies. *J. Alloys Compd.* **2017**, *723*, 1155–1161. [CrossRef]
66. Yan, S.; Salley, S.O.; Ng, K.S. Simultaneous transesterification and esterification of unrefined or waste oils over $ZnO{-}La_2O_3$ catalysts. *Appl. Catal. A: Gen.* **2009**, *353*, 203–212. [CrossRef]
67. Khan, S.; Shah, S.S.; Anjum, M.A.R.; Khan, M.R.; Janjua, N.K. Electro-oxidation of ammonia over copper oxide impregnated $\gamma{-}Al_2O_3$ nanocatalysts. *Coatings* **2021**, *11*, 313. [CrossRef]
68. Ashna, R.; Yulizar, Y.; Apriandanu, D. *Strobilanthes crispus* (B.) leaf extract-assisted green synthesis of $ZnO{-}La_2O_3$ composite and preliminary study of its photocatalytic activity. In *IOP Conference Series: Materials Science and Engineering*; IOP Publishing: Bristol, UK, 2020.
69. Klingenberg, B.; Vannice, M.A. Influence of pretreatment on lanthanum nitrate, carbonate, and oxide powders. *Chem. Mater.* **1996**, *8*, 2755–2768. [CrossRef]
70. Subhan, M.A.; Fahim, A.M.M.; Saha, P.C.; Rahman, M.M.; Begum, K.; Azad, A.K. Structural study, photoluminescence and photocatalytic properties of La_2O_3 Fe_3O_4 ZnO, AgO NiO ZnO and La_2O_3 AgO ZnO nanocomposites. *Nano-Struct. Nano-Objects* **2017**, *10*, 30–41. [CrossRef]
71. Saif, M.; Hafez, H.; Nabeel, A. Photo-induced self-cleaning and sterilizing activity of Sm^{3+} doped ZnO nanomaterials. *Chemosphere* **2013**, *90*, 840–847. [CrossRef] [PubMed]
72. Mu, Q.; Wang, Y. Synthesis, characterization, shape-preserved transformation, and optical properties of $La(OH)_3$, $La_2O_2CO_3$, and La_2O_3 nanorods. *J. Alloys Compd.* **2011**, *509*, 396–401. [CrossRef]
73. Mandlimath, T.R.; Gopal, B. Catalytic activity of first row transition metal oxides in the conversion of p-nitrophenol to p-aminophenol. *J. Mol. Catal. A Chem.* **2011**, *350*, 9–15. [CrossRef]
74. Zoromba, M.S.; Abdel-Aziz, M. Ecofriendly method to synthesize poly (o-aminophenol) based on solid state polymerization and fabrication of nanostructured semiconductor thin film. *Polymer* **2017**, *120*, 20–29. [CrossRef]
75. Guan, H.; Chao, C.; Lu, Y.; Shang, H.; Zhao, Y.; Yuan, S.; Zhang, B. PtNi nanoparticles embedded in porous silica microspheres as highly active catalysts for p-nitrophenol hydrogenation to p-aminophenol. *J. Chem. Sci.* **2016**, *128*, 1355–1365. [CrossRef]

coatings

Article

Electro-Oxidation of Ammonia over Copper Oxide Impregnated γ-Al_2O_3 Nanocatalysts

Safia Khan [1], Syed Sakhawat Shah [1], Mohsin Ali Raza Anjum [2,*], Mohammad Rizwan Khan [3] and Naveed Kausar Janjua [1,*]

1 Department of Chemistry, Quaid-i-Azam University, Islamabad 45320, Pakistan; safiakhan@chem.qau.edu.pk (S.K.); sakhawat_shah@yahoo.com (S.S.S.)
2 Chemistry Division, Directorate of Science, Pakistan Institute of Nuclear Science and Technology, Nilore, Islamabad 45650, Pakistan
3 Advanced Materials Research Chair, Chemistry Department, College of Science, King Saud University, Riyadh 11451, Saudi Arabia; mrkhan@ksu.edu.sa
* Correspondence: mohsinpieas@hotmail.com (M.A.R.A.); nkausarjanjua@yahoo.com (N.K.J.)

Abstract: Ammonia electro-oxidation (AEO) is a zero carbon-emitting sustainable means for the generation of hydrogen fuel, but its commercialization is deterred due to sluggish reaction kinetics and the poisoning of expensive metal electrocatalysts. With this perspective, CuO impregnated γ-Al_2O_3 (CuO/γ-Al_2O_3) hybrid materials were synthesized as effective and affordable electrocatalysts and investigated for AEO in alkaline media. Structural investigations were performed via different characterization techniques, i.e., X-ray diffraction (XRD), Fourier transformed infrared spectroscopy (FTIR), scanning electron microscopy (SEM), transmission electron microscopy (TEM), and electrochemical impedance spectroscopy (EIS). The morphology of γ-Al_2O_3 support as interconnected porous structures rendered the CuO/γ-Al_2O_3 nanocatalysts with robust activity. The additional CuO impregnation resulted in the enhanced electrochemical active surface area (ECSAs) and diffusion coefficient and spiked the electrocatalytic performance for NH_3 electrolysis. Owing to good values of diffusion coefficient for AEO, low bandgap, and availability of ample ECSA at higher CuO to γ-Al_2O_3 ratio, these proposed electrocatalysts were proved to be effective in AEO. Due to good reproducibility, electrochemical stability, and higher activity for ammonia electro-oxidation, CuO/γ-Al_2O_3 nanomaterials are proposed as efficient promoters, electrode materials, or catalysts in ammonia electrocatalysis.

Keywords: CuO/γ-Al_2O_3; ammonia electro-oxidation (AEO); electrochemical surface area (ECSA); electrocatalysts

Citation: Khan, S.; Shah, S.S.; Anjum, M.A.R.; Khan, M.R.; Janjua, N.K. Electro-Oxidation of Ammonia over Copper Oxide Impregnated γ-Al_2O_3 Nanocatalysts. *Coatings* **2021**, *11*, 313. https://doi.org/10.3390/coatings11030313

Received: 31 January 2021
Accepted: 5 March 2021
Published: 9 March 2021

1. Introduction

Ammonia (NH_3) is a corrosive, pungent and carcinogenic inorganic gaseous pollutant and is being produced from both biogenic and anthropogenic sources, i.e., livestock waste, animal agriculture, refrigeration, nitrogen fertilizer and petroleum refining industries [1,2]. Many efforts have been devoted to removing NH_3 from gaseous and waste streams through chemical, biological, and physical methods, but all have their limitations [2–4]. Recently, electrocatalytic oxidation of ammonia (AEO) has attracted much attention of researchers and scientists because, as a hydrogen-rich carrier, it possesses 70% higher volumetric hydrogen content than pure liquid hydrogen [5,6]. Theoretical and experimental investigations reveal that hydrogen generation via ammonia electro-oxidation (AEO) is a cost-effective approach compared to water electrolysis because it requires much lower oxidation potential (0.06 V) than water (−1.23 V), as described in Equations (1–3) [7,8]. Theoretical ammonia electrolysis consumes energy of about ~1.55 $Wh \cdot g^{-1}$ H_2 at standard conditions, which is 95% less than water electrolysis [9]. Due to fast reaction kinetics and modest operating conditions, ammonia electro-oxidation is considered a promising future technology to produce hydrogen from ammonia polluted wastewaters [10,11]. At 25 °C, the ammonia electrooxidation potential is −0.77 V vs. Standard Hydrogen Electrode (SHE), merely 0.06 V less negative than the oxidation potential of −0.83 V vs. SHE for hydrogen evolution in alkaline solution [12].

$$\textit{Anode reaction: } NH_3(aq) + 3OH^- \rightarrow 1/2N_2 + 3H_2O + 3e^- \rightarrow E_{anode} = -0.77 \text{ V vs. SHE} \quad (1)$$

$$\textit{Cathode reaction: } 3H_2O + 3e^- \rightarrow 3/2H_2 + 3OH^- \rightarrow E_{cathode} = -0.83 \text{ V vs. SHE} \quad (2)$$

$$\textit{Overall reaction: } NH_3(aq) \rightarrow 1/2N_2 + 3/2H_2 \rightarrow \Delta E = 0.06 \text{ V} \quad (3)$$

Electro-oxidation of ammonia prompts the inhibition of the oxygen evolution reaction (OER) because of adsorption of the products in result of ammonia oxidation on the electrode surface. The OER inhibition boosts the AEO reaction, which is the main reaction. The amino radicals as intermediates, formed in the course of ammonia oxidation, trigger a reaction chain where dissolved molecular oxygen is involved in the AEO [13]. Nitrogen present as ammonia or ionized ammonium or both forms deplete dissolved oxygen through oxidation in aqueous solutions, which boost the AEO reaction [14].

Moreover, innovation in the design and fabrication of reasonable and proficient electrocatalysts for AEO are essential for its successful implementation. To commercialize this direct ammonia fuel cell (DAFC) technology, an efficient, stable, and economical electrocatalyst is required. Thus far, expensive noble metal-based (Pt, Ru, Ir, Rh, Pd) alloys are considered as the best performing AEO catalysts [5–8,15,16]. Sluggish reaction kinetics of AEO, high cost and low resistance to poisoning by reaction intermediates of these catalysts hinder this technology in commercialization. Recently, some effort has been devoted to developing non-noble metal electrocatalysts including NiO-TiO_2 [17] and CuO-TiO_2 [1] for AEO, but they require higher overpotential. Similarly, Ni oxides and hydroxides-based catalysts have displayed better performance towards AEO; however, they get easily corroded and deactivated in ammonia solution [9,17,18]. Although it is proved theoretically that copper displays comparable activity to Pt, it forms too weak a bond with Ni than Pt, leading to a very high overpotential [19]. Therefore, it displays poor catalytic performance toward AOR experimentally [20]. Cupric oxide (CuO) is a promising p-type semiconductor and has been investigated in various applications such as photocatalytic hydrogen production [21], dye degradation [22], electro-oxidation of hydrazine [23–25], methanol oxidation [26], and so on. Recently, the oxidation behavior of CuO-modified TiO_2 was investigated for ammonia (0.5 M H_2SO_4 + 0.1 M NH_3) by using linear sweep voltammetry [1]. However, extensive research is required to explore the cost-effective and efficient electrocatalysts to replace the incumbent Pt-based alloys [27]. Porous materials behave as potential electrodes owing to their high conductivity [28].

Furthermore, catalyst support materials exhibit great influence on the cost, performance, and durability of polymer electrolyte membrane (PEM) fuel cells. Due to high chemical/electrochemical stability, surface area and versatility, the high surface area metal oxides including alumina (γ-Al_2O_3), silica (SiO_2), titania (TiO_2) and zirconia (ZrO_2) are considered as better catalyst supports than conventional carbon materials [29–32]. Besides, multiple types of transition alumina are recurrently used as supports for developing heterogeneous catalysts which consist of an active phase dispersed on a carrier or support. Alpha (α), beta (β) and gamma (γ) are three different types of alumina. The α-Al_2O_3, also known as nano-alumina, is a white puffy powder. It has lower specific surface area, limited high temperature resistance and it is inert, therefore it does not belong to activated alumina, and hence displays almost no catalytic activity. β-Al_2O_3 is hexagonal, with lamellar structure and the unit cell contains two alumina spinel-based block. It also exhibits low catalytic and strengthening properties. Among the different known alumina types, γ-Al_2O_3 is perhaps the most important with direct application as a catalyst support in the several industries. γ-Al_2O_3 possesses high purity and provides excellent dispersion and high specific surface area, offering commendable resistance to high temperature and high activity. It is porous; hence, it is said to be activated alumina and used as catalyst support as well as adsorbent [33]. The efficacy of γ-Al_2O_3 can be regarded to an auspicious combination of its textural possessions, e.g., surface area, pore volume, pore-size distribution and acid/base characteristics, which are mainly related to local microstructure, surface chemical composition and phase composition [34]. The highly porous γ-Al_2O_3 can be synthesized at moderate temperatures; therefore, it is widely used as support in many applications [22,26,35,36].

Herein, cost-effective and easily synthesized electrocatalysts (CuO/γ-Al_2O_3) were investigated for efficient ammonia electro-oxidation reaction in alkaline media. The CuO impregnated γ-Al_2O_3 nanomaterials were prepared by a simple impregnation-annealing process [26]. In such systems, impregnation of γ-Al_2O_3 with an aqueous active metal salt, preceded by drying and calcination, typically results in diffusion of active metal ions into the support surface, forming interaction species. Depending on the calcination temperature and time, only a finite amount of metal ions can be accommodated in the vacant lattice sites of the support. Once all of the available lattice sites are saturated, further addition of metal ions can be accommodated only by segregation of a separate metal oxide phase. The morphology of γ-Al_2O_3 support was modulated from plate-type to network-like by altering the CuO contents in compositions [37]. The electrocatalytic performance of as-synthesized CuO/γ-Al_2O_3 was observed by using cyclic voltammetry as the investigation tool. The prepared nanomaterials displayed high electrochemical active surface areas (ECSA), diffusion coefficients, and electrocatalytic activity for NH_3 electro-oxidation. Also, the conductive and stable catalytic performance towards ammonia electro-oxidation is indebted to their low bandgaps. Increasing the CuO contents in CuO/γ-Al_2O_3 nanomaterials enhanced the catalytic performance because of the suppressed formation of active reaction intermediates as observed via CV profiles. The electrochemical stability and higher performance towards the ammonia electro-oxidation rendered the CuO/γ-Al_2O_3 nanomaterials as efficient electrocatalysts.

2. Experimental

2.1. Preparation of CuO/γ-Al_2O_3 Electrocatalysts

All chemicals were purchased from Sigma Aldrich, St. Louis, MO, USA. and were used without further purification. Firstly, the catalyst's support alumina (γ-Al_2O_3) was prepared by calcination of pre-precipitated aluminum hydroxide ($Al(OH)_3$). Typically, 36.2 g of aluminum nitrate nonahydrate ($Al_2(NO_3)_3{\cdot}9H_2O$) was dissolved in distilled water (50 mL) and aqueous ammonia (35%) was added dropwise to obtain the white precipitates of aluminum hydroxide. Finally, as-prepared vacuum dried (80 °C) powder was calcined in NEY 2-525 furnace at 800 °C in the air for two hours to obtain γ-Al_2O_3. The copper oxide coated alumina (CuO/γ-Al_2O_3) nanomaterials were synthesized in two-steps; in the first step, as-synthesized γ-Al_2O_3 was impregnated in the required amount of aqueous solution of copper nitrate trihydrate ($Cu(NO_3)_2{\cdot}3H_2O$) for 48 h. and dried in an oven at 200 °C. The X-CuO/Al_2O_3 (X = 4, 8, 12, 16 and 20 wt.% of CuO) impregnated catalysts were obtained by calcination at 500 °C in the air for two hours as shown in Figure 1. The gradual variation in color of synthesized catalysts from bluish-white to bluish green was observed with incremental CuO content.

2.2. Electrochemical Investigations

Electrochemical analysis of all prepared electrocatalysts was carried out by using Gamry potentiostat interface 1000 and three-electrode system in which silver/silver chloride (Ag/AgCl), silver wire and modified glass carbon electrode are used as a reference, counter and the working electrode, respectively [38–40]. Ag/AgCl does not interfere with the system under investigation as it has been used as a reference electrode in ammonia and other basic media [41–43]. Silver wire is also used as a conductive counter electrode without affecting the system performance [44]. The glassy carbon electrode was polished with alumina slurry and cleaned with ethanol before dropping the catalyst ink that was prepared in ethanol while 5% Nafion (2.0 μL) solution was poured on the powder catalyst as a binder. The electrochemical active surface areas (ECSAs) of all electrocatalysts were determined in a standard redox solution (5.0 mM $K_4[Fe(CN)_6]$ + 3.0 M KCl) at a scan rate of 100 mVs^{-1}. Ammonia electro-oxidation was conducted by cyclic voltammetry (CV) in 1.0 M solution of NH_3 prepared in 0.1 M KOH solution.

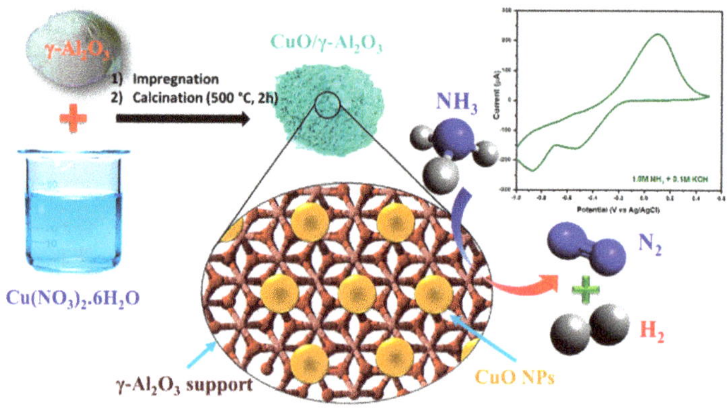

Figure 1. Graphical abstract for the synthesis of CuO/γ-Al_2O_3 nanomaterials for ammonia electro-oxidation.

3. Results and Discussion

3.1. Structural Characterization

The phase purity and crystal structure of all prepared samples was determined by powder X-ray diffraction (PXRD) with a scan rate 0.4° per minute in 2θ window of 10°–70° via continuous scan type using PANalytical X'PERT High Score's diffractometer, Malvern, UK. As shown in Figure 2a, the diffraction peaks at 32.4° and 36.7° are assigned to CuO which become more prominent as the loading of CuO is increased from 4% to 20%, while 37°, 61.5° and 68.1° corresponds to Al_2O_3. All these peaks have been indexed for pure cubic system and corresponded to JCPDS card No. 45-0937 [26]. There is no visible diffraction peak shift due to addition of Cu contents which indicates CuO species on the alumina surface exist as highly dispersed species [45]. Moreover, if 20%-CuO/Al_2O_3 is compared with pure alumina and 4%-CuO/Al_2O_3, an increase in peak intensity can be analyzed. Particle size increased from 2.5 in γ-Al_2O_3 to 16.2 nm in 20%-CuO/γ-Al_2O_3. However, this increment in particle size is not too much to affect the crystalline integrity of γ-Al_2O_3 matrix. The additional broadness of diffraction peaks at lower copper percentages is dedicated to more amorphous phase of composites [46].

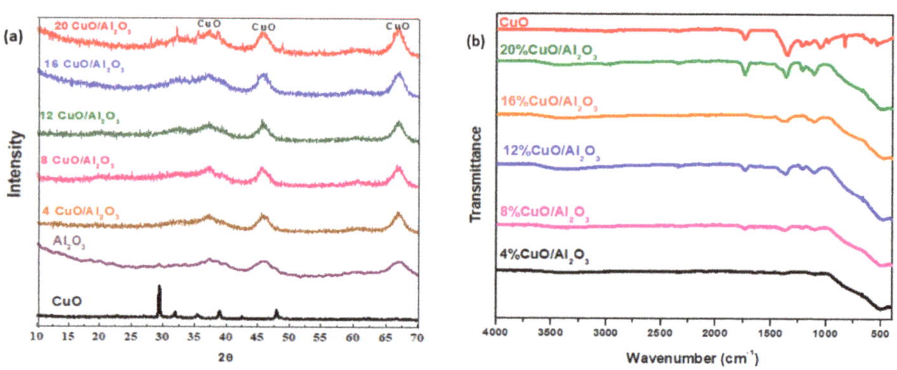

Figure 2. X-ray diffraction patterns (**a**) and Fourier-transform infrared (FTIR) spectra (**b**) of pure γ-Al_2O_3 and CuO/Al_2O_3 nanomaterials.

Debye-Scherrer formula was used to estimate the average crystallite size, Equation (4):

$$D_{av} = K \cdot \lambda / (\beta \cdot \cos \theta) \quad (4)$$

Here, D_{av} is the average crystallite size in nm, λ is the wavelength of X-ray (1.5418 nm) while using Cu as an anode, θ is Bragg's angle and β is the full width of the diffraction peak at half of the maximum. The calculated crystallite size for CuO is 14.6 nm which lies in the range between 10 to 30 nm as already reported [47–50]. The crystallite sizes of alumina increase from 2.5 nm (for pure γ-Al_2O_3) to 16.2 nm (20%-CuO/γ-Al_2O_3) as shown in Table 1. This increment in crystallite size was already observed in CuO/Al_2O_3 materials, which might be due to the insertion of CuO items into the crystal lattice of γ-Al_2O_3 [44]. In all samples, the CuO impregnation and calcination steps resulted in the increased crystallite size.

Table 1. Average crystallite size calculated from Scherrer formula.

Electrocatalyst	D_{av} (nm)
CuO	14.6
γ-Al_2O_3	2.5
4%-CuO/γ-Al_2O_3	7.1
8%-CuO/γ-Al_2O_3	9.1
12%-CuO/γ-Al_2O_3	11.5
16%-CuO/γ-Al_2O_3	15.1
20%-CuO/γ-Al_2O_3	16.2

Fourier-transform infrared (FTIR) spectra of all materials were obtained by using Nicolet 5PC, Nicolet Analytical Instrument Protea, Cambridgeshire, UK and compared with that of the bare CuO in Figure 2b. The absorption signals at ~500, ~1300 and ~1670 cm^{-1} indicate the stretching vibration of the Cu–O bond [51]. This analysis further confirms the successful impregnation of CuO into γ-Al_2O_3 support. Our results indicate that, at low copper loadings (below 10%), Cu^{2+} ions form a well-dispersed interaction species with the support. At high copper loadings (after saturation of the support, i.e., >16% Cu), segregation of bulk-like CuO occurred. The crystallite size of this CuO phase increases with metal loading [52].

3.2. Surface Characterization

The surface morphology, porosity and effect of CuO impregnation on the textural nature of Al_2O_3 in all synthesized materials were investigated by scanning electron microscopy (SEM) using TESCAN (Brno, Czech Republic) MAIA3, i.e., an ultra-high-resolution SEM. As displayed in Figure 3, γ-Al_2O_3 showed plate-like morphology with large pores (Figure 3a–c at different magnifications) and significant dendrites distributed over the surface similar to the earlier reports [53,54]. Interestingly, these plates were converted to a porous network-like structure when the contents of CuO were increased in materials as shown in Figure 3d–h). The CuO nanoparticles are heterogeneously distributed on the surface of Al_2O_3 nanostructures. The energy dispersive spectrum corresponds to the presence and elemental purity of Al, Cu and O, as displayed in Figure 3h.

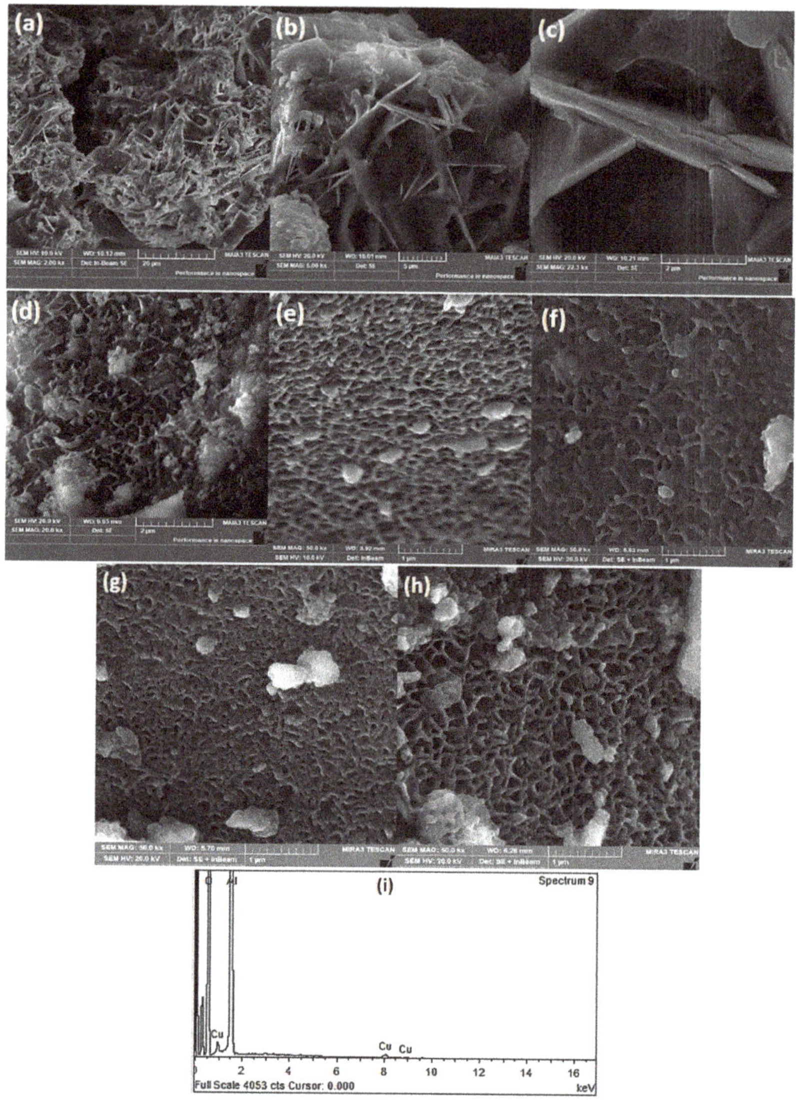

Figure 3. SEM micrographs for pure Al_2O_3 at 20 μm (**a**), 5 μm (**b**) and 2 μm (**c**) and for 4%–20%-CuO/γ-Al_2O_3 materials respectively at 1 μm (**d**–**h**) and (**i**) Energy dispersive spectrum for 4%-CuO/γ-Al_2O_3.

Low-magnification TEM nanostructures shown in Figure 4a–f are taken using TEM, JEOL 2100F (Tokyo, Japan), 200 kV to assess the surface picture of as-synthesized γ-Al_2O_3 and 16%-CuO/γ-Al_2O_3 matrices and if any agglomeration observed in the nanoparticles. TEM images for γ-Al_2O_3 shown in Figure 4 a and b displayed the globular structure of particles and the images corresponds with the already reported TEM nanostructures [55,56]. Figure 4c–f represents the TEM images for 16%-CuO/γ-Al_2O_3 and it is evident that incorporation of CuO into γ-Al_2O_3 matrix alters the surface structure. The modification in the γ-Al_2O_3 structure and a reconstructed composite surface is associated with dispersion of fine metal particles in catalysts [56]. Also, it can be perceived that the composite oxide is assembled from nearly uniform particles exhibiting similar shape, and significant porosity, and no big blocks are observed [57]. In such catalysts, CuO dispersed on the surface of the flake-like γ-Al_2O_3 generate synergy and coupling effects for ammonia adsorption [58].

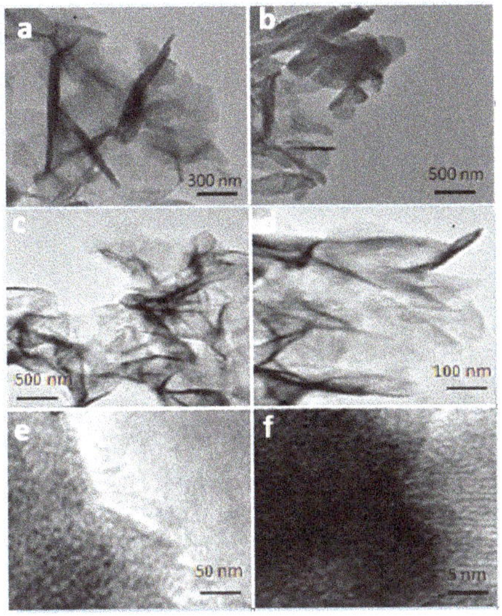

Figure 4. TEM images for pure Al_2O_3 at 300 nm (**a**), 500 nm (**b**), and for 16%-CuO/γ-Al_2O_3 materials at 500 nm (**c**), 100 nm (**d**), 50 nm (**e**) and 5 nm (**f**).

3.3. *Electrochemical Impedance Spectroscopy EIS*

The electron transfer properties of all as-prepared nano-electrodes were studied via EIS in 1 M NH_3 and 0.1 M KOH. The Nyquist plots witnessed for CuO/γ-Al_2O_3 modified glassy carbon electrodes are displayed in Figure 5, and associated EIS parameters are tabulated in Table 2. Systematically, the electron transfer resistance decreased with increase in copper content while again undergoing a decrease after an optimum composition. It endorsed that 16% CuO/γ-Al_2O_3 is optimal for electrochemical catalysis. A significantly lower value of R_{ct} referred to the superior conductivity and much efficient electrocatalytic activity of 16%-CuO/γ-Al_2O_3, comparative to other composites of the series. The differences in electrochemical behavior of the as-synthesized electrocatalysts depend upon the relative feasibility of electron transferal [26].

The nature of electrodes exhibits no influence on solution resistance (R_s) and Warburg resistance (R_w) because these are features of electrolyte and diffusion of electroactive specie that are common in all observations. However, the charge-transfer resistance (R_{ct}) and constant phase element (CPE) are influenced by modification of electrodes, as they are associated with conductive properties of the active material. α represents capacitance and surface roughness, respectively and its value varies from 0 to 1. Herein, currently modified electrode systems have α value ranging from 0.56 to 0.87, revealing that catalysts depicted enough surface roughness, which also correlates with the SEM and TEM observations. The electron-transfer rate constant k_{app} for as-proposed catalysts was calculated by Equation (5) [59].

$$k_{app} = RT/F^2 \cdot R_{ct} \cdot C \quad (5)$$

Here, F represents the Faraday's constant, C corresponds to concentration of analyte and R is universal constant in SI units. It is obvious from Table 2 that the value of k_{app} for 16%-CuO/γ-Al_2O_3 is greatest among the series of nanomaterials, referring to its highest capacity to assist the AEO reaction.

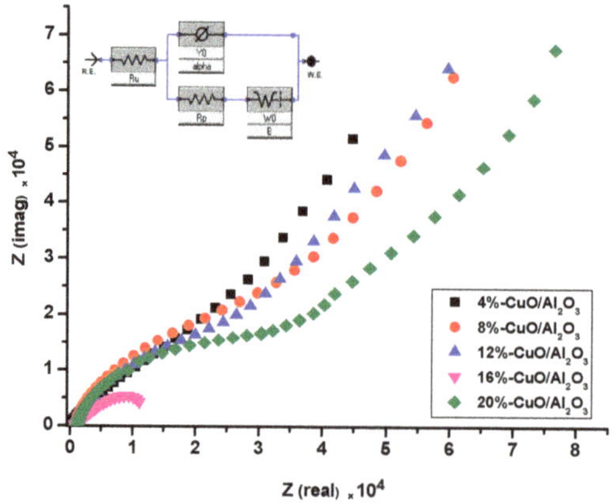

Figure 5. Electrochemical impedance spectroscopy (EIS) spectra for 4%–20%-$CuO/\gamma\text{-}Al_2O_3$ electrodes recorded in 1 M NH_3 + 0.1 M KOH aqueous solutions. Inset represents the model fit for calculating parameters.

Table 2. Parameters estimated from EIS analysis of electrocatalysts.

Electrocatalysts	R_s (Ω)	R_{ct} (kΩ)	CPE (μF)	α	Wo (Ω)	$k_{app}/10^{-9}$ (cms^{-1})
4%-$CuO/\gamma\text{-}Al_2O_3$	816.0	23.92	2.57	0.80	17.0	10.4
8%-$CuO/\gamma\text{-}Al_2O_3$	770.6	22.7	2.12	0.87	16.7	8.81
12%-$CuO/\gamma\text{-}Al_2O_3$	968.0	22.2	2.26	0.82	16.0	10.8
16%-$CuO/\gamma\text{-}Al_2O_3$	**736.0**	**3.92**	**55.4**	**0.56**	**13.9**	**61.2**
20%-$CuO/\gamma\text{-}Al_2O_3$	934.2	27.9	4.77	0.84	23.6	8.73

3.4. Active Surface Area of the $CuO/\gamma\text{-}Al_2O_3$ Modified Electrodes

The electrochemical active surface area (ECSA) of a catalyst is an important performance indicator of any electrochemical reaction; therefore, cyclic voltammograms of all prepared electrocatalysts were recorded in a standard redox solution (5.0 mM $K_4[Fe(CN)_6]$ + 3 M KCl) at 100 mVs^{-1} for ECSA estimation. Peak current (i_p) increment with CuO contents in the observed CV profile corresponded to a reversible one-electron transfer process using the synthesized nanomaterials as electron mediators in modified electrodes in $K_4[Fe(CN)_6]$ electrolyte (Figure 6a). This observation of a reversible CV profile points to the facile electro kinetics in the model redox couple, which correlates the electrocatalytic behavior of the used materials. The peak currents corresponding to $[Fe(CN)_6]^{4-}$ oxidation and peak current of $[Fe(CN)_6]^{3-}$ reduction increase with the increase in the concentration of active CuO thus overall a diffusion-controlled process [60]. The ECSA of electrodes was calculated by applying the Randles-Sevcik Equation (6) [26].

$$i_p = 2.69 \times 10^5 \cdot n^{3/2} \cdot A \cdot D^{1/2} \cdot \upsilon^{\frac{1}{2}} \cdot C \quad (6)$$

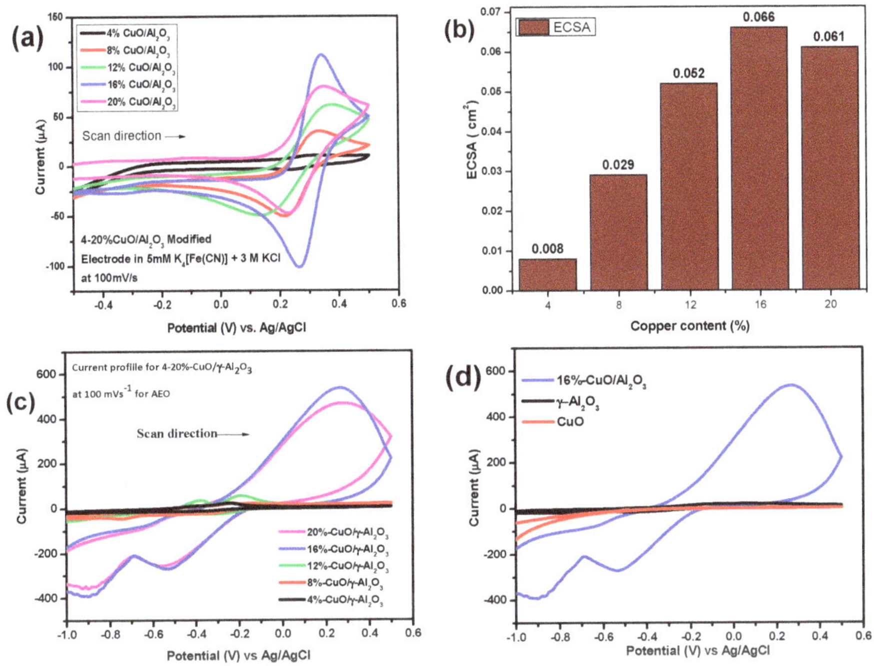

Figure 6. Cyclic voltammetric profiles on 4%–20%-CuO/γ-Al_2O_3 modified electrodes materials in (5.0 mM $K_4[Fe(CN)_6]$ + 3.0 M KCl) solution at scan ate of 100 $mV{\cdot}s^{-1}$ (**a**) and comparison of ECSA (**b**), Comparison of AEO in 0.1 M KOH + 1.0 M NH_3 at 100 $mV{\cdot}s^{-1}$ (**c**) and 16%-CuO/γ-Al_2O_3 with bare CuO and γ-Al_2O_3 (**d**).

Here, i_p is the peak current, n is the number of electrons transferred, A is the electrochemical active surface area (cm^2), D is the diffusion coefficient (0.76×10^{-5} $cm^2{\cdot}s^{-1}$ at 25 °C), [61] υ is the scan rate ($V{\cdot}s^{-1}$) and C is the concentration of the analyte. ECSA of catalysts are compared in Figure 6b that increases in the following order: 4%-CuO/γ-Al_2O_3 < 8%-CuO/γ-Al_2O_3 < 12%-CuO/γ-Al_2O_3 < 20%-CuO/γ-Al_2O_3 < 16%-CuO/γ-Al_2O_3. This little drop in ECSA of 20%-CuO/γ-Al_2O_3 may affect from the agglomeration and consequent phase-out of CuO nanoparticles at its higher loading.

3.5. Electrochemical Studies

Electrocatalytic responses of all CuO/γ-Al_2O_3 materials towards ammonia electro-oxidation (AEO) was investigated by cyclic voltammetry in 1.0 M NH_3 and 0.1 M KOH as displayed in Figure 6c,d. The peaks observed in both the forward (anodic) and reverse (cathodic) scans correspond to oxidative and reductive removal of chemisorbed species of ammonia, respectively [62]. The height of the anodic peak represents the electro-oxidation of ammonia on the surface of the electrodes. The anodic peak current increases with CuO contents in CuO/γ-Al_2O_3 nanomaterials; however, 16%-CuO/γ-Al_2O_3 shows the best performance towards AEO compared to other serial materials (Figure 6c), bare CuO and γ-Al_2O_3 (Figure 6d). The onset potential for AEO is found to be about −0.35 V vs. Ag/AgCl (i.e., −0.2 V vs. NHE) which is comparable to the Pt electrode [5,63–65].

To examine the electrode kinetics and the I–V responses towards AEO, voltammograms were also recorded for each catalyst at various scan rates from 10 to 100 mVs^{-1}. As shown in Figure 7a–e, a linear increase in peak current of AEO at ~0.2 V is observed with a scan rate that indicates a facilitated electron transfer process of ammonia electro-oxidation. Therefore, the proposed nanocatalysts behave like the adsorptive species on the electrode surface [9]. The electrochemical response of each material towards AEO was enhanced with varying contents of active component, i.e., CuO in CuO/γ-Al_2O_3. This further confirms that CuO species present at the surface of catalysts play important role in AEO, more active sites available at the surface will maximize the ammonia adsorption thus leading to AEO high performance. The AEO strongly depends on the adsorption/desorption of NH_3 species and the number of available active sites of active components at the surface of the electrode [66]. Similarly, two peaks in the cathodic curve are attributed to desorption of ammonia and proton at the surface of the catalysts, Figures 7 and 8 [67]. The first cathodic peak is attributed to the reduction of reaction intermediates. Besides, the anodic and cathodic peak currents are also increased with the number of available active sites due to CuO. An anodic shoulder peak at −0.4 V can be attributed to the structural sensitivity of catalysts towards oxidation of pre-adsorbed hydrogen or nitrogen-containing intermediates at the surface of the electrode [68]. However, the anodic shoulder peak is suppressed by the ammonia oxidation peak when the contents of CuO are increased from 4% to 20% while the cathodic shoulder peaks increase because of the conversion/reduction of reaction intermediates. In this way, it can be said that irreversibility character increases with an increase in copper content up to an optimum level. As observed in Randles-Sevcik plots (Figure 7f), the peak current (i_p) exhibits a linear relationship with the square root of scan rate ($\upsilon^{1/2}$), which is an indication of a diffusion-controlled process for AEO [67,68].

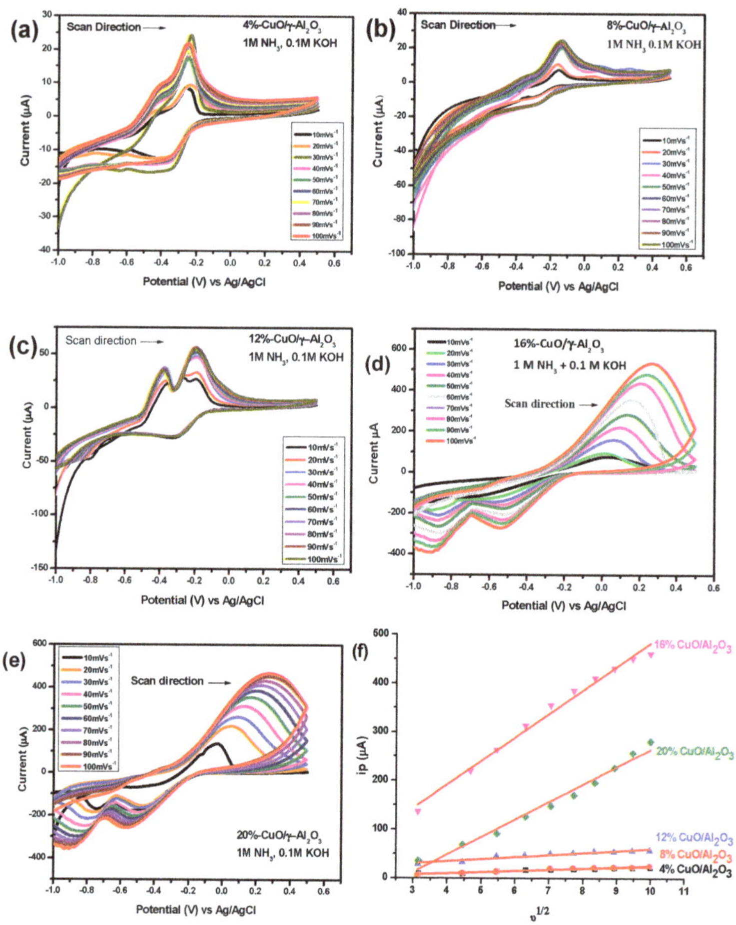

Figure 7. Cyclic voltammograms for 4%–20%-CuO/γ-Al_2O_3 recorded in (1.0 M NH_3 + 0.1 M KOH) solution; scan rate from 10–100 mV·s^{-1} (**a**–**e**) and Randles-Sevcik plots derived from cyclic voltammetry (CV) curves of each catalyst (**f**).

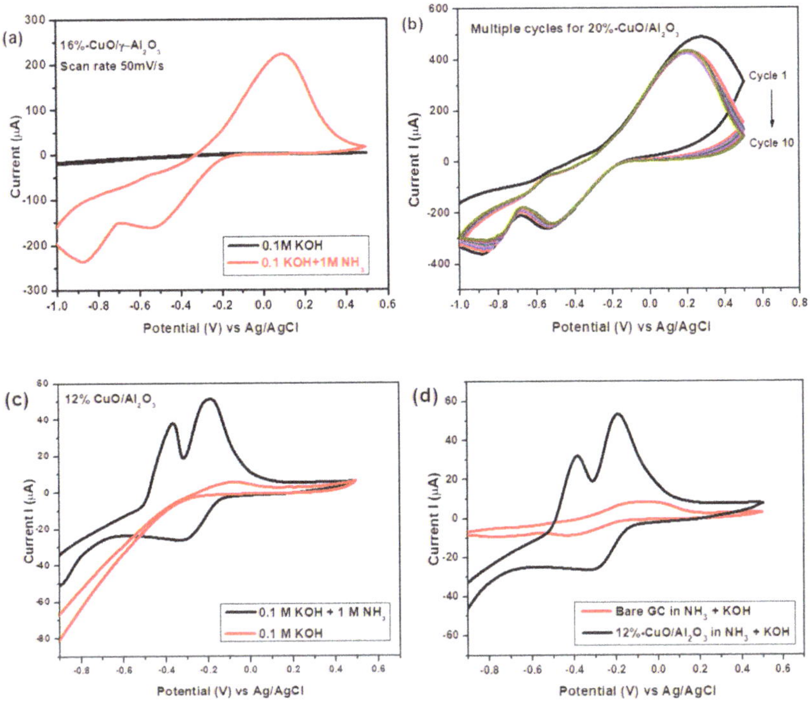

Figure 8. Comparison of CV curves for 16%-CuO/γ-Al_2O_3 recorded pure 0.1 M KOH and solution of 1.0 M NH_3 in 0.1 M KOH (**a**), Multiple cycles for 20%-CuO/γ-Al_2O_3 (**b**), Comparison of NH_3 free voltammetry for 12%-CuO/γ-Al_2O_3 (**c**) and Comparison of 12%-CuO/γ-Al_2O_3 with bare glassy carbon electrode (**d**).

ECSA can be affected by the change of electrode layer structure; thus, 16%-CuO/γ-Al_2O_3 shows a higher peak current output than 20%-CuO/γ-Al_2O_3. After a certain limit, a further increase in CuO percentage leads towards agglomeration. Also, more ammonia molecules lead to an increase in diffusion-layer thickness, which results in a lower catalytic response in 20%-CuO/γ-Al_2O_3 [69]. The 16%-CuO/γ-Al_2O_3 is inferred to be the optimal loading of CuO for AEO under the optimal conditions. Subsequently, compositions above 20%-CuO/γ-Al_2O_3 were not studied to avoid the agglomerated material, as depicted by ECSA. Also, the same composition has been proved as the promising optimal loading of CuO onto γ-Al_2O_3 support for glucose sensing and methanol electro-oxidation [26].

To further assess the electrochemical activity of catalysts towards AEO, the electrochemical response of 16%CuO/γ-Al_2O_3 electrode was profiled by recording the cyclic voltammograms in pure 0.1 M KOH and 1.0 M NH_3 solution. As displayed in Figure 8, no significant oxidation peak current is observed in 0.1 M KOH compared to 1.0 M NH_3 as an analyte. This endorses the theory that the major peak corresponds to ammonia electro-oxidation rather than the electrolyte at the electrode surface [70]. The optimal onset potential for AEO is −0.35 V and the maximum oxidation peak current is observed at ~0.1 V. The electrochemical stability is a vital element in the commercialization of an electrocatalyst; therefore, it was determined by repeating CV cycles in 1.0 M NH_3 + 0.1 M KOH at 100 mV·s^{-1}. No significant current loss was observed after 10 CV cycles in Figure 8b, which indicates that the used catalysts give the stable performance towards AEO and can be commercialized for industrial applications. Concerning low CuO percentages, it is observed from Figure 8c that it also oxidizes ammonia to an extent, although current response is lower than that for high copper contents. Figure 8d gives a validation that a bare glassy carbon electrode when modified with 12%-CuO/γ-Al_2O_3, gave an oxidation output. Hence, it can be inferred that even a small amount of CuO can electro-catalyze ammonia oxidation.

3.6. Diffusion Coefficients for AEO on $CuO/\gamma\text{-}Al_2O_3$ Modified Electrodes

Diffusion coefficient (D) is an important parameter to study mass transfer kinetics, and it depends on various factors including analyte size and the concentration of electrolyte. Thus, the diffusion coefficients for ammonia (D_{NH3}) in 0.1 M KOH were determined by applying the Randles-Sevcik formula (Equation (5)), which shows a linear relationship between the peak current and square root of the scan rate ($\upsilon^{1/2}$). The adsorption of ammonia occurs until the surface achieves saturation. Adsorbed nitrogen species cover and hence block the surface, whereas NH_2(ads) and NH(ads) seem to be the active species, which recombine to form N_2H_x (x = 2–4) species as reaction intermediates. These finally dehydrogenate to form N_2. The recombination of NH_x species is proposed to be the rate-determining step [71]. Generalized from overall ammonia electro-oxidation reaction (Equation (3)), three electrons (i.e., n = 3) are required for AEO. Moreover, quite similar ammonia oxidation voltammograms are reported earlier, describing it as a three-electron process [72]. The D_{NH3} using all the catalysts can be estimated from the slope of the Randles-Sevcik (i_p vs. $\upsilon^{1/2}$) plot at a constant scan rate (100 $mV{\cdot}s^{-1}$) at 25 °C [73]. Resultantly, 16%-$CuO/\gamma\text{-}Al_2O_3$ gives the highest value of the diffusion coefficient as compared to the other materials Ni/Pt [65] and Ni/Ni$(OH)_2$ [18], as enlisted in Table 3. Usually, the diffusion coefficient for AEO in KOH electrolyte (10^{-9}) is much lower than that of water (2.4×10^{-5}) [66] due to the presence of hydroxyl (OH^-) ions in the electrolyte [20].

Table 3. Diffusion coefficients values for AEO using $CuO/\gamma\text{-}Al_2O_3$ modified electrodes in 0.1 M KOH at 25 °C.

Electrocatalyst	D_{NH3} 10^{-9} (cm^2s^{-1})
4%-$CuO/\gamma\text{-}Al_2O_3$	1.0
8%-$CuO/\gamma\text{-}Al_2O_3$	2.6
12%-$CuO/\gamma\text{-}Al_2O_3$	2.7
16%-$CuO/\gamma\text{-}Al_2O_3$	**4.1**
20%-$CuO/\gamma\text{-}Al_2O_3$	3.8
Ni/Pt [55]	1.2
Ni/Ni$(OH)_2$ [16]	2.8
Ionic liquids [16]	0.1

3.7. Estimation of Bandgap Values of $CuO/\gamma\text{-}Al_2O_3$ for Ammonia Electro-Oxidation

The electrochemical bandgap (E_g) and frontier orbitals energy levels (E_{HOMO} and E_{LUMO}) are important factors to understand the electrical and electrochemical properties of any material [74]. Although it is difficult to quantify the exact values for bandgaps, it can be estimated from onset potentials, as is widely done in the literature [75–77]. As described in mathematical expressions (Equations (7) and (8)), the onset potentials of oxidation (anodic) curve, i.e., $(E_{onset})_{ox}$ and reduction (cathodic) curve, i.e., $(E_{onset})_{red}$ linearly correlate with energies of frontier orbitals, HOMO (E_{HOMO}) and LUMO (E_{LUMO}), respectively [78].

$$E_{HOMO} = -[(E_{onset})_{ox} + 4.4]\ eV \quad (7)$$

$$E_{LUMO} = -[(E_{onset})_{red} + 4.4]\ eV \quad (8)$$

Here, the onset potentials are calibrated regarding the saturated calomel electrode [79]. As presented in Table 4, the estimated bandgap narrows by increasing the copper contents in materials from 0.98 eV (4%-$CuO/\gamma\text{-}Al_2O_3$) to 0.22 eV (16%-$CuO/\gamma\text{-}Al_2O_3$), which results in maximizing the conducting properties. Also, the bandgap decreases with the increase of crystallite size due to the quantum confinement effect [80,81]. Furthermore, the conductive electrocatalysts let ample electrons speed up the electrochemical reactions [82,83]. Therefore, in general, the materials containing higher CuO contents give better AEO performance due to the rapid transfer of electrons from the conduction band. Accordingly, the most conductive response is provided by the catalyst of optimal composition, i.e., 16%-CuO/γ-Al_2O_3.

Table 4. Electrochemical bandgap values derived from onset potentials vs. sat'd calomel electrode for AEO on the CuO/γ-Al_2O_3 materials.

Catalyst	$E_{(ons)oxi}$ (V)	$E_{(ons)red}$ (V)	E_{HOMO} (eV)	E_{LUMO} (eV)	E_g (eV)
4%-CuO/γ-Al_2O_3	−0.64	−0.25	−3.16	−4.14	0.98
8%-CuO/γ-Al_2O_3	−0.43	−0.13	−3.56	−4.31	0.75
12%-CuO/γ-Al_2O_3	−0.53	−0.21	−3.98	−4.69	0.71
16%-CuO/γ-Al_2O_3	**−0.41**	**−0.18**	**−3.90**	**−4.20**	**0.22**
20%-CuO/γ-Al_2O_3	−0.32	−0.06	−3.89	−4.30	0.41

4. Conclusions

The copper oxide modified γ-Al_2O_3 electrocatalysts (CuO/γ-Al_2O_3) were synthesized by the facile co-impregnation and calcination process. The CuO nanocrystals were successfully grown on γ-Al_2O_3 supports, as depicted by X-ray diffraction and FTIR. The shape of γ-Al_2O_3 support was changed from irregular to network-like by increasing the CuO contents, as observed in SEM images. The network-like structures coincided with the electrochemical active surface areas (ECSAs), which also varied with the CuO contents. The surface coarseness and homogeneity were seen in TEM images. Electrochemical characterization involved cyclic voltammetry and EIS that conferred to the absolute catalytic behavior of as-proposed electrocatalysts. The maximum ECSA and minimum charge-transfer resistance have been displayed by 16%-CuO/γ-Al_2O_3, which makes it the optimal composition. The effect of CuO contents was investigated in the catalytic performance of ammonia electro-oxidation (AEO) in alkaline media. It was observed that the AEO is an irreversible and diffusion-controlled process under alkaline conditions on the surface of CuO/γ-Al_2O_3 electrodes. Additionally, the suppressing of catalysts by nitrogenous species can be significantly reduced by increasing the CuO contents, as displayed by the continuous disappearance of anodic shoulder peak from lower to higher CuO loading. The successive increment in diffusion coefficients for NH_3 with increasing CuO to γ-Al_2O_3 ratio showed that these materials can effectively electro-oxidize the NH_3 due to facile electron transfer. Correspondingly, the prepared electrocatalysts demonstrated good electrocatalytic activity, reproducibility, and stability towards the ammonia electro-oxidation. Recent investigation of CuO/γ-Al_2O_3 electrocatalysts has revealed a large synergistic effect towards AEO at ambient temperatures. Therefore, these nanomaterials could be used as efficient electrocatalysts or promoters in AEO due to the higher diffusion coefficient for NH_3, low bandgap, and chemical/electrochemical characteristics.

Author Contributions: S.K. and N.K.J. envisioned the study scheme and S.K. performed the experiments and the electrochemical measurements at Department of Chemistry, Quaid-I-Azam University Islamabad. N.K.J., M.A.R.A. and S.K. analyzed, discussed the results and wrote the manuscript. S.S.S. and N.K.J. are the supervisor and co-supervisor of S.K., respectively and did proof reading. SEM and EDS mapping were performed by M.A.R.A. while N.K.J. and M.A.R.A. are the corresponding authors of this paper. M.R.K. performed the TEM analysis. All authors have read and agreed to the published version of the manuscript.

Funding: The authors acknowledge the Fuel Cell Lab, Department of Chemistry, Quaid-i-Azam University, Islamabad that furnished this research. Also grateful to the Deanship of Scientific Research, King Saud University for this collaborative project through Vice Deanship of Scientific Research Chairs.

Institutional Review Board Statement: Not applicable.

Informed Consent Statement: Not applicable.

Data Availability Statement: Data is contained within article.

Conflicts of Interest: The authors declare no conflict of interest.

References

1. Dong, C.-D.; Chen, C.-W.; Kao, C.-M.; Hung, C.-M. Synthesis, characterization, and application of CuO-modified TiO_2 electrode exemplified for ammonia electro-oxidation. *Process Saf. Environ. Prot.* **2017**, *112*, 243–253. [CrossRef]
2. Yuan, M.-H.; Chen, Y.-H.; Tsai, J.-Y.; Chang, C.-Y. Ammonia removal from ammonia-rich wastewater by air stripping using a rotating packed bed. *Process Saf. Environ. Prot.* **2016**, *102*, 777–785. [CrossRef]
3. Sheng, Y.; Fang, L.; Zhang, L.; Wang, Y. Aimed at building a healthy living environment: An analysis of performance of Clean-Air Heat Pump system for ammonia removal. *Build. Environ.* **2020**, *171*, 106639. [CrossRef]
4. Ulu, F.; Kobya, M. Ammonia removal from wastewater by air stripping and recovery struvite and calcium sulphate precipitations from anesthetic gases manufacturing wastewater. *J. Water Process. Eng.* **2020**, *38*, 101641. [CrossRef]
5. Barbosa, J.R.; Leon, M.N.; Fernandes, C.M.; Antoniassi, R.M.; Alves, Q.C.; Ponzio, E.A.; Silva, J.C.M. $PtSnO_2$/C and Pt/C with preferential (100) orientation: High active electrocatalysts for ammonia electro-oxidation reaction. *Appl. Catal. B Environ.* **2020**, *264*, 118458. [CrossRef]
6. Katsounaros, I.; Figueiredo, M.C.; Calle-Vallejo, F.; Li, H.; Gewirth, A.A.; Markovic, N.M.; Koper, M.T. On the mechanism of the electrochemical conversion of ammonia to dinitrogen on Pt (1 0 0) in alkaline environment. *J. Catal.* **2018**, *359*, 82–91. [CrossRef]
7. Diaz, L.A.; Botte, G.G. Mathematical modeling of ammonia electrooxidation kinetics in a Polycrystalline Pt rotating disk electrode. *Electrochimica Acta* **2015**, *179*, 519–528. [CrossRef]
8. Bayati, M.; Liu, X.; Abellan, P.; Pocock, D.; Dixon, M.; Scott, K. Synergistic Coupling of a Molybdenum Carbide Nanosphere with Pt Nanoparticles for Enhanced Ammonia Electro-Oxidation Activity in Alkaline Media. *ACS Appl. Energy Mater.* **2020**, *3*, 843–851. [CrossRef]
9. Zhang, H.; Wang, Y.; Wu, Z.; Leung, D.Y. An ammonia electrolytic cell with NiCu/C as anode catalyst for hydrogen production. *Energy Procedia* **2017**, *142*, 1539–1544. [CrossRef]
10. Li, L.; Liu, Y. Ammonia removal in electrochemical oxidation: Mechanism and pseudo-kinetics. *J. Hazard. Mater.* **2009**, *161*, 1010–1016. [CrossRef]
11. Marinčić, L.; Leitz, F.B. Electro-oxidation of ammonia in waste water. *J. Appl. Electrochem.* **1978**, *8*, 333–345. [CrossRef]
12. Botte, G.G.; Benedetti, L.; Gonzalez, J. Electrolysis of ammonia: An in situ hydrogen production process. In Proceedings of the 2005 AIChE Annual Meeting, Cincinnati, OH, USA, 30 October–4 November 2005.
13. Michels, N.-L.; Kapałka, A.; Abd-El-Latif, A.; Baltruschat, H.; Comninellis, C. Enhanced ammonia oxidation on BDD induced by inhibition of oxygen evolution reaction. *Electrochem. Commun.* **2010**, *12*, 1199–1202. [CrossRef]
14. Mahvi, A.H.; Ebrahimi, S.J.A.D.; Mesdaghinia, A.; Gharibi, H.; Sowlat, M.H. Performance evaluation of a continuous bipolar electrocoagulation/electrooxidation–electroflotation (ECEO–EF) reactor designed for simultaneous removal of ammonia and phosphate from wastewater effluent. *J. Hazard. Mater.* **2011**, *192*, 1267–1274. [CrossRef] [PubMed]
15. Allagui, A.; Sarfraz, S.; Ntais, S.; Al Momani, F.; Baranova, E.A. Electrochemical behavior of ammonia on Ni98Pd2 nano-structured catalyst. *Int. J. Hydrog. Energy* **2014**, *39*, 41–48. [CrossRef]
16. Zhang, Q.; Zhang, T.; Xia, F.; Zhang, Y.; Wang, H.; Ning, P. Promoting effects of acid enhancing on N_2 selectivity for selectivity catalytic oxidation of NH3 over RuOx/TiO2: The mechanism study. *Appl. Surf. Sci.* **2020**, *500*, 144044. [CrossRef]
17. Almomani, F.; Bhosale, R.; Khraisheh, M.; Kumar, A.; Tawalbeh, M. Electrochemical oxidation of ammonia on nickel oxide nanoparticles. *Int. J. Hydrog. Energy* **2020**, *45*, 10398–10408. [CrossRef]
18. Kapałka, A.; Cally, A.; Neodo, S.; Comninellis, C.; Wächter, M.; Udert, K.M. Electrochemical behavior of ammonia at Ni/$Ni(OH)_2$ electrode. *Electrochem. Commun.* **2010**, *12*, 18–21. [CrossRef]
19. Xu, W.; Lan, R.; Du, D.; Humphreys, J.; Walker, M.; Wu, Z.; Wang, H.; Tao, S. Directly growing hierarchical nickel-copper hydroxide nanowires on carbon fibre cloth for efficient electrooxidation of ammonia. *Appl. Catal. B Environ.* **2017**, *218*, 470–479. [CrossRef]
20. HHerron, J.A.; Ferrin, P.; Mavrikakis, M. Electrocatalytic oxidation of ammonia on transition-metal surfaces: A first-principles study. *J. Phys. Chem. C* **2015**, *119*, 14692–14701. [CrossRef]
21. Kum, J.M.; Yoo, S.H.; Ali, G.; Cho, S.O. Photocatalytic hydrogen production over CuO and TiO_2 nanoparticles mixture. *Int. J. Hydrog. Energy* **2013**, *38*, 13541–13546. [CrossRef]
22. Hua, L.; Ma, H.; Zhang, L. Degradation process analysis of the azo dyes by catalytic wet air oxidation with catalyst CuO/γ-Al_2O_3. *Chemosphere* **2013**, *90*, 143–149. [CrossRef]
23. Hosseini, S.R.; Kamali-Rousta, M. Preparation of electro-spun CuO nanoparticle and its application for hydrazine hydrate electro-oxidation. *Electrochim. Acta* **2016**, *189*, 45–53. [CrossRef]
24. Firdous, N.; Janjua, N.K. $CoPt_x$/γ-Al_2O_3 bimetallic nanoalloys as promising catalysts for hydrazine electrooxidation. *Heliyon* **2019**, *5*, e01380. [CrossRef]
25. Ma, Y.; Li, H.; Wang, R.; Wang, H.; Lv, W.; Ji, S. Ultrathin willow-like CuO nanoflakes as an efficient catalyst for electro-oxidation of hydrazine. *J. Power Sources* **2015**, *289*, 22–25. [CrossRef]
26. Mujtaba, A.; Janjua, N.K. Fabrication and electrocatalytic application of CuO@ Al_2O_3 hybrids. *J. Electrochem. Soc.* **2015**, *162*, H328–H337. [CrossRef]

27. Firdous, N.; Janjua, N.K.; Qazi, I.; Wattoo, M.H.S. Optimal Co–Ir bimetallic catalysts supported on γ-Al_2O_3 for hydrogen generation from hydrous hydrazine. *Int. J. Hydrog. Energy* **2016**, *41*, 984–995. [CrossRef]
28. Zhang, S.; Feng, D.; Shi, L.; Wang, L.; Jin, Y.; Tian, L.; Li, Z.; Wang, G.; Zhao, L.; Yan, Y. A review of phase change heat transfer in shape-stabilized phase change materials (ss-PCMs) based on porous supports for thermal energy storage. *Renew. Sustain. Energy Rev.* **2021**, *135*, 110127. [CrossRef]
29. López, M.; Palacio, R.; Mamede, A.-S.; Fernández, J.J.; Royer, S. Hydrodeoxygenation of guaiacol into cyclohexane over mesoporous silica supported Ni–ZrO_2 catalyst. *Microporous Mesoporous Mater.* **2020**, *309*, 110452. [CrossRef]
30. Ramesh, A.; Tamizhdurai, P.; Krishnan, P.S.; Ponnusamy, V.K.; Sakthinathan, S.; Shanthi, K. Catalytic transformation of non-edible oils to biofuels through hydrodeoxygenation using Mo-Ni/mesoporous alumina-silica catalysts. *Fuel* **2020**, *262*, 116494. [CrossRef]
31. Thou, C.Z.; Khan, F.S.A.; Mubarak, N.M.; Ahmad, A.; Khalid, M.; Jagadish, P.; Walvekar, R.; Abdullah, E.C.; Khan, S.; Khan, M.; et al. Surface charge on chitosan/cellulose nanowhiskers composite via functionalized and untreated carbon nanotube. *Arab. J. Chem.* **2021**, *14*, 103022. [CrossRef]
32. Kashif, M.; Ngaini, Z.; Harry, A.V.; Vekariya, R.L.; Ahmad, A.; Zuo, Z.; Sahari, S.K.; Hussain, S.; Khan, Z.A.; Alarifi, A. An experimental and DFT study on novel dyes incorporated with natural dyes on titanium dioxide (TiO_2) towards solar cell application. *Appl. Phys. A* **2020**, *126*, 1–13. [CrossRef]
33. Paranjpe, K.Y. Alpha, Beta and Gamma Alumina as Catalyst. *Pharma. Innov. J.* **2017**, *6*, 236–238.
34. Trueba, M.; Trasatti, S.P. γ-Alumina as a support for catalysts: A review of fundamental aspects. *Eur. J. Inorg. Chem.* **2005**, 3393–3403. [CrossRef]
35. Badri, M.T.; Barati, M.; Rasa, S.H. A kinetic study of acetone acidic oxidation with $KMnO_4$ in the absence and presence of $CuO/$-Al_2O_3 as a heterogeneous nano-catalyst. *Sci. Iran.* **2020**, *27*, 1234–1242.
36. Bradu, C.; Frunza, L.; Mihalche, N.; Avramescu, S.-M.; Neaţă, M.; Udrea, I. Removal of Reactive Black 5 azo dye from aqueous solutions by catalytic oxidation using CuO/Al_2O_3 and NiO/Al_2O_3. *Appl. Catal. B Environ.* **2010**, *96*, 548–556. [CrossRef]
37. Firdous, N.; Janjua, N.K.; Wattoo, M.H.S. Promoting effect of ruthenium, platinum and palladium on alumina supported cobalt catalysts for ultimate generation of hydrogen from hydrazine. *Int. J. Hydrog. Energy* **2020**, *45*, 21573–21587. [CrossRef]
38. Minakshi; Pundir, C.S. Construction of an amperometric enzymic sensor for triglyceride determination. *Sens. Actuators B Chem.* **2008**, *133*, 251–255. [CrossRef]
39. Hooda, V.; Gahlaut, A. Amperometric cholesterol determination using HRP incorporated carbon paste electrode. *Biosci. Biotechnol. Res. Asia* **2020**, *17*, 53–64. [CrossRef]
40. Odeyemi, C.S.; Awodugba, A.O. Improving the efficiency of titanium dioxide based dye sensitized solar cell (DSSC) using silver surface counter electrode. *Futa J. Eng. Eng. Technol.* **2018**, *12*, 1–6.
41. Xu, W.; Du, D.; Lan, R.; Humphreys, J.; Miller, D.N.; Walker, M.; Wu, Z.; Irvine, J.T.; Tao, S. Electrodeposited NiCu bimetal on carbon paper as stable non-noble anode for efficient electrooxidation of ammonia. *Appl. Catal. B Environ.* **2018**, *237*, 1101–1109. [CrossRef]
42. Strehlitz, B.; Gründig, B.; Kopinke, H. Sensor for amperometric determination of ammonia and ammonia-forming enzyme reactions. *Anal. Chim. Acta* **2000**, *403*, 11–23. [CrossRef]
43. Kang, Y.; Wang, W.; Li, J.; Li, Q.; Liu, S.; Lei, Z. A highly efficient Pt-NiO/C electrocatalyst for ammonia electro-oxidation. *J. Electrochem. Soc.* **2017**, *164*, F958. [CrossRef]
44. Schechner, P.; Kroll, E.; Bubis, E.; Chervinsky, S.; Zussman, E. Silver-plated electrospun fibrous anode for glucose alkaline fuel cells. *J. Electrochem. Soc.* **2007**, *154*, B942–B948. [CrossRef]
45. Luo, M.F.; Fang, P.; He, M.; Xie, Y.L. In situ XRD, Raman, and TPR studies of CuO/Al_2O_3 catalysts for CO oxidation. *J. Mol. Catal. A Chem.* **2005**, *239*, 243–248. [CrossRef]
46. Bagtache, R.; Saib, F.; Abdmeziem, K.; Trari, M. A new hetero-junction p-CuO/Al_2O_3 for the H_2 evolution under visible light. *Int. J. Hydrog. Energy* **2019**, *44*, 22419–22424. [CrossRef]
47. Azam, A. Size-dependent antimicrobial properties of CuO nanoparticles against Gram-positive and-negative bacterial strains. *Int. J. Nanomed.* **2012**, *7*, 3527–3535. [CrossRef]
48. Li, Y.; Liang, J.; Tao, Z.; Chen, J. CuO particles and plates: Synthesis and gas-sensor application. *Mater. Res. Bull.* **2008**, *43*, 2380–2385. [CrossRef]
49. Ren, G.; Hu, D.; Cheng, E.W.; Vargas-Reus, M.A.; Reip, P.; Allaker, R.P. Characterisation of copper oxide nanoparticles for antimicrobial applications. *Int. J. Antimicrob. Agents* **2009**, *33*, 587–590. [CrossRef] [PubMed]
50. Baqer, A.A.; Matori, K.A.; Al-Hada, N.M.; Shaari, A.H.; Kamari, H.M.; Saion, E.; Chyi, J.L.Y.; Abdullah, C.A.C. Synthesis and characterization of binary $(CuO)0.6(CeO_2)0.4$ nanoparticles via a simple heat treatment method. *Results Phys.* **2018**, *9*, 471–478. [CrossRef]
51. Gholami, T.; Salavati-Niasari, M. Effects of copper: Aluminum ratio in CuO/Al_2O_3 nanocomposite: Electrochemical hydrogen storage capacity, band gap and morphology. *Int. J. Hydrog. Energy* **2016**, *41*, 15141–15148. [CrossRef]
52. Strohmeier, B.R.; Levden, D.E.; Field, R.S.; Hercules, D.M. Surface spectroscopic characterization of $CuAl_2O_3$ catalysts. *J. Catal.* **1985**, *94*, 514–530. [CrossRef]
53. Keshavarz, A.R.; Rezaei, M.; Yaripour, F. Nanocrystalline gamma-alumina: A highly active catalyst for dimethyl ether synthesis. *Powder Technol.* **2010**, *199*, 176–179. [CrossRef]

54. Liu, Y.; Ma, D.; Han, X.; Bao, X.; Frandsen, W.; Wang, D.; Su, D. Hydrothermal synthesis of microscale boehmite and gamma nanoleaves alumina. *Mater. Lett.* **2008**, *62*, 1297–1301. [CrossRef]
55. Schmücker, M.; Albers, W.; Schneider, H. Mullite formation by reaction sintering of quartz and α-Al_2O_3—A TEM study. *J. Eur. Ceram. Soc.* **1994**, *14*, 511–515. [CrossRef]
56. Rozita, Y.; Brydson, R.; Scott, A.J. An investigation of commercial gamma-Al_2O_3 nanoparticles. *J. Phys. Conf. Ser.* **2010**, *241*, 012096. [CrossRef]
57. Nie, M.-X.; Li, X.-Z.; Liu, S.-R.; Guo, Y. ZnO/CuO/Al_2O_3 composites for chloroform detection. *Sens. Actuators B Chem.* **2015**, *210*, 211–217. [CrossRef]
58. Duan, N.; Liu, J.; Li, Q.; Xiao, H. Enhanced adsorption performance of CuO-Al_2O_3 composite derived from cotton template. *Can. J. Chem. Eng.* **2015**, *93*, 2015–2023. [CrossRef]
59. Maximiano, E.M.; de Lima, F.; Cardoso, C.A.L.; Arruda, G.J. Incorporation of thermally activated zeolite into carbon paste electrodes for voltammetric detection of carbendazim traces in milk samples. *J. Appl. Electrochem.* **2016**, *46*, 713–723. [CrossRef]
60. Muhammad, S.; Banin Zahra, U.; Ahmad, A.; Shah, L.A.; Muhammad, A. Understanding the basics of electron transfer and cyclic voltammetry of potassium ferricyanide-an outer sphere heterogeneous electrode reaction. *J. Chem. Soc. Pak.* **2020**, *42*, 813.
61. Cossar, E.; Houache, M.S.; Zhang, Z.; Baranova, E.A. Comparison of electrochemical active surface area methods for various nickel nanostructures. *J. Electroanal. Chem.* **2020**, *870*, 114246. [CrossRef]
62. Zhan, T.; Sun, X.; Wang, X.; Sun, W.; Hou, W. Application of ionic liquid modified carbon ceramic electrode for the sensitive voltammetric detection of rutin. *Talanta* **2010**, *82*, 1853–1857. [CrossRef]
63. Zhang, X.-B.; Han, S.; Yan, J.-M.; Shioyama, H.; Kuriyama, N.; Kobayashi, T.; Xu, Q. Electrochemical oxidation of ammonia borane on gold electrode. *Int. J. Hydrog. Energy* **2009**, *34*, 174–179. [CrossRef]
64. Le Vot, S.; Roué, L.; Bélanger, D. Synthesis of Pt–Ir catalysts by coelectrodeposition: Application to ammonia electrooxidation in alkaline media. *J. Power Sources* **2013**, *223*, 221–231. [CrossRef]
65. Li, Z.-F.; Wang, Y.; Botte, G.G. Revisiting the electrochemical oxidation of ammonia on carbon-supported metal nanoparticle catalysts. *Electrochim. Acta* **2017**, *228*, 351–360. [CrossRef]
66. Matinise, N.; Mayedwa, N.; Ikpo, C.O.; Hlongwa, N.W.; Ndipingwi, M.M.; Molefe, L.; Dywili, N.; Yonkeu, A.L.D.; Waryo, T.; Baker, P.G.L.; et al. Bimetallic nanocomposites of palladium (100) and ruthenium for electrooxidation of ammonia. *J. Nano Res.* **2016**, *44*, 100–113. [CrossRef]
67. Álvarez-Ruiz, B.; Gómez, R.; Orts, J.M.; Feliu, J.M. Role of the metal and surface structure in the electro-oxidation of hydrazine in acidic media. *J. Electrochem. Soc.* **2002**, *149*, D35–D45. [CrossRef]
68. Zhou, L.; Cheng, Y. Catalytic electrolysis of ammonia on platinum in alkaline solution for hydrogen generation. *Int. J. Hydrog. Energy* **2008**, *33*, 5897–5904. [CrossRef]
69. Zhu, P.; Zhao, Y. Effects of electrochemical reaction and surface morphology on electroactive surface area of porous copper manufactured by Lost Carbonate Sintering. *RSC Adv.* **2017**, *7*, 26392–26400. [CrossRef]
70. da Silva, S.W.; Navarro, E.M.O.; Rodrigues, M.A.S.; Bernardes, A.M.; Pérez-Herranz, V. The role of the anode material and water matrix in the electrochemical oxidation of norfloxacin. *Chemosphere* **2018**, *210*, 615–623. [CrossRef] [PubMed]
71. Moran, E.; Cattaneo, C.; Mishima, H.; De Mishima, B.A.L.; Silvetti, S.P.; Rodriguez, J.L.; Pastor, E. Ammonia oxidation on electrodeposited Pt–Ir alloys. *J. Solid State Electrochem.* **2008**, *12*, 583–589. [CrossRef]
72. Yao, K.; Cheng, Y. Electrodeposited Ni–Pt binary alloys as electrocatalysts for oxidation of ammonia. *J. Power Sources* **2007**, *173*, 96–101. [CrossRef]
73. Khan, M.; Janjua, N.K.; Khan, S.; Qazi, I.; Ali, S.; Saad Algarni, T. Electro-oxidation of ammonia at novel Ag_2O-PrO_2/γ-Al_2O_3 Catalysts. *Coatings* **2021**, *11*, 257. [CrossRef]
74. Diaz, L.A.; Valenzuela-Muñiz, A.; Muthuvel, M.; Botte, G.G. Analysis of ammonia electro-oxidation kinetics using a rotating disk electrode. *Electrochim. Acta* **2013**, *89*, 413–421. [CrossRef]
75. Cheng, Y.-J.; Yang, S.-H.; Hsu, C.-S. Synthesis of conjugated polymers for organic solar cell applications. *Chem. Rev.* **2009**, *109*, 5868–5923. [CrossRef] [PubMed]
76. Admassie, S.; Inganäs, O.; Mammo, W.; Perzon, E.; Andersson, M.R. Electrochemical and optical studies of the band gaps of alternating polyfluorene copolymers. *Synth. Met.* **2006**, *156*, 614–623. [CrossRef]
77. Loganathan, K.; Pickup, P.G. Poly(Δ4,4′-dicyclopenta[2,1-b:3,4-b′] dithiophene–co-3,4-ethylenedioxythiophene): Electrochemically generated low band gap conducting copolymers. *Electrochim. Acta* **2005**, *51*, 41–46. [CrossRef]
78. Jessop, I.; Zamora, P.P.; Díaz, F.R.; del Valle, M.A.; Leiva, A.; Cattin, L.; Makha, M.; Bernède, J.C. New polymers based on 2, 6-di (thiophen-2-yl) aniline and 2, 2′-(thiophen-2, 5-diyl) dianiline monomers. preparation, characterization and thermal, optical, electronic and photovoltaic properties. *Int. J. Electrochem. Sci.* **2012**, *7*, 9502–9517.
79. Lohrman, J.; Zhang, C.; Zhang, W.; Ren, S. Semiconducting single-wall carbon nanotube and covalent organic polyhedron-C60 nanohybrids for light harvesting. *Chem. Commun.* **2012**, *48*, 8377–8379. [CrossRef]
80. Rehman, S.; Mumtaz, A.; Hasanain, S.K. Size effects on the magnetic and optical properties of CuO nanoparticles. *J. Nanoparticle Res.* **2011**, *13*, 2497–2507. [CrossRef]
81. Phoka, S.; Laokul, P.; Swatsitang, E.; Promarak, V.; Seraphin, S.; Maensiri, S. Synthesis, structural and optical properties of CeO_2 nanoparticles synthesized by a simple polyvinyl pyrrolidone (PVP) solution route. *Mater. Chem. Phys.* **2009**, *115*, 423–428. [CrossRef]

82. Nasir, M.H.; Janjua, N.K.; Santoki, J. Electrochemical performance of carbon modified $LiNiPO_4$ as Li-ion battery cathode: A combined experimental and theoretical study. *J. Electrochem. Soc.* **2020**, *167*, 130526. [CrossRef]
83. Butt, T.M.; Janjua, N.K.; Mujtaba, A.; Zaman, S.A.; Ansir, R.; Rafique, A.; Sumreen, P.; Mukhtar, M.; Pervaiz, M.; Yaqub, A.; et al. B-site doping in lanthanum cerate nanomaterials for water electrocatalysis. *J. Electrochem. Soc.* **2020**, *167*, 026503. [CrossRef]

coatings

MDPI

Article

Plant Extract Induced Biogenic Preparation of Silver Nanoparticles and Their Potential as Catalyst for Degradation of Toxic Dyes

Khalida Naseem [1,*], Muhammad Zia Ur Rehman [2], Awais Ahmad [3,*], Deepak Dubal [4,*] and Tahani Saad AlGarni [5]

1 Department of Chemistry, Faculty of Sciences, University of Central Punjab, Lahore 54000, Pakistan
2 Department of Chemical Engineering, University of Engineering and Technology, Lahore 54890, Pakistan; timsheet123@gmail.com
3 Department of Chemistry, The University of Lahore, Lahore 54590, Pakistan
4 School of Chemistry, Physics and Mechanical Engineering, Queensland University of Technology, O Block, Level 4, Room O-406, Gardens Point Campus, Brisbane, QLD 4001, Australia
5 Chemistry Department, College of Science, King Saud University, Riyadh 11451, Saudi Arabia; Tahanis@ksu.edu.sa
* Correspondence: khalida.naseem@ucp.edu.pk (K.N.); awaisahmed@gcuf.edu.pk (A.A.); deepak.dubal@qut.edu.au (D.D.)

Received: 30 October 2020; Accepted: 14 December 2020; Published: 16 December 2020

Abstract: This study focusses on the synthesis of silver nanoparticles (Ag-nPs) by citrus fruit (*Citrus paradisi*) peel extract as reductant while using $AgNO_3$ salt as source of silver ions. Successful preparation of biogenic CAg-nPs catalyst was confirmed by turning the colorless reaction mixture to light brown. The appearance of surface Plasmon resonance (SPR) band in UV-Vis spectra further assured the successful fabrication of nPs. Different techniques such as FTIR, TGA and DLS were adopted to characterize the CAg-nPs. CAg-nPs particles were found to excellent catalysts for reduction of Congo red (CR), methylene blue (MB), malachite green (MG), Rhodamine B (RhB) and 4-nitrophenol (4-NP). Reduction of CR was also performed by varying the contents of $NaBH_4$, CR and catalyst to optimize the catalyst activity. The pseudo first order kinetic model was used to explore the value of rate constants for reduction reactions. Results also interpret that the catalytic reduction of dyes followed the *Langmuir–Hinshelwood* (LH) mechanism. According to the LH mechanism, the CAg-nPs role in catalysis was explained by way of electrons transfer from donor ($NaBH_4$) to acceptor (dyes). Due to reusability and green synthesis of the CAg-nPs catalyst, it can be a promising candidate for the treatment of water sources contaminated with toxic dyes.

Keywords: nanoparticles; dyes; catalysis; reduction

1. Introduction

Recently, different environmental pollutants such as toxic dyes have been identified as threats because these dyes are harmful for humans and aquatic life [1,2]. Discharge of these dyes containing effluents in the environment causes the natural ecosystem to become unbalanced [3]. Therefore, the effective removal of toxic dyes from wastewater or their conversation in usable sources before its discharge in the water system is a primary global issue [4]. Different methods like biodegradation, electrochemical, physicochemical and photochemical treatment including the advanced oxidation process via photo-catalysis or chemical reduction, adsorption and ultra-filtration have been adopted to address this problem [5,6]. Most of these methods are not useful for the treatment of dyes' polluted wastewater. Methods such as the physicochemical method are inefficient for degradation of dyes due to their high stability. The biodegradation method, however, is cost effective, yet it is a very slow method, and dyes are also harmful for microorganisms. Adsorption transfers the toxic dyes from one medium to another rather than eliminating them. However, catalytic degradation of dyes has emerged as the best method for their treatment [1]. Harmful substances can be distrusted into less toxic or non-toxic substances via photo-catalysis [7,8] or chemical catalysis [9]. Various metal nanoparticles like Au, Pt, Pd, Fe and Ag-nPs are reported as catalysts for degradation of various toxic dyes [10–13]. One of the most commonly used metal nPs for direct reduction reactions of pollutants is iron nPs. These nPs are less toxic and cost effective. However, Ag-nPs are more stable and highly active catalysts for reduction reactions.

Among these nPs, silver nanoparticles (Ag-nPs) have been considered as interesting candidates over the last decade due to exceptional properties such as high catalytic activity with controlled surface area and low cost [14–18].

Several synthetic methods such as chemical, photochemical and electrochemical methods have been adopted to prepare Ag-nPs by treatment of silver salt [17,19]. However, these methods involve the use of toxic chemicals along with drastic reaction conditions and induce severe environmental pollutions. Due to negative impacts of these reported methods, a new method for preparation of Ag-nPs by plant extracts has been introduced [20]. Plant extract mediated synthesis of nPs tunes their size, shape and size distribution. The biological method for preparation of nPs exhibits several merits over other reported physiochemical methods such as use of non-toxic solvent (water), no utilization of toxic or harmful chemicals, mild reaction conditions and cost effectiveness. Mostly, plant extracts pose different organic compounds with different functional groups such as amino acid and carboxylic acid that act as stabilizing and reducing agents. Plant extracts also act as reducing and stabilizing agents due to the presence of polyphenols [21]. Size and shape of nPs can also be tuned by changing the contents of plant extract utilized during their preparation. Thus, alternatively, catalytic properties of nPs can also be tuned according to the requirement as catalysis is the surface phenomenon [22]. Due to these reasons, many researchers turned toward the utilization of biological units such as plant extract for preparation and stabilization of metal nPs [21,23]. Gardea-Torresdey at al. successfully fabricated Ag-nPs by growing live alfalafa plants in $AuCl_4$ rich media and concluded that bio items can be used efficiently for preparation of inorganic nanoparticles [24]. Kasthuri et al. prepared Au-nPs by using the chloroauric acid as source of gold ions in the presence of phyllanthin extract at room temperature [22].

Citrus is one of the major fruit crops that is widely consumed as fresh fruit or juice by removing the peel. Peel of fruits is discarded as waste material. Citrus fruit peel has polyphenolic compounds, flavonoids, ascorbic acid etc. These phytochemical components have bioactive properties like antiproliferation, antibacterial, antifungal, antioxidant and antiviral activities [25]. Due to these functional groups, citrus peel extract also acts as a capping agent. Thus, *Citrus paradisi* peel extract induced nPs remain stable for a prolonged amount of time due to the formation of interaction between these functional groups and nPs [21]. Values of apparent rate constant and reaction completion time for reduction of dyes and aromatic compounds in the presence of different metal nanoparticles stabilized by a variety of plant extract for comparative purpose are given in Table 1. Activity of reported catalysts was high for reduction of different dyes as compared to previously used biogenic metal nanoparticles stabilized by plant extracts.

Many research papers used the *Citrus paradisi* peel extract for preparation of different metal nPs like Fe_3O_4 [26,27], ZnO [28] and Ag-nPs [21,25,28–30]. It acts as a reducing as well as a capping agent during preparation of Ag-nPs. Fabricated CAg-nPs find applications in synthesis of durable cotton and silk fabric due to their antibacterial and antimicrobial activity and essential oil effect due to the use of citrus plant extracts [31]. These nanoparticles also find use as potential bio-pesticides to control different types of pathogens in aqueous medium [25,32].

Table 1. Comparative analysis kinetic parameters for catalytic reduction of different dyes using different plant extract induced metal nanoparticles.

Dye	Catalyst	Plant Extract	k_{app} (min^{-1})	Reaction Completion Time (min)	References
CR	Au-nPs	Salmalia malabarica gum	0.236	10	[10]
	Ag-nPs	Gum tragacanth	0.148	15	[33]
	Ag-nPs	Thunbergia grandiflora	0.099	18	[34]
	Ag-nPs	Citrus paradise	0.591	5	This work
MB	Au-nPs	Salmalia malabarica gum	0.241	9	[10]
	Ag-nPs	Gum tragacanth	0.182	12	[33]
	Ag-nPs	*Gmelina arborea*	-	10	[35]
	Ag-nPs	Citrus paradise	0.613	4	This work
4-NP	Ag-nPs	Coleus forskohlii root extract	0.101	24	[36]
	Au-nPs	Prunus domestica (plum) fruit extract	0.114	9	[37]
	Ag-nPs	Dolichos lablab	-	40	[38]
	Ag-nPs	Citrus paradise	0.247	9	This work

However, no one used these metal nanoparticles as catalysts for degradation of toxic dyes. Thus, we reported here for the first time the preparation of citrus peel induced biogenic CAg-nPs particles and their use as catalysts for degradation of different toxic dyes like CR, MB, MG, RhB and 4-NP. Recyclability and reusability of catalysts were also performed for reduction of CR. Prepared biogenic CAg-nPs particles were analyzed by UV-Vis, FTIR, DLS and TGA.

2. Materials and Methods

2.1. Materials

Congo red (CR) (98%), methylene blue (MB) (98%), malachite green (MG) (98%), Rhodamine B (RhB) (98%), 4-nitrophenol (4-NP) (98%), sodium borohydride ($NaBH_4$) (98%) and silver nitrate ($AgNO_3$) (98%) were purchased from Scharlau (Barcelona, Spain) and used as such without further treatment. Deionized water was used throughout the experimentation. Filtration of plant extract was done by Whatmann No. 1 filter paper (Merck Darmstadt, Germany).

2.2. Synthesis of CAg-nPs

Citrus paradisi peels were washed and completely dried in shade. The dried peels were grinded into fine powder using mortar. Afterward, 0.5 g of fine peel powder was stirred with 80 mL deionized water for 3 h in a flask at 60 °C on hotplate and filtered for further use. After that, a 30 mL peel extract was treated with 30 mL of 1.0 mM $AgNO_3$ solution at 70 °C for 50 min on a hotplate in a round bottom flask under constant stirring and nitrogen supply. Reaction mixture turned light brown on treatment with plant extract. Afterward, a light brown emulsion type mixture was filtered and saved in a sample bottle covered with aluminum foil for analysis. It was also used as a catalyst for catalytic reduction of toxic dyes. Prepared CAg-nPs were also dried in powdered form for pursuing other analyses. Diagrammatic representation of preparation of biogenic CAg-nPs and their use as catalysts for reduction of CR are shown in Figure 1.

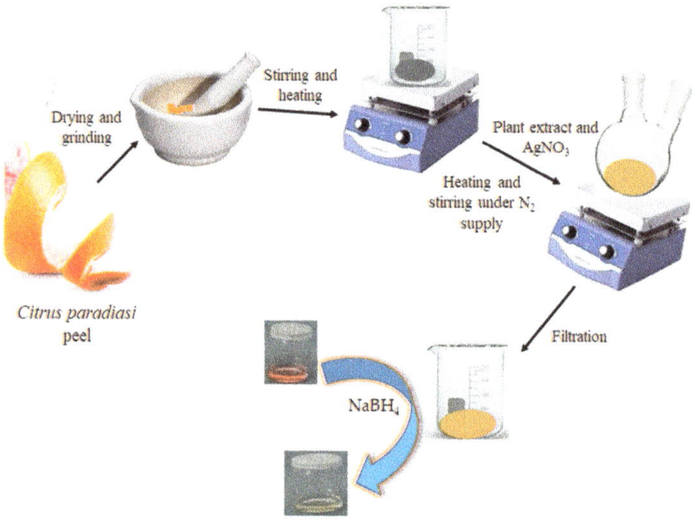

Figure 1. Scheme for preparation of Ag-nPs by the *Citrus paradisi* peel extract and its use as catalyst for reduction of CR.

2.3. Characterization of CAg-nPs

Functional groups of plant extract and biosynthesized CAg-nPs particles were evaluated by scanning FTIR spectra. For this aim, FTIR spectra of powdered samples were scanned on RXI FTIR spectrometer (Perkin ELMER, Waltham, MA, USA). UV-Vis analysis of plant extract, aqueous solution of $AgNO_3$ and biogenic CAg-nPs particles was also performed on a UV VIS Spectrophotometer (Stalwart, Germany) in the wavelength range from 250 to 750 nm at 25 °C. DLS analysis was done to check the particle size distribution of CAg-nPs particles. For this purpose, analysis was performed on B1-200SM (Brookhaven Instrument Corp, Holtsville, NY, USA at 90° (angle of scattering)) while the He–Ne laser was used as a light source with a wavelength of 637 nm. TGA analysis of plant extract and biogenic CAg-nPs particles was also done on Thermal analyzer (Model: SDT, Q-600, TA Shanghai, China) in the presence of N_2.

2.4. Catalytic Reduction of Toxic Dyes

In addition, 0.060 mM CR and 7.88 mM $NaBH_4$ were taken in a cuvette along with 1.8 mg/mL of CAg-nPs catalyst and spectra were recorded in the range of 380 to 680 nm with a one minute interval on a UV-Vis spectrophotometer (Stalwart, Dehli India) to check the reaction progress. CR reduction was also performed by varying the contents of reaction mixture such as $NaBH_4$ (4.50 to 12.38 mM), CR (0.057 to 0.078 mM) and catalyst dose (1.2 to 2.7 mg/mL). Various other toxic dyes like MB, MG, RhB and 4-NP were also reduced successfully in the presence of biogenic CAg-nPs catalysts and $NaBH_4$ reductants.

3. Results and Discussion

3.1. Analysis of Biosynthesized CAg-nPs

FTIR analysis of plant extract and CAg-nPs particles was done to show the interaction between different constituents of plant extract and silver nanoparticles as shown in Figure 2. Peaks of different functional groups appeared at almost the same position in the case of both plant extract and CAg-nPs particles. Peak appeared in the range of 2900 to 3400 cm^{-1} was due to stretching vibrations of O-H and NH_2 groups present in plant extracts in the form of alcohols, amides, amines, esters, ethers and carboxylic acids [25]. Broadened peaks that appeared in the case of CAg-nPs particles as compared to plant extracts, relevant to stretching vibrations of these groups, are an indication of the involvement of these groups in the stabilization of fabricated silver nPs. A slight change in the peak positions of these groups in the case of CAg-nPs particles also illustrates their function as ligation agents. Bands appearing at 694.23 cm^{-1} in both plant extracts and CAg-nPs particles are an indication of the presence of aromatic hydrocarbons [39]. A peak seen at 1517.71 cm^{-1} is a characteristic signal of aromatic C=C bonds [25]. Peaks appearing at 1199.27 cm^{-1} are an indication of C–O stretching vibrations [17,40]. A peak appearing at 1638.85 cm^{-1} unveiled the vibrations of carbonyl group (C=O). Kasthuri et al. prepared biosynthesized gold and silver nanoparticle by using phyllanthin extract at room temperature and concluded that there is a formation of some sort of interaction between moieties of plant extract and metal nanoparticles by FTIR analysis [22].

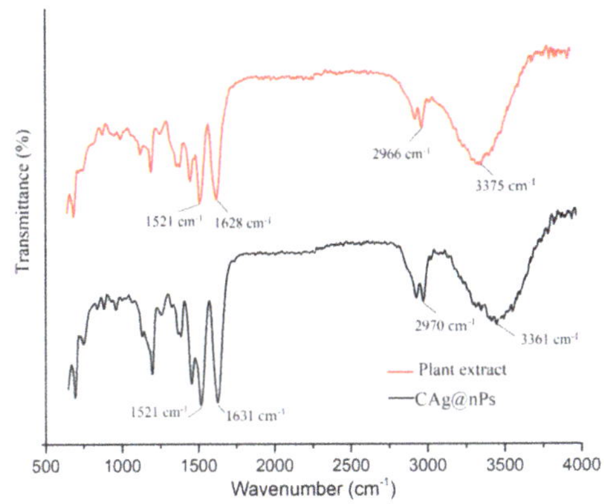

Figure 2. FTIR spectra of plant extract and biosynthesized CAg-nPs.

UV-Vis spectra of CAg-nPs particles were also obtained at different time intervals during their preparation as shown in Figure 3. At the beginning, a mixture of plant extract and silver salt showed no peak. However, a small peak with low absorbance intensity appeared 16 min after the start of reaction. This peak became sharp, distinct and less broad with progress of reaction. An absorbance intensity of the peak also increased along with its shifting toward a high wavelength (red shifting). Actually, an increase in reaction time induces nucleation of metal nanoparticles to a large amount. Free electrons present on large sized Ag-nPs oscillate with electromagnetic radiations of low energy or high wavelength. As a result, the SPR band was red shifted. The change of appearance of suspension from light yellow to light brown also illustrates the successful fabrication of CAg-nPs particles.

UV-Vis spectra of plant extracts, aqueous $AgNO_3$ salt and CAg-nPs particles were also recorded for comparison purposes as shown in Figure 4. No peak was seen in the case of plant extracts of aqueous solution of $AgNO_3$. This shows the transparent nature of moieties of plant extract and silver nitrate salt to UV-Vis radiations. However, a sharp peak appeared at 405 nm in the case of CAg-nPs particles, which illustrates their successful fabrication. A single, sharp and prominent peak unveils the spherical shaped and narrow size distributed nPs [39]. This peak appears in the visible region due to the surface plasmon resonance (SPR) phenomenon of Ag-nPs. Actually, electrons present on the surface of nanoparticles exhibit oscillation. Oscillating electrons resonate with electromagnetic radiations of specific frequency and result in the appearance of SPR band in the visible region [41].

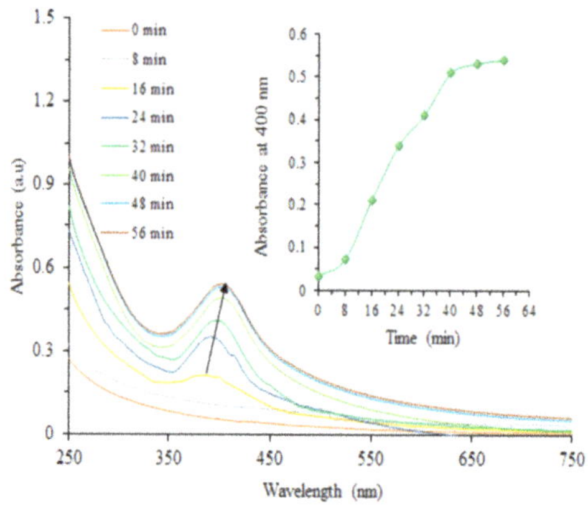

Figure 3. Time based UV-Vis spectra during preparation of biogenic CAg-nPs.

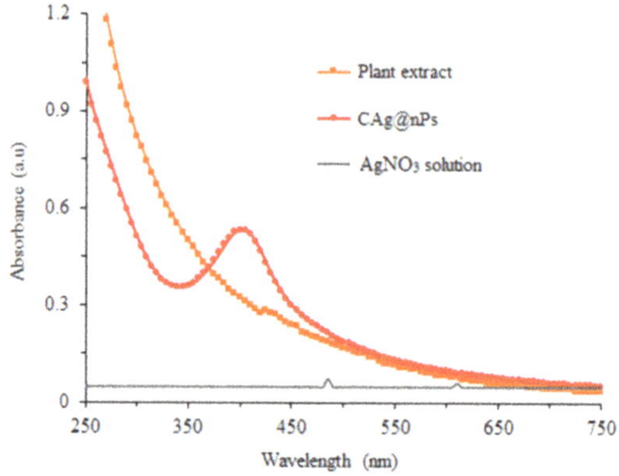

Figure 4. UV-Vis spectra of plant extract, $AgNO_3$ solution and CAg-nPs at room temperature.

Time based stability of biogenic CAg-nPs particles was also checked. For this purpose, spectra of freshly prepared and 15-day-old CAg-nPs particles were obtained (Figure 5a). The SPR band appeared at the same position (405 nm) with a slight decrease in the absorbance value in both cases. These results show the stability of biosynthesized CAg-nPs particles. Actually, biomolecules present in plant extracts play a role as capping/stabilizing agents and enhance the life span of silver nanoparticles [1]. Particle size distribution of CAg-nPs was evaluated by DLS analysis and spectra are shown in Figure 5b.

The average diameter of CAg-nPs particles was found as 28 nm. TGA analysis of plant extract and biosynthesized CAg-nPs was done by changing temperature from 0 to 480 °C as shown in Figure 6. TGA curves for CAg-nPs and plant extract can be divided into three stages. The first stage in TGA curve of plant extract illustrates the weight loss with temperature increase up to 100 °C because of evaporation of water contents present in plant extracts [42]. In the second stage, due to an increase of temperature up to 250 °C, there was almost no change in weight loss. During this stage, most of the heat is absorbed by the biomacromolecules to decrease the strength of intermolecular forces. In the third stage, there was a sharp decrease in weight loss due to the decomposition of plant extract contents. However, in the case of biogenic CAg-nPs catalyst, there was a steady and sharp weight loss as temperature was increased from 0 to 350 °C with respect to plant extract thermal behavior. This sharp weight loss of biogenic CAg-nPs particles may be associated with assembling Ag-nPs that promote the decomposition of plant extract contents [22]. However, weight loss was not approached to zero due to the presence of silver nanoparticles in CAg-nPs particles. Thus, less weight loss was seen in the case of CAg-nPs particles as compared to pure plant extract at high temperature. Percentage content of Ag-nPs was also calculated by TGA analysis. It was seen that 15% silver nanoparticles were present in biosynthesized CAg-nPs as calculated from the difference of weight loss curves of CAg-nPs and of plant extract at 450 °C. Mata et al. also performed TGA analysis to study the thermal behavior of *Plumeria alba* extract treated gold nanoparticles and observed the same temperature induced weight loss trend [42]. Ayinde and coworkers investigated the size of *Citrus paradisi* peel extract induced Ag-nPs by SEM analysis [25]. The size of spherical shaped and bio-synthesized Ag-nPs was found as 14.84 nm. Kalia et al. also illuminated the shape of peel extract induced prepared Ag-nPs vis SEM analysis and observed the rod shape metal nanoparticles [43].

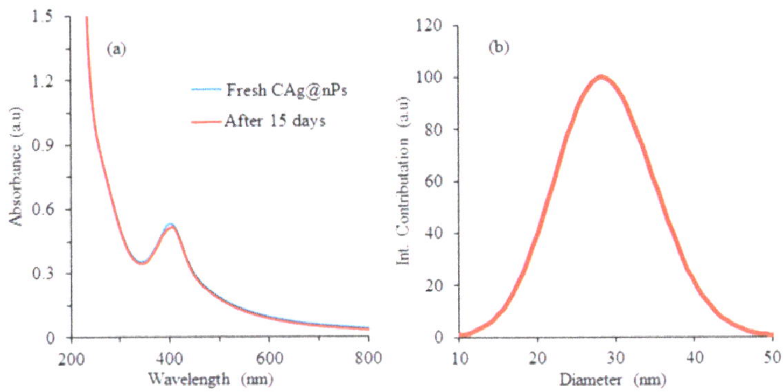

Figure 5. (**a**) Time dependent stability of biosynthesized CAg-nPs and (**b**) particle size distribution of biosynthesized CAg-nPs particles.

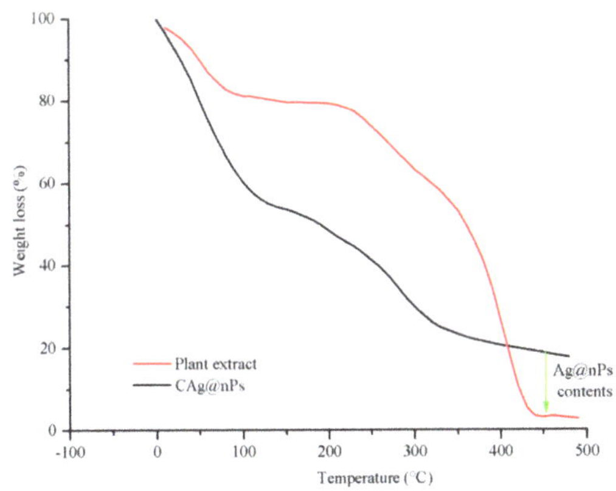

Figure 6. TGA analysis of plant extract and CAg-nPs nanoparticles.

3.2. Catalytic Reduction of Toxic Dyes

Various toxic dyes are released into water sources due to massive industrialization [44]. These dyes are stable and highly dangerous for the environment. About 10–20% of dyes are lost in wastewater streams and make their removal a major concern [45]. The best way to reduce toxic effects of these dyes is to convert them into less toxic products by using $NaBH_4$ as reductant. Thus, these products are alternatively reduced to further non-toxic products [46]. However, the reduction of dyes by BH_4^{-1} ions is a thermodynamically favorable process. However, the conversion of dyes by BH_4^{-1} ions in the absence of catalyst is kinetically unfavorable and such reactions proceed at a slow speed. This may be due to the energy barrier present between BH_4^{-1} ions and substrate dyes [42]. Thus, non-catalytic reduction of dyes requires more time. However, the presence of metal nanoparticles (mNPs) in reaction mixture speed up the rate of reduction of dyes which result in increased reaction efficiency. mNPs exhibit a large surface area for reactant adsorption and result in high catalytic activity for dye reduction. Actually, mNPs provide a new path for the reactants with a low level energy barrier and convert them into products easily. Thus, mNPs catalysts act as conveyor belts for electron transfer from reductant to substrate and facilitate the reduction of toxic dyes.

CR is azo dye with carcinogenic and mutagenic nature [1]. It is red in color and shows maximum absorbance at 495 nm in UV-Vis spectra. In addition, 0.060 mM CR and 7.88 mM $NaBH_4$ were added in quartz cells along with 1.8 mg/mL CAg-nPs catalyst and spectra were recorded (380 to 680 nm) with a time interval of one min at room temperature (Figure 7). The addition of CAg-nPs in mixture of CR and $NaBH_4$ leads to a decrease in absorbance intensity of the peak. Thus, the decrease in absorption intensity shows the decrease in CR concentration with the passage of time according to Beer–Lambert law [45]. Actually, the catalyst acts as a carrier to transfer hydride and electrons from borohydride to dye.

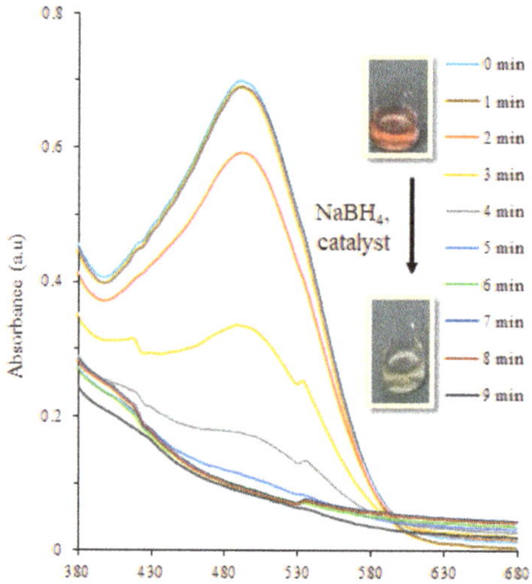

Figure 7. CR reduction in the presence of biosynthesized CAg-nPs catalyst ([CR] = 0.060 mM, [$NaBH_4$] = 7.88 mM, catalyst = 1.8 mg/mL) at ambient temperature.

The catalytic bleaching of red color of CR to white color was obtained in 9 min. The presence of CAg-nPs particles did not interfere in the monitoring of CR reduction by the spectrophotometer because the surface plasmon resonance (SPR) band of CAg-nPs particles appeared at 405 nm (Figure 4). The same reaction was also carried out as control reaction with $NaBH_4$ (absence of catalyst) (Figure 8a), with plant extract (in the absence of $NaBH_4$) (Figure 8b) and with catalyst (absence of $NaBH_4$) (Figure 8) to confirm whether the fading of the red color of CR was due to its degradation induced by biosynthesized CAg-nPs catalyst or adsorption by plant extract or catalyst.

Absorbance intensity of peak was slightly decreased in the presence of $NaBH_4$ and absence of CAg-nPs catalyst (Figure 8a). It explains the thermodynamic feasibility of reaction but the presence of large kinetic barrier between the reactants. No change in absorbance intensity of reaction mixture was observed in the presence of plant extract or Ag-nPs. These results explain that a decrease in absorbance intensity of CR or fading of its color was due to inclusion of a biogenic CAg-nPs catalyst in a reaction mixture that speeds up the rate of CR reduction in the presence of $NaBH_4$. Results also explain that CR was reduced sharply in less time in the presence of catalyst rather than being adsorbed by the plant extract or Ag-nPs. Actually, reduction of CR was completed in a short time rather than being adsorbed by the surface. Thus, this result illustrates that the decrease in absorbance intensity of dye was due to catalytic reduction rather than adsorption phenomenon.

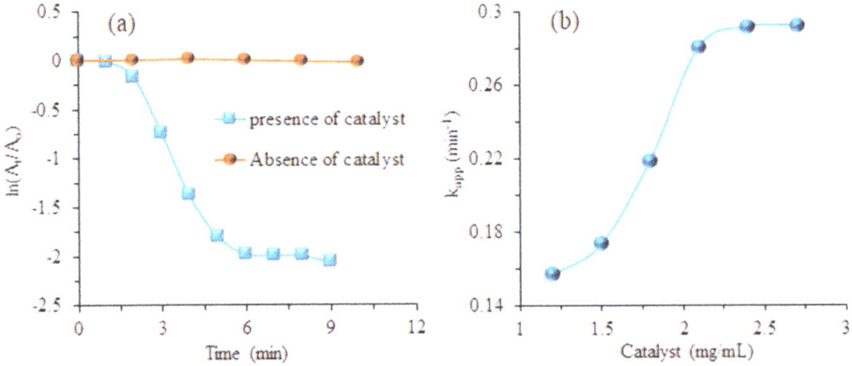

Figure 8. (**a**) Plot of $\ln(A_t/A_o)$ vs. time and (**b**) plot of catalyst dose vs. k_{app} for reduction of CR [conditions: CR = 0.045 mM, $NaBH_4$ = 7.88 mM] at room temperature.

3.3. Kinetic Study

Monitoring of reaction was continued until the change in absorbance of reaction mixture became constant. Due to excessive use of $NaBH_4$ as compared to CR ($NaBH_4$/CR ≥ 100), the kinetic aspect of reduction reactions was explained by pseudo first order [$\ln(A_t/A_o)$ against time] [46]. CR was not reduced immediately after the addition of catalyst in the reaction mixture in the presence of $NaBH_4$ reductant. This delay in reduction of CR was due to the presence of oxygen in reaction mixtures that prevents the immediate reduction of CR. Time between adding of catalyst in reaction mixture and start of reaction is called delay time. After delay time, CR was rapidly reduced to the product in the presence of CAg-nPs catalyst and reducing agent. As the Beer–Lambert law states, the A_t (absorbance at any time) and A_o (absorbance at zero time) are directly proportional to C_t (concentration at any time) and C_o (concentration at zero time), respectively. Thus, the change in absorbance at λ_{max} of CR actually explains its concentration at that specific time. Pseudo first order plots for CR degradation with and without CAg-nPs biogenic catalyst are shown in Figure 8a. Values of apparent rate constant (k_{app}) for CR reduction were calculated from linear regions of plots shown in Figure 8a. k_{app} for CR degradation was found as 0.468 and 0.003 min^{-1} with and without biogenic catalyst, respectively.

Percentage conversion of CR in product was also calculated for catalytic and un-catalyzed reaction [44]. Value of percentage conversion of CR with and without CAg-nPs biogenic catalyst was found to be 87.28% and 2.62%, respectively.

Varadayenkatesan et al. prepared Ag-nPs by using the flower extract of *Thunbergia grandiflora* and employ these particles as a catalyst for reduction of CR [34]. The value of apparent rate constant (k_{app}) for reduction of CR was found as 0.0999 min^{-1}. Indana and co-workers utilized gum tragacanth fabricated Ag-nPs as a catalyst for reduction of CR using $NaBH_4$ as reductant [33]. The value of k_{app} was found as 0.148 min^{-1} and reaction was completed in 15 min. Thus, results showed that our reported biogenic Ag-nPs catalyst showed high activity with less reaction time and high value of k_{app} as compared to previously reported work.

3.4. *Effect of Reaction Conditions*

Reduction of CR was also performed by changing the concentrations of CR (0.057 to 0.078 mM) while keeping $NaBH_4$ and catalyst as 7.88 mM and 2.1 mg/mL, respectively. Values k_{app}, half life and reaction completion time for reduction of CR under its different concentrations are given in Table 2. It can be seen from Table 2 that the value of k_{app} was initially increased, attained the maxima and then was decreased as CR concentration was increased. At low contents of CR, both CR molecules and BH_4^{-1} ions were adsorbed on the surface of catalyst simultaneously and results in a high value of k_{app}. However, at high concentrations of CR, most of the active sites of the catalyst are occupied by CR molecules as compared to BH_4^{-1} ions. Due to insufficient adsorbed BH_4^{-1} ions compared to CR molecules, the rate of catalytic reduction of CR was decreased. This leads to a low value of k_{app}. The plot of k_{app} vs. CR concentration for its reduction in the presence of reducing agent is shown in Figure 9a. The curve obtained for CR dependent k_{app} value shows that CR reduction followed the LH mechanism. According to this mechanism, reacting species like CR and BH_4^{-1} ions were first adsorbed on a fixed number of active sites of biogenic CAg-nPs catalyst. Then, adsorbed CR molecules and BH_4^{-1} ions reacted with each other on the surface of the catalyst, and CR was converted into environmental benign products. In the next step, adsorbed product was desorbed and diffused to bulk, making the availability of active sites for adsorption of more reacting species. Actually, biogenic CAg-nPs catalyst acts as a conveyer belt for speedy electrons transfer from BH_4^{-1} ions to CR molecules due to the large surface area of nanoparticles. Thus, a high activity of CAg-nPs catalyst lies in their efficiency during electrons transfer process [42].

Table 2. Effect of CR, $NaBH_4$ and catalyst contents for reduction of CR using biogenic CAg-nPs catalyst at ambient temperature.

Contents	CR (mM)	$NaBH_4$ (mM)	Catalyst (mg/mL)	k_{app} (min^{-1})	Half Life (min)	R^2	Induction Time (min)	Reaction Completion Time (min)
CR	0.057	7.88	2.1	0.229	3.026	0.988	2	13
	0.060	7.88	2.1	0.294	2.357	0.992	1	9
	0.063	7.88	2.1	0.544	1.274	0.951	0	4
	0.066	7.88	2.1	0.591	1.173	0.963	0	5
	0.069	7.88	2.1	0.469	1.478	0.982	1	7
	0.072	7.88	2.1	0.432	1.604	0.967	2	8
	0.075	7.88	2.1	0.404	1.715	0.995	2	9
	0.078	7.88	2.1	0.321	2.159	0.935	2	9
$NaBH_4$	0.072	4.50	2.1	0.156	4.442	0.972	3	27
	0.072	5.63	2.1	0.204	3.397	0.955	3	21
	0.072	6.75	2.1	0.255	2.718	0.933	1	12
	0.072	7.88	2.1	0.448	1.547	0.945	1	09
	0.072	8.33	2.1	0.390	1.777	0.9664	0	09
	0.072	9.00	2.1	0.281	2.466	0.9655	2	14
	0.072	10.13	2.1	0.249	2.783	0.9372	2	16
	0.072	11.25	2.1	0.194	3.572	0.978	2	20
	0.072	12.38	2.1	0.160	4.331	0.965	3	22
Catalyst	0.072	7.88	1.2	0.157	4.414	0.997	3	26
	0.072	7.88	1.5	0.173	4.006	0.994	2	23
	0.072	7.88	1.8	0.218	3.179	0.982	1	19
	0.072	7.88	2.1	0.281	2.466	0.978	1	15
	0.072	7.88	2.4	0.292	2.373	0.995	0	14
	0.072	7.88	2.7	0.293	2.365	0.999	0	12

CR reduction was also done by varying the concentration of $NaBH_4$ (4.50 to 12.38 mM) while CR and catalysts were taken as 0.072 mM and 2.1 mg/mL, respectively. Values of k_{app}, reaction completion time and half life for CR reduction by using different concentrations of CR are given in Table 2. Reaction completion time for CR reduction was first decreased and then was increased as the contents of $NaBH_4$ in reaction mixture were increased. A trend observed for $NaBH_4$ dependent values of k_{app} for CR reduction was similar to that of CR concentration depending on k_{app} value. A plot of k_{app} against the concentration of $NaBH_4$ for CR reduction is shown in Figure 9b. k_{app} was first increased, approaching the highest values and then was increased as $NaBH_4$ content in reaction mixture was increased.

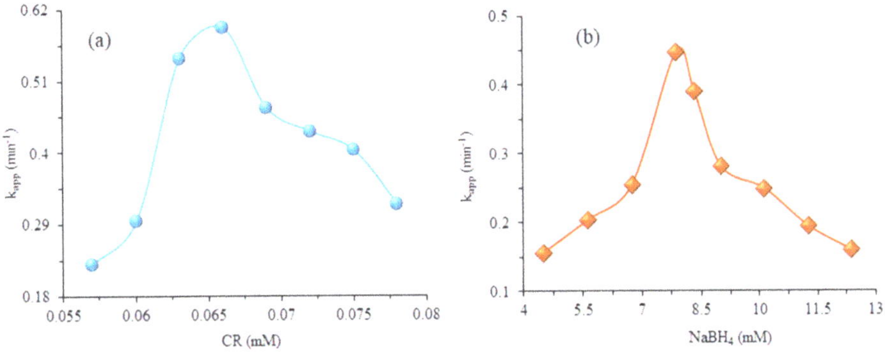

Figure 9. (**a**) CR concentration, (**b**) Sodium borohydride concentration dependent k_{app} relation for degradation of CR at ambient temperature (Reaction conditions: $[NaBH_4]$ = 7.88 mM, [CR] = 0.072 mM and catalyst = 2.1 mg/mL).

The amount of catalyst was changed from 1.2 to 2.7 mg/mL in reaction mixture while the amount of $NaBH_4$ and CR was kept as 7.88 and 0.072 mM, respectively, for reduction of CR. Values of k_{app}, induction time, reaction completion time and half life for CR degradation at different amounts of catalyst are given in Table 2. The value of k_{app} was increased with the increase of catalyst amount up to a limit, and then it became constant. Actually, a high catalyst dose offers a large number of active sites for the adsorption of reacting species. A high catalyst dose induced a high k_{app} value for CR reduction. At a very large amount of catalyst, all reacting species are adsorbed on the surface of the catalyst and lead to the saturation of catalyst surface. Thus, a further increase of the catalyst amount has no effect on the rate of reaction or on the value of k_{app}. As a result, the catalyst dependent value of k_{app} becomes constant at a high amount of catalyst as shown in Figure 8b. Previous studies show that catalytic reduction of dyes depends on the available active site on the surface of catalyst for adsorption of reactants as well as the number of nPs per volume. Thus, our results were found to be in agreement with the previous literature.

3.5. Reduction of Other Dyes and Nitroarenes

Various other toxic organic compounds like MG, MB, RhB and 4-NP having concentrations of 0.072 mM were also reduced individually in the presence of $NaBH_4$ reducing agent and CAg-nPs catalyst as shown in Figure 10. Reactions were completed in feasible time intervals and monitoring of reaction was easy while using 0.072 mM solution of all dyes individually.

Various parameters for catalytic reduction of MG, MB, RhB and p-NP are given in Table 3. It was concluded that RhB was not reduced completely in the presence of catalyst as compared to MB and RhB. The decreasing order of reduction of dyes in terms of k_{app} was MB > MG > 4-NP > RhB. A high reduction efficiency of biocatalysts for MB was due to its nature. Mb was degraded easily while the lowest reduction efficiency for RhB was due to its complex structure.

Table 3. Reduction of various dyes in the presence of CAg-nPs using $NaBH_4$ as reducing agent.

Dyes	k_{app} (min^{-1})	Intrinsic Rate Constant ($mL \cdot mg^{-1} \cdot min^{-1}$)	Reaction Completion Time (min)	Reduction Efficiency (%)
MB	0.613	0.292	4	93.29
MG	0.451	0.215	7	83.73
4-NP	0.247	0.118	9	88.90
RhB	0.085	0.041	18	60.53

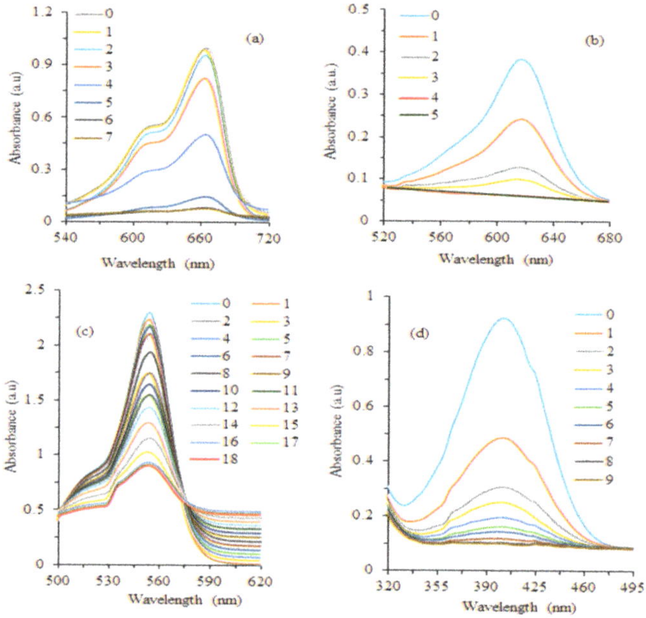

Figure 10. Catalytic reduction of various dyes (0.072 mM) such as (**a**) Mb, (**b**) MG, (**c**) RhB and (**d**) 4-Np in the presence of biogenic CAg-nPs catalyst (2.1 mg/mL) and $NaBH_4$ (7.88 mM) at an ambient temperature.

Catalyst was recycled from the reaction mixture by performing centrifugation at a high speed and reused for CR reduction. There was not a remarkable decrease in the reduction efficiency of the catalyst for up to three consecutive cycles (reduction efficiency ≈ 87% to 82% from the 1st to 3rd cycle). Recoverability and reusability of catalyst were also investigated after three cycles. A sharp reduction in the value of percentage removal was observed. This may be due to coagulation of Ag-nPs due to repeated usage. Reduction efficiency of the biogenic CAg-nPs catalyst for different reusability cycles for CR reduction is shown in Figure 11.

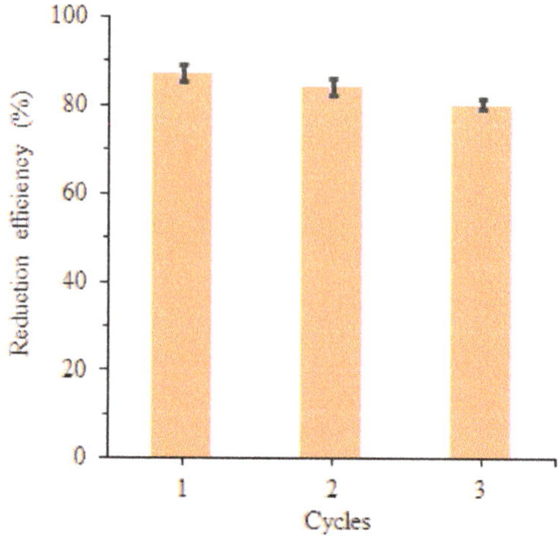

Figure 11. Recyclability of biogenic CAg-nPs catalyst for reduction of CR (0.216 mM) in the presence of $NaBH_4$ (24.03 mM) and catalyst (8.42 mg/mL) at an ambient temperature.

4. Conclusions

This study presents the ecofriendly biogenic synthesis of CAg-nPs by an aqueous peel extract of *Citrus paradise*. FTIR and UV-Vis analysis demonstrates the involvement of different metabolites of plant extract in the bio-reduction process and stabilization of silver nPs. UV-Vis analysis also confirms the successful fabrication of silver nanoparticles along with their spherical shape and narrow size distribution. TGA analysis shows that the prepared biogenic CAg-nPs catalyst is incorporated with 15% silver nanoparticle contents. Catalytic activity of biogenic CAg-nPs was explored against the reduction of various toxic dyes such as CR, MB, MG, RhB and 4-NP using BH_4^{-1} ions as hydrogen/electrons source. Results prove the remarkable catalytic efficiency of biogenic CAg-nPs nano-catalysts for reduction of reported dyes. The CAg-nPs catalyst was conveniently recovered from a reaction mixture by high speed centrifugation and applied for subsequent reaction without a remarkable decrease in its activity.

The CAg-nPs catalyst was applied to speed up the rate of reduction of selective dyes. However, textile industry wastewater contains a lot of other toxic chemicals. The effect of those chemicals on the activity of catalysts can also be addressed to make it effective for treatment of industrial wastewater.

Author Contributions: Conceptualization, K.N. and A.A.; Methodology, A.A.; Software, M.Z.U.R.; Validation, K.N. and D.D.; Formal Analysis, T.S.A.; Investigation, A.A.; Resources, T.S.A.; Data Curation, K.N.; Writing—Original Draft Preparation, K.N.; Writing—Review & Editing, A.A.; Visualization, D.D.; Supervision, D.D.; Project Administration, M.Z.U.R.; Funding Acquisition, T.S.A. All authors have read and agreed to the published version of the manuscript.

Funding: This work was funded by the Researchers Supporting Project Number (RSP-2020/254) King Saud University, Riyadh, Saudi Arabia.

Conflicts of Interest: The authors declare no conflict of interest.

References

1. Naseem, K.; Farooqi, Z.H.; Begum, R.; Irfan, A. Removal of Congo red dye from aqueous medium by its catalytic reduction using sodium borohydride in the presence of various inorganic nano-catalysts: A review. *J. Clean. Prod.* **2018**, *187*, 296–307. [CrossRef]
2. Zhang, X.; Xu, P.; Liu, G.; Ahmad, A.; Chen, X.; Zhu, Y.; Alothman, A.; Hussain, S.; Qiao, G. Synthesis, characterization and wettability of Cu–Sn alloy on the Si-implanted 6H–SiC. *Coatings* **2020**, *9*, 906. [CrossRef]
3. Albukhari, S.M.; Ismail, M.; Akhtar, K.; Danish, E.Y. Catalytic reduction of nitrophenols and dyes using silver nanoparticles@ cellulose polymer paper for the resolution of waste water treatment challenges. *Colloids Surf. Physicochem. Eng. Asp.* **2019**, *577*, 548–561. [CrossRef]
4. Aravind, M.; Ahmad, A.; Ahmad, I.; Amalanathan, M.; Naseem, K.; Mary, S.M.M.; Parvathiraja, C.; Hussain, S.; Algarni, T.S.; Pervaiz, M.; et al. Critical green routing synthesis of silver NPs using jasmine flower extract for biological activities and photocatalytical degradation of methylene blue. *J. Environ. Chem. Eng.* **2020**, 104877. [CrossRef]
5. Hussain, S.; Khan, A.J.; Arshad, M.; Javed, M.S.; Ahmad, A.; Shah, S.S.A.; Khan, M.R.; Akram, S.; Zulfiqar; Ali, S.; et al. Charge storage in binder-free 2D-hexagonal $CoMoO_4$ nanosheets as a redox active material for pseudocapacitors. *Ceram. Int.* **2020**. [CrossRef]
6. Naseem, K.; Farooqi, Z.H.; Begum, R.; Ghufran, M.; Rehman, M.Z.U.; Najeeb, J.; Irfan, A.; Al-Sehemi, A.G. Poly (N-isopropylmethacrylamide-acrylic acid) microgels as adsorbent for removal of toxic dyes from aqueous medium. *J. Mol. Liq.* **2018**, *268*, 229–238. [CrossRef]
7. Ahmad, A.; Jini, D.; Aravind, M.; Parvathiraja, C.; Ali, R.; Kiyani, M.Z.; Alothman, A. A novel study on synthesis of egg shell based activated carbon for degradation of methylene blue via photocatalysis. *Arab. J. Chem.* **2020**. [CrossRef]
8. Ahmad, A.; Mubharak, N.M.; Naseem, K.; Tabassum, H.; Rizwan, M.; Najda, A.; Kashif, M.; Bin-Jumah, M.; Hussain, A.; Shaheen, A.; et al. Recent advancement and development of chitin and chitosan-based nanocomposite for drug delivery: Critical approach to clinical research. *Arab. J. Chem.* **2020**, *13*, 8935–8964. [CrossRef]

9. Suresh, M.; Sivasamy, A. Fabrication of graphene nanosheets decorated by nitrogen-doped ZnO nanoparticles with enhanced visible photocatalytic activity for the degradation of Methylene Blue dye. *J. Mol. Liq.* **2020**, *317*, 114112. [CrossRef]
10. Ganapuram, B.R.; Alle, M.; Dadigala, R.; Dasari, A.; Maragoni, V.; Guttena, V. Catalytic reduction of methylene blue and Congo red dyes using green synthesized gold nanoparticles capped by salmalia malabarica gum. *Int. Nano Lett.* **2015**, *5*, 215–222. [CrossRef]
11. Naseem, K.; Begum, R.; Wu, W.; Irfan, A.; Al-Sehemi, A.G.; Farooqi, Z.H. Catalytic reduction of toxic dyes in the presence of silver nanoparticles impregnated core-shell composite microgels. *J. Clean. Prod.* **2019**, *211*, 855–864. [CrossRef]
12. Lu, Y.; Proch, S.; Schrinner, M.; Drechsler, M.; Kempe, R.; Ballauff, M. Thermosensitive core-shell microgel as a "nanoreactor" for catalytic active metal nanoparticles. *J. Mater. Chem.* **2009**, *19*, 3955–3961. [CrossRef]
13. Graça, C.A.; Mendes, M.A.; Teixeira, A.C.S.; de Velosa, A.C. Anoxic degradation of chlorpyrifos by zerovalent monometallic and bimetallic particles in solution. *Chemosphere* **2020**, *244*, 125461. [CrossRef]
14. Kashif, M.; Ngaini, Z.; Harry, A.V.; Vekariya, R.L.; Ahmad, A.; Zuo, Z.; Sahari, S.K.; Hussain, S.; Khan, Z.A.; Alarifi, A. An experimental and DFT study on novel dyes incorporated with natural dyes on titanium dioxide (TiO_2) towards solar cell application. *Appl. Phys. A* **2020**, *126*, 1–13. [CrossRef]
15. Zhang, J.; Ma, N.; Tang, F.; Cui, Q.; He, F.; Li, L. pH-and glucose-responsive core–shell hybrid nanoparticles with controllable metal-enhanced fluorescence effects. *ACS Appl. Mater. Interfaces* **2012**, *4*, 1747–1751. [CrossRef]
16. Shameli, K.; Ahmad, M.B.; Zamanian, A.; Sangpour, P.; Shabanzadeh, P.; Abdollahi, Y.; Zargar, M. Green biosynthesis of silver nanoparticles using Curcuma longa tuber powder. *Int. J. Nanomed.* **2012**, *7*, 5603. [CrossRef]
17. Rostami-Vartooni, A.; Nasrollahzadeh, M.; Alizadeh, M. Green synthesis of seashell supported silver nanoparticles using Bunium persicum seeds extract: Application of the particles for catalytic reduction of organic dyes. *J. Colloid Interface Sci.* **2016**, *470*, 268–275. [CrossRef]
18. Naseem, K.; Farooqi, Z.H.; Begum, R.; Wu, W.; Irfan, A.; Ajmal, M. Systematic study for catalytic degradation of nitrobenzene derivatives using core@ shell composite micro particles as catalyst. *Colloids Surf. Physicochem. Eng. Asp.* **2020**, *594*, 124646. [CrossRef]
19. Tang, F.; Ma, N.; Tong, L.; He, F.; Li, L. Control of metal-enhanced fluorescence with pH-and thermoresponsive hybrid microgels. *Langmuir* **2012**, *28*, 883–888. [CrossRef]
20. Abbasi, E.; Milani, M.; Fekri Aval, S.; Kouhi, M.; Akbarzadeh, A.; Tayefi Nasrabadi, H.; Nikasa, P.; Joo, S.W.; Hanifehpour, Y.; Nejati-Koshki, K. Silver nanoparticles: Synthesis methods, bio-applications and properties. *Crit. Rev. Microbiol.* **2016**, *42*, 173–180. [CrossRef]
21. Nava, O.; Soto-Robles, C.; Gómez-Gutiérrez, C.; Vilchis-Nestor, A.; Castro-Beltrán, A.; Olivas, A.; Luque, P. Fruit peel extract mediated green synthesis of zinc oxide nanoparticles. *J. Mol. Struct.* **2017**, *1147*, 1–6. [CrossRef]
22. Kasthuri, J.; Kathiravan, K.; Rajendiran, N. Phyllanthin-assisted biosynthesis of silver and gold nanoparticles: A novel biological approach. *J. Nanopart. Res.* **2009**, *11*, 1075–1085. [CrossRef]
23. Shankar, S.S.; Rai, A.; Ahmad, A.; Sastry, M. Rapid synthesis of Au, Ag, and bimetallic Au core–Ag shell nanoparticles using Neem (Azadirachta indica) leaf broth. *J. Colloid Interface Sci.* **2004**, *275*, 496–502. [CrossRef]
24. Gardea-Torresdey, J.; Parsons, J.; Gomez, E.; Peralta-Videa, J.; Troiani, H.; Santiago, P.; Yacaman, M.J. Formation and growth of Au nanoparticles inside live alfalfa plants. *Nano Lett.* **2002**, *2*, 397–401. [CrossRef]
25. Ayinde, W.; Gitari, W.; Samie, A. Optimization of microwave-assisted synthesis of silver nanoparticle by Citrus paradisi peel and its application against pathogenic water strain. *Green Chem. Lett. Rev.* **2019**, *12*, 225–234. [CrossRef]
26. Najafinejad, M.S.; Mohammadi, P.; Afsahi, M.M.; Sheibani, H. Green synthesis of the Fe_3O_4@ polythiophen-Ag magnetic nanocatalyst using grapefruit peel extract: Application of the catalyst for reduction of organic dyes in water. *J. Mol. Liq.* **2018**, *262*, 248–254. [CrossRef]
27. Kumar, B.; Smita, K.; Galeas, S.; Sharma, V.; Guerrero, V.H.; Debut, A.; Cumbal, L. Characterization and application of biosynthesized iron oxide nanoparticles using Citrus paradisi peel: A sustainable approach. *Inorg. Chem. Commun.* **2020**, *119*, 108116. [CrossRef]
28. Kumar, B.; Smita, K.; Cumbal, L.; Debut, A. Green approach for fabrication and applications of zinc oxide nanoparticles. *Bioinorg. Chem. Appl.* **2014**, *2014*, 523869. [CrossRef]
29. Soto, K.M.; Quezada-Cervantes, C.T.; Hernández-Iturriaga, M.; Luna-Bárcenas, G.; Vazquez-Duhalt, R.; Mendoza, S. Fruit peels waste for the green synthesis of silver nanoparticles with antimicrobial activity against foodborne pathogens. *LWT* **2019**, *103*, 293–300. [CrossRef]

30. Jackson, T.C.; Uwah, T.O.; Ifekpolugo, N.L.; Emmanuel, N.A. Comparison of antimicrobial activities of silver nanoparticles biosynthesized from some Citrus species. *Am. J. Nano Res. Appl.* **2018**, *6*, 54–59.
31. Vankar, P.S.; Shukla, D. Biosynthesis of silver nanoparticles using lemon leaves extract and its application for antimicrobial finish on fabric. *Appl. Nanosci.* **2012**, *2*, 163–168. [CrossRef]
32. Barbinta-Patrascu, M.E.; Badea, N.; Ungureanu, C.; Iordache, S.M.; Constantin, M.; Purcar, V.; Rau, I.; Pirvu, C. Ecobiophysical aspects on nanosilver biogenerated from citrus reticulata peels, as potential biopesticide for controlling pathogens and wetland plants in aquatic media. *J. Nanomater.* **2017**, *2017*, 4214017. [CrossRef]
33. Indana, M.K.; Gangapuram, B.R.; Dadigala, R.; Bandi, R.; Guttena, V. A novel green synthesis and characterization of silver nanoparticles using gum tragacanth and evaluation of their potential catalytic reduction activities with methylene blue and Congo red dyes. *J. Anal. Sci. Technol.* **2016**, *7*, 19. [CrossRef]
34. Varadavenkatesan, T.; Selvaraj, R.; Vinayagam, R. Green synthesis of silver nanoparticles using Thunbergia grandiflora flower extract and its catalytic action in reduction of Congo red dye. *Mater. Today Proc.* **2020**, *23*, 39–42. [CrossRef]
35. Saha, J.; Begum, A.; Mukherjee, A.; Kumar, S. A novel green synthesis of silver nanoparticles and their catalytic action in reduction of Methylene Blue dye. *Sustain. Environ. Res.* **2017**, *27*, 245–250. [CrossRef]
36. Naraginti, S.; Sivakumar, A. Eco-friendly synthesis of silver and gold nanoparticles with enhanced bactericidal activity and study of silver catalyzed reduction of 4-nitrophenol. *Spectrochim. Acta Part A Mol. Biomol. Spectrosc.* **2014**, *128*, 357–362. [CrossRef]
37. Dauthal, P.; Mukhopadhyay, M. Prunus domestica fruit extract-mediated synthesis of gold nanoparticles and its catalytic activity for 4-nitrophenol reduction. *Ind. Eng. Chem. Res.* **2012**, *51*, 13014–13020. [CrossRef]
38. Kahsay, M.H.; RamaDevi, D.; Kumar, Y.P.; Mohan, B.S.; Tadesse, A.; Battu, G.; Basavaiah, K. Synthesis of silver nanoparticles using aqueous extract of Dolichos lablab for reduction of 4-Nitrophenol, antimicrobial and anticancer activities. *OpenNano* **2018**, *3*, 28–37. [CrossRef]
39. Kumar, P.; Govindaraju, M.; Senthamilselvi, S.; Premkumar, K. Photocatalytic degradation of methyl orange dye using silver (Ag) nanoparticles synthesized from Ulva lactuca. *Colloids Surf. B Biointerfaces* **2013**, *103*, 658–661. [CrossRef]
40. Begum, R.; Naseem, K.; Ahmed, E.; Sharif, A.; Farooqi, Z.H. Simultaneous catalytic reduction of nitroarenes using silver nanoparticles fabricated in poly (N-isopropylacrylamide-acrylic acid-acrylamide) microgels. *Colloids Surf. Physicochem. Eng. Asp.* **2016**, *511*, 17–26. [CrossRef]
41. Farooqi, Z.H.; Naseem, K.; Begum, R.; Ijaz, A. Catalytic reduction of 2-nitroaniline in aqueous medium using silver nanoparticles functionalized polymer microgels. *J. Inorg. Organomet. Polym. Mater.* **2015**, *25*, 1554–1568. [CrossRef]
42. Mata, R.; Bhaskaran, A.; Sadras, S.R. Green-synthesized gold nanoparticles from Plumeria alba flower extract to augment catalytic degradation of organic dyes and inhibit bacterial growth. *Particuology* **2016**, *24*, 78–86. [CrossRef]
43. Kalia, A.; Manchanda, P.; Bhardwaj, S.; Singh, G. Biosynthesized silver nanoparticles from aqueous extracts of sweet lime fruit and callus tissues possess variable antioxidant and antimicrobial potentials. *Inorg. Nano Met. Chem.* **2020**, *50*, 1–10. [CrossRef]
44. Naseem, K.; Farooqi, Z.H.; Begum, R.; Wu, W.; Irfan, A.; Al-Sehemi, A.G. Silver Nanoparticles Engineered Polystyrene-Poly (N-isopropylmethacrylamide-acrylic acid) Core Shell Hybrid Polymer Microgels for Catalytic Reduction of Congo Red. *Macromol. Chem. Phys.* **2018**, *219*, 1800211. [CrossRef]
45. Alzahrani, H.A.; Buckingham, M.A.; Wardley, W.P.; Tilley, R.D.; Ariotti, N.; Aldous, L. Gold nanoparticles immobilised in a superabsorbent hydrogel matrix: Facile synthesis and application for the catalytic reduction of toxic compounds. *Chem. Commun.* **2020**, *56*, 1263–1266. [CrossRef]
46. Ismail, M.; Khan, M.; Khan, M.A.; Akhtar, K.; Asiri, A.M.; Khan, S.B. Plant-supported silver nanoparticles: Efficient, economically viable and easily recoverable catalyst for the reduction of organic pollutants. *Appl. Organomet. Chem.* **2019**, *33*, e4971. [CrossRef]

coatings

MDPI

Article

Weight Loss during Calcination and Sintering Process of $Na_{0.5}Bi_{0.5}TiO_3$–$Bi_{1/2}(Mg_{2/3}Nb_{1/3})O_3$ Composite Lead-Free Piezoelectric Ceramics

Syed Ali Afza [1,2], Fayaz Hussain [1,2], Sajid Hussain Siyal [1], Muhammad Sufyan Javed [3,4,*], Muhammad Saleem [5], Muhammad Imran [6], Mohammed A. Assiri [6], Aboud Ahmed Awadh Bahajjaj [7], Ayman A. Ghfar [7], Murefah Mana AL-Anazy [8], Mohamed Ouladsmane [7], Saad Al-Tamrah [7] and Shafaqat Ali [9,10,*]

1 Metallurgy & Materials Engineering Department, Dawood University of Engineering and Technology, Karachi 74800, Pakistan; syedaliafzal@duet.edu.pk (S.A.A.); fhussain@neduet.edu.pk (F.H.); Sajid.Hussain@duet.edu.pk (S.H.S.)
2 Department of Materials Engineering, NED University of Engineering & Technology, Karachi 75270, Pakistan
3 School of Physical Science and Technology, Lanzhou University, Lanzhou 730000, China
4 Department of Physics, Lahore Campus, COMSATS University Islamabad, Lahore 54000, Pakistan
5 Institute of Physics, The Islamia University of Bahawalpur, Bahawalpur 63100, Pakistan; saleem.malikape@gmail.com
6 Department of Chemistry, Faculty of Science, King Khalid University, P.O. Box 9004, Abha 61413, Saudi Arabia; miahmad@kku.edu.sa (M.I.); maassiri@kku.edu.sa (M.A.A.)
7 Advanced Materials Research Chair, Chemistry Department, College of Science, King Saud University, Riyadh 11451, Saudi Arabia; aawadh@ksu.edu.sa (A.A.A.B.); Aghafr@ksu.edu.sa (A.A.G.); mouladsmane@ksu.edu.sa (M.O.); altamrah@ksu.edu.sa (S.A.-T.)
8 Department of Chemistry, College of Science, Princess Nourah bint Abdulrahman University, Riyadh 11671, Saudi Arabia; mmalanazy@pnu.edu.sa
9 Department of Environmental Science and Engineering, Government College University, Faisalabad 38000, Pakistan
10 Department of Biological Sciences and Technology, China Medical University, Taichung 40402, Taiwan
* Correspondence: safisabri@gmail.com (M.S.J.); shafaqataligill@yahoo.com (S.A.)

Citation: Afzal, S.A.; Hussain, F.; Siyal, S.H.; Javed, M.S.; Saleem, M.; Imran, M.; Assiri, M.A.; Bahajjaj, A.A.A.; Ghfar, A.A.; AL-Anazy, M.M.; et al. Weight Loss during Calcination and Sintering Process of $Na_{0.5}Bi_{0.5}TiO_3$–$Bi_{1/2}(Mg_{2/3}Nb_{1/3})O_3$ Composite Lead-Free Piezoelectric Ceramics. *Coatings* **2021**, *11*, 676. https://doi.org/10.3390/coatings11060676

Received: 11 May 2021
Accepted: 31 May 2021
Published: 3 June 2021

Abstract: $(1-x)(Na_{0.5}Bi_{0.5})TiO_3$–$xBi(Mg_{2/3}Nb_{1/3})O_3$ ceramics with x = 0.00 mol.% (0BMN), 0.01 mol.% (1BMN), 0.03 mol.% (3BMN), and 0.05 mol.% (5BMN) were synthesized using a solid-state processing technique. The thermogravimetric analysis (TGA) of uncalcined samples up to 730 °C showed that the maximum weight loss was observed for 3BMN, whereas the minimum weight loss was attributed to the 0BMN sample. After that, calcination was performed at 800 °C for 4 h. The XRD of calcined samples showed the successful formation of the perovskite phase with no impurity phases. 1BMN and 3BMN samples showed some of the lattice strain; however, a morphotropic phase boundary (MPB) existed around x = 0.03 between the rhombohedral and tetragonal structure. The TGA of the green pellets showed weight loss up to the sintering temperature (1100 °C) and during the 3 h holding period. 5BMN showed the maximum weight loss up to sintering temperature, as well as during the holding period, whereas 0BMN displayed the minimum weight loss up to sintering temperature, as well as some weight gain during the holding period. The relative permittivity (ε_r) was maximum at low frequencies, but the addition of BMN improved the ε_r. The frequency dependence of dielectric loss (tanδ) showed that the maximum loss was observed for 3BMN at lower frequencies, and 5BMN showed the maximum loss at higher frequency among all samples.

Keywords: lead-free; NBT–BMN; weight loss; dielectric; piezoelectric ceramics

1. Introduction

Piezoelectric ceramics primarily based on lead oxide (PbO) are commonly used as piezoelectric actuators, sensors, and transducers for their exceptional piezoelectric properties [1]. Nonetheless, evaporation related to the hazardous nature of PbO during high-temperature sintering not only leads to ecological pollution, but also triggers variability of the composition and electrical characteristics of material [2,3]. Consequently, researchers around the world are focused on the synthesis of green lead-free piezoelectric ceramics to eliminate the hazardous lead oxide-based ceramics [4]. The research on lead-free piezoceramics has already started a new journey to replace lead zirconium titanate (PZT) with nonharmful choices [5]. Moreover, the European Union (EU) passed legislation on Restriction of Hazardous Substances (RoHS), Waste from Electrical and Electronic Equipment (WEEE), and End of Life Vehicles (ELV), propelling scientists to take out poisonous substances from electrical and electronic devices to lessen their effect on the surroundings and wellbeing of people [6]. Sodium bismuth titanate ($Na_{0.5}Bi_{0.5}TiO_3$ denoted as NBT) is one of the emerging lead-free ceramics, and it was discovered in 1960 [7,8]. It has a distorted perovskite structure (ABO_3) with rhombohedral R3c symmetry at room temperature (RT). It exhibits superior dielectric, piezoelectric (d_{33} = 73 pC/N), and electromechanical properties and could potentially replace PZT; however, it has some major drawbacks, such as the vaporization of Na^{1+} and Bi^{3+} at high temperatures, the low piezoelectric constant (d_{33}), high coercive field (E_c), low depolarization temperature (T_d), and high conductivity [9,10]. It has been established that Bi vaporizes at high sintering temperatures (>1150 °C). Moreover, Hiruma et al. revealed that Bi vaporizes when sintered at 1100 °C due to the high dielectric loss of undoped BNT ceramics at $T > T_d$, which makes poling treatment difficult due to high conductivity and pinning of domains. The difficulties related to bismuth (Bi) volatilization can be reduced by doping stable oxide-containing compounds in NBT [11]. Kimura et al. observed that the excess Bi_2O_3 increased the sintered density because of a decrease in grain growth. It was reported that excess Bi_2O_3 enhances texture development and improves piezoelectric properties [12]. In another study, stoichiometric and nonstoichiometric BNT ceramics were prepared to clarify the effects of Bi and Na vaporizations on phase transitions and electrical properties. It was concluded that the dielectric losses were decreased and d_{33} increased when the Bi/Na ratio increased. Furthermore, the resistivity was increased with excess Bi and the leakage current was very small above T_d for BNT-Bi(0.01) because of low conductivity [11]. The CeO_2-doped NBT also showed a superior piezoelectric constant (d_{33}) and other electronic properties [13]. Recently, a new NBT-based relaxor ferroelectric was prepared by combining NBT with $Bi_{1/2}(Mg_{2/3}Nb_{1/3})$ O3 (BMN), which gave d_{33} = 94 pC/N at x = 0.07. Furthermore, the values of dielectric constant and electromechanical coupling factor also increased when BMN was added to NBT [14]. Another study showed the effects of excess Bi on $0.99Bi_x(Na_{0.8}K_{0.2})_{0.5}$–$0.01SrTiO_3$ (x = 0.5–0.535). The excess Bi increased RT strain properties and gave a d_{33}^* value of 440 pm/V [15]. Furthermore, a study on (1 − x) ($0.8Bi_{1/2}Na_{1/2}TiO_3$–$0.2Bi_{1/2}K_{1/2}TiO_3$)–$xBiMg_{2/3}Nb_{1/3}O_3$ (100xBMN) (BNT–BKT–BMN) ceramics revealed that the introduction of BMN gave rise to strain properties and aided the transformation from ferroelectric to relaxor phase. The maximum value of d_{33}^* was also found to be 784 pm/V [16]. Another study showed the improvement in ferroelectric and piezoelectric properties of $0.8Bi_{1/2}Na_{1/2}TiO_3$–$0.2Bi_{1/2}K_{1/2}TiO_3$ (BNT–BKT, BNKT) by doping $Bi_{x/3}Mg_{y/3}Nb_{z/3}O_3$ (BMN) [17]. Another study revealed that the $(1-x)[0.85BaTiO_3$–$0.15Bi(Mg_{2/3}Nb_{1/3})O_3]$–$_xNa_{0.5}Bi_{0.5}TiO_3$ (BT–BMN–NBT) system gives an improved dielectric constant and dielectric loss [18]. Therefore, in this study, we prepared a binary solid solution of $(1-x)Na_{0.5}Bi_{0.5}TiO_3$-$xBi(Mg_{2/3}Nb_{1/3})$ ceramics to observe the effect of BMN on Na and Bi volatility during the calcination and sintering process. Another goal of this study was to study the effect of different concentrations of BMN on the dielectric properties of NBT–BMN piezoelectric ceramics.

2. Experimental Procedure

The $(1-x)Na_{0.5}Bi_{0.5}TiO_3$–$xBi(Mg_{2/3}Nb_{1/3})O_3$ (NBT–BMN) ceramics with x = 0.00 mol.% (0BMN), 0.01 mol.% (1BMN), 0.03 mol.% (3BMN), and 0.05 mol.% (5BMN) were prepared using a conventional solid-state sintering technique. Na_2CO_3, Bi_2O_3, TiO_2, MgO, and Nb_2O_5 were used as raw materials with 99.9% purity (Daejung Chemicals & Metals Co., Ltd, Shiheung-city, Korea) and were stoichiometrically weighed in hot conditions to avoid any moisture pickup. The powders were subjected to planetary ball milling at 300 rpm in ethanol solution for 2 h. Zirconia jars and zirconia balls of diameter ranging from 5 to 20 mm were used as grinding media. After the ball milling process, the slurry was dried in a drying oven at 80 °C to let the ethanol evaporate. Then, the mixed powders were crushed in a mortar and pestle to break the agglomerates. Finally, the powder was passed through a sieve of mesh size 149 µm. Thermogravimetric analysis (TGA)/differential scanning calorimetry (DSC) was performed on the sample using an SDT Q600 Thermal Analyzer (TA instruments, New Castle, DE, USA) to determine the weight losses and perovskite phase formation temperature. The temperature was raised from 30 °C to 950 °C at a rate of 10 °C/min. According to the results, the calcination temperature was determined and finalized. All the compositions were calcined at 800 °C for 4 h in an alumina crucible by a chamber furnace (PLF 130/18, Protherm Furnaces, Ankara, Turkey). Calcination is an important step to remove the carbonates from the powders and for perovskite phase formation. The heating rate and cooling rate were set to be 5 °C/min. A small sample from the calcined batch was crushed using a mortar and pestle, and then the X-ray Diffraction (XRD) of these samples was performed on an X'Pert PRO (PANalytical, Almelo, The Netherland), to verify the perovskite phase formation. The diffraction patterns were recorded over the angular range (2θ) from 15° to 80° at room temperature. The whole batch of the calcined sample was again ball-milled in ethanol for 1 h, dried, crushed using mortar and pestle, and sieved over the mesh size 149 µm. The powders were then die-pressed into discs of 12.8 mm diameter in a pressing machine. Polyvinyl alcohol (PVA) was used as the binder, and a pressure of 2000 psi (140 bar) was uniaxially applied on the powder for 45–60 s to get the compacted green pellet. The TGA/DSC of green pellets was performed on an SDT Q600 Thermal Analyzer (TA instruments) to understand the weight loss under the sintering conditions. The heating rate was set to be 5 °C/min and, after reaching 1100 °C, the samples were held for 3 h in air at this temperature to complete the sintering cycle. The green pellets were sintered in an energy-saving box furnace at 1100 °C for 3 h. The pellets were first heated at 600 °C at a rate of 5 °C/min for 2 h to remove the binder and other organic materials. Then, the temperature was raised to 1100 °C at a heating rate of 5 °C/min. After sintering at 1100 °C, the samples were allowed to cool down at a rate of 5 °C/min until 400 °C. Below 400 °C, the furnace was turned off and the samples were allowed to furnace cool until room temperature. The dielectric measurements were carried out using an LCR meter TH2826 by Tonghui (Changzhou Tonghui Electronic Co., Ltd., Changzhou, China). The silver coating was applied on the sintered samples to measure the frequency dependence of relative permittivity (ε_r). The current was kept constant at 100 mV, and the frequency was increased from 1 kHz to 1 MHz.

3. Results and Discussion

3.1. *TGA/DSC of Uncalcined Samples*

Figure 1 shows the TGA/DSC results of the uncalcined samples. There were three distinct/major weight loss steps, as shown in Figure 1b; therefore, the weight loss during TGA was divided into three stages. Furthermore, the weight losses for all the batches are summarized in Table 1.

Table 1. Weight loss during TGA of uncalcined samples.

Batch	Weight Loss due to Moisture	Weight Loss during Crystallization	Weight Loss during Calcination	Total Weight Loss (%)
0BMN	2.03	5.35	7.62	10.57
1BMN	2.52	5	7.76	12.42
3BMN	3.57	7.16	10.68	16.52
5BMN	3.78	3.71	6.83	12.51

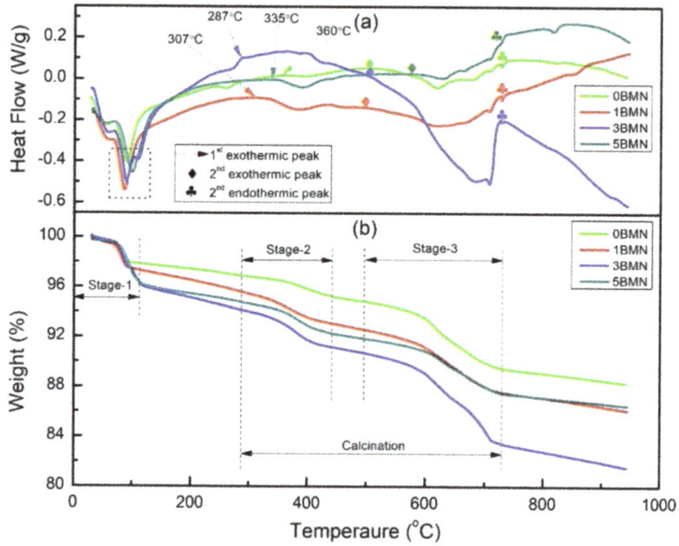

Figure 1. (**a**) DSC and (**b**) TGA of uncalcined samples.

Weight loss during stage-1 was associated with the evaporation of moisture [19]. This was confirmed by the first endothermic peaks in the DSC curve, which appeared around 100 °C and are represented inside the dashed square in Figure 1a. Furthermore, the maximum weight loss due to moisture was observed for the 5BMN sample. In addition, as the amount of BMN increased, the moisture loss also increased.

Weight loss during stage-2 was due to the decomposition of carbonates and the escape of CO_2 gas [20]. The start of stage-2 was attributed to the first exothermic peak in the DSC curve, represented by dashed arrows in Figure 1a. However, the starting temperature of stage-2 was different for all the batches ranging from 287 °C for 3BMN to 360 °C for 0BMN. The last weight loss step, i.e., Stage-3, was attributed to the weight loss during perovskite phase formation and crystallization of (1 − x) NBT–xBMN. In the DSC curve (Figure 1a), there were two peaks in the temperature range of 500 °C and 750 °C. The second exothermic peak marks the start of stage-3 (crystallization) and the second endothermic peak marks its end. A maximum weight loss of 7.16% was observed for 3BMN during stage-3. No further significant weight loss was observed beyond 730 °C, showing the complete crystallization of perovskite structure [21].

To evaluate the weight loss during calcination due to the removal of CO and CO_2, the weight loss from the start of stage-2 to the end of stage-3 was considered, as shown in Figure 1b. The maximum weight loss of 10.68% was observed for the 3BMN sample. 5BMN displayed the minimum weight loss during calcination (6.83%). The total weight loss was the total weight reduced from the material from the beginning of stage-1 to the end of stage-3 (~730 °C). The maximum total weight loss was observed for the 3BMN sample, which was around 16.52%. The reason for this could be the presence of rhombohedral–tetragonal MPB that exists around x = 0.03, as discussed in the XRD results in the next section. The distortion in the crystal structure may have been the reason for the high volatility of the sample as shown in Table 1. The minimum total weight loss of 10.57% was shown by 0BMN. This shows that the addition of BMN promoted the weight loss in NBT as compared to pure NBT during calcination. The TGA curve even showed a slight weight loss beyond stage-3, indicating the volatility of Na and Bi at higher temperatures. Therefore, to minimize further weight loss, the calcination temperature was set to 800 °C.

3.2. XRD of Calcined Samples

The XRD of calcined samples was performed to assess the perovskite phase formation and the effect of BMN on the crystal structure of NBT. Figure 2a shows the XRD spectra in the 2θ range from 15° to 75° of all the samples calcined at 800 °C. A perovskite structure without any impurity peaks was confirmed for all the samples, showing that BMN completely diffused into the structure of NBT [16,22].

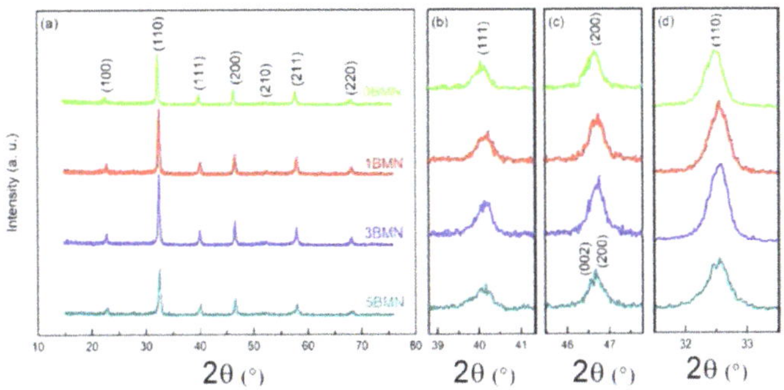

Figure 2. (**a**) XRD patterns of calcined (1 − x)NBT–xBMN samples; (**b**–**d**) show enlarged XRD patterns of (111), (200), and (110) diffraction peaks, respectively.

The (111), (200), and (110) peaks are magnified in Figure 2b–d, respectively. In Figure 2c, the (200) peak started to split, and a new peak started to emerge from the left in the 3BMN sample. This new (002) peak can be seen in the 5BMN sample. This means that the structures for 0BMN and 1BMN were rhombohedral with R3c symmetry [15,16]; however, with the increasing amount of BMN, the structure became tetragonal with P4bm symmetry [23,24]. Consequently, it could be said that there is a rhombohedral–tetragonal MPB that exists around x = 0.03 for the NBT–BMN system.

As shown in Figure 2d, the height of the (110) peak was low for 0BMN; however, the height increased for 1BMN and 3BMN samples. This increase in peak intensity indicated a more ordered and highly crystallized structure [25]. Nevertheless, the intensity sharply decreased in the case of 5BMN, suggesting strain relaxation in the structure as it transformed from rhombohedral to tetragonal. Since the effective radii of $(Mg_{2/3}Nb_{1/3})^{3+}$ and Ti^{4+} are 0.069 and 0.061 nm respectively, the broadening of the (111) peak in Figure 2b also indicated the lattice strains in the crystal induced by the addition of BMN due to the substitution of $(Mg_{2/3}Nb_{1/3})^{3+}$ for Ti^{4+} [26].

3.3. TGA during the Sintering Process

The TGA/DSC of green pellets for all four samples was done to evaluate the weight loss up to and at the sintering temperature. Figure 3 shows the TGA curves of all the samples during sintering. The weight loss observations are summarized in Table 2. It has been established by previous researchers that Na^{1+} and Bi^{3+} cations start evaporating at high sintering temperatures (>1130 °C) [24]. This causes vacancies at the A-site in the crystal lattice, which affects the structural homogeneity, stoichiometry, and properties of NBT [13].

Figure 3 shows that 5BMN displayed the maximum weight loss of 5.3% and 0BMN showed the minimum weight loss of 2.88% up until sintering temperature. The major portion of this weight loss could be attributed to the moisture and binder loss, which was added during the compaction of green pellets. Moreover, upon holding the samples at 1100 °C for 3 h, there was a notable increase in the weight of 0BMN, whereas 5BMN displayed further weight loss of 1.18% during the same holding period. Moreover, 1BMN and 3BMN showed very slight weight loss. 0BMN displayed a weight gain of (+)0.4% until the end of the sintering period. It is interesting to note that, at such a high temperature, the weight increased instead of decreasing.

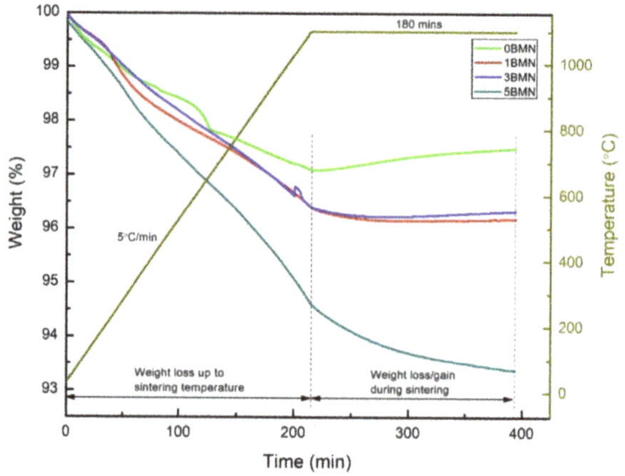

Figure 3. TGA curves for 0BMN, 1BMN, 3BMN, and 5BMN during sintering at 1100 °C for 3 h in the air.

Table 2. Weight loss/gain (+) during sintering.

Batch	Weight Loss up until Sintering Temperature (%)	Weight Loss/Gain (+) during Sintering (%)
0BMN	2.88	(+)0.4
1BMN	3.61	0.18
3BMN	3.57	0.06
5BMN	5.3	1.18

In order to study the phenomenon behind the weight gain, it is important to determine the crystal structure changes and/or new compound formation during sintering. For this purpose, the TGA/DSC curve during sintering was also plotted for 0BMN, as shown in Figure 4.

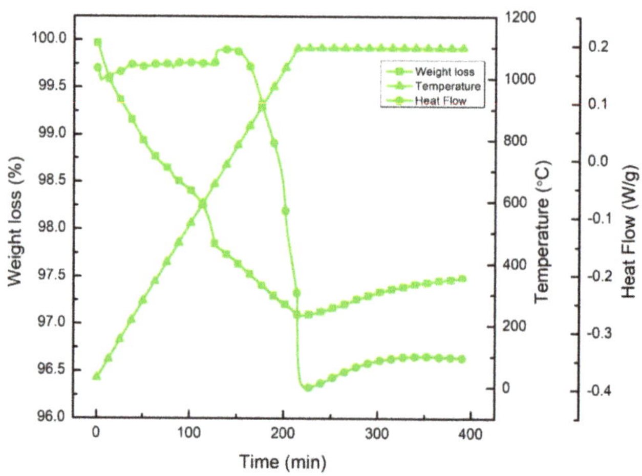

Figure 4. TGA/DSC curves of 0BMN sample during sintering at 1100 °C for 3 h in the air.

Nevertheless, there were no visible exothermic and endothermic peaks during the 3 h holding at 1100 °C. This represents no major crystal structure changes. However, sintering in the air reduced the number of oxygen vacancies and led to the decomposition of the material to keep charge neutrality. Further research could be done for 0BMN to establish the reason for weight gain during sintering.

3.4. Frequency Dependence of Relative Permittivity (ε_r)

Figure 5 shows the frequency dependence of relative permittivity (ε_r) of 0BMN, 1BMN, 3BMN, and 5BMN samples at room temperature. The ε_r was maximum at low frequencies for all four samples. The ε_r of the 5BMN sample (750) was the maximum among all the samples, whereas the ε_r of the undoped sample i.e., 0BMN was the minimum among all at all frequencies. This showed the increase in ε_r as BMN increased in NBT [17,18]. Moreover, the ε_r decreased when measuring frequency increased due to a decrease in polarization [27]. At low frequencies, all the polarization mechanisms were activated, whereas, at higher frequencies, polarization mechanisms ceased to function. Therefore, the net polarization of the material decreased, which led to the decrease in ε_r. It is proven that relative permittivity (ε_r) was a function of frequency and was very prominent in the above results. The lattice distortions caused by the A or B site disorder caused the abovementioned behaviors. The Bi^{3+} ions substituted the A site and $(Mg_{2/3}Nb_{1/3})^{3+}$ ions substituted the B site of the perovskite crystal structure and caused lattice distortions in the material [10]. The polarization switching required a high dielectric constant as BMN increased in NBT [16]. Furthermore, the higher ε_r of 5BMN could be due to better densification during sintering [22]. The difference in the decrease in relative permittivity (ε_r) as the frequency increased was maximum in the 5BMN sample and minimum in the 0BMN sample. This trend shows the difference in the decrease in the rate of relative permittivity (ε_r) increase as BMN doping increased.

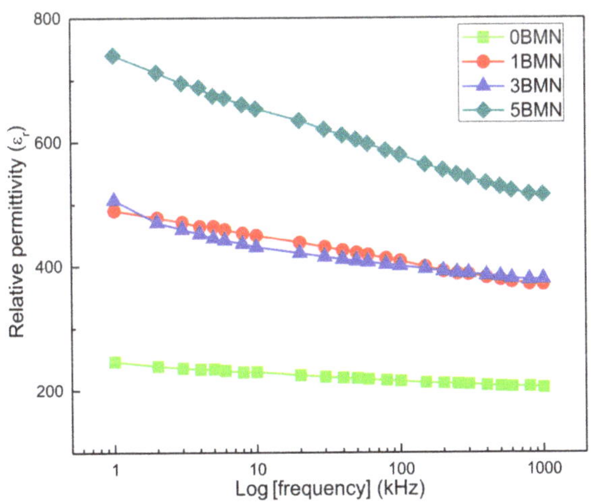

Figure 5. Frequency dependence of relative permittivity (ε_r).

3.5. Frequency Dependence of Dielectric Loss (tanδ)

The frequency dependence of dielectric loss (tanδ) for sintered 0BMN, 1BMN, 3BMN, and 5BMN samples at room temperature was determined, as shown in Figure 6. For 3BMN, the steep decrease in tanδ in the frequency range of 1 kHz and 100 kHz indicated leakage current [28]. During polarization (below 10 kHz), 3BMN showed the maximum loss. As the frequency started to increase, the dielectric loss slightly decreased for 0BMN. For 1BMN, it started to increase with the increase in frequency, whereas, for 5BMN, it stayed within a certain range. At lower frequencies, the friction due to dipole rotation increased dielectric losses. There was a slight hump in the losses of all samples in the range of 100 kHz to 1000 kHz, showing relaxor behavior [29]. This could be attributed to the change in the polarization mechanism due to the increase in frequency [9,30]. At higher frequencies (i.e., 1000 kHz), 5BMN displayed the highest tanδ, whereas 0BMN displayed the lowest tanδ among all samples.

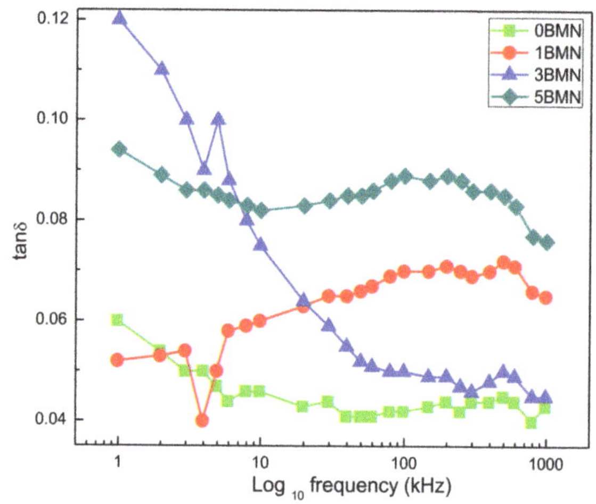

Figure 6. Frequency dependence of dielectric loss (tanδ).

4. Conclusions

In this study, $(1 - x)Na_{0.5}Bi_{0.5}TiO_3$–$xBi(Mg_{2/3}Nb_{1/3})O_3$ ceramics (x = 0 mol.%, 1 mol.%, 3 mol.%, and 5 mol.%) were synthesized by conventional solid-state processing route to evaluate their thermal behavior during calcination and sintering. The calcination was performed at 800 °C for 4 h, and sintering was performed at 1100 °C for 3 h. TGA/DSC of the as-milled powder was performed to understand the weight loss during calcination. 3BMN displayed the maximum and 5BMN displayed the minimum weight loss during the calcination process; however, the minimum total weight loss of 10.57% was shown by 0BMN. The XRD spectra showed that all samples had a single-phase perovskite structure, but there was a structural change from rhombohedral to tetragonal symmetry around x = 0.03. Moreover, 1BMN and 3BMN samples also showed lattice strain due to the addition of BMN. TGA of the green pellets was also done to evaluate the weight loss during sintering. 5BMN showed the maximum weight loss up until sintering temperature, as well as during the holding period; however, 0BMN displayed the minimum weight loss up until sintering temperature, along with some weight gain during the holding period. The DSC curve for 0BMN showed no major peaks that could indicate structural changes during the holding period. Further research could be done to evaluate the reason behind this weight gain. The frequency dependence of relative permittivity (ε_r) showed that the ε_r was maximum at low frequencies for all the samples; however, the increase in BMN content increased ε_r. Furthermore, increasing the frequency resulted in a decrease in ε_r due to the lattice distortions of dopant ions. The frequency dependence of dielectric loss (tanδ) showed that the maximum loss was observed for 3BMN at lower frequencies (<10 kHz). However, 5BMN showed the maximum loss at higher frequency among all samples.

Author Contributions: Conceptualization, S.A.A. and F.H.; methodology, M.I.; software, S.H.S.; validation, M.S., formal analysis, M.A.A.; investigation, A.A.A.B.; resources, A.A.G.; data curation, M.M.A.-A.; writing—original draft preparation, M.S.J.; writing—review and editing, S.A.; visualization, S.A.-T.; supervision, F.H.; project administration, S.A.; funding acquisition, M.O. All authors have read and agreed to the published version of the manuscript.

Funding: The authors are grateful to the Deanship of Scientific Research, King Saud University for funding through Vice Deanship of Scientific Research Chairs and this research was funded by the Dean-ship of Scientific Research at Princess Nourah bint Abdulrahman University through the Fast-track Research Funding Program.

Institutional Review Board Statement: Not applicable.

Informed Consent Statement: Not applicable.

Conflicts of Interest: The authors declare no conflict of interest.

References

1. Shibata, K.; Wang, R.; Tou, T.; Koruza, J. Applications of lead-free piezoelectric materials. *MRS Bull.* **2018**, *43*, 612–616. [CrossRef]
2. Xie, K.; Hu, H.; Xu, S.; Chen, T.; Huang, Y.; Yang, Y.; Yang, F.; Yao, H. Fate of heavy metals during molten salts thermal treatment of municipal solid waste incineration fly ashes. *Waste Manag.* **2020**, *103*, 334–341. [CrossRef] [PubMed]
3. Tiwari, S.; Tripathi, I.; Tiwari, H. Effects of lead on Environment. *Int. J. Emerg. Res. Manag. Technol.* **2013**, *2*, 204–212.
4. Akça, E.; Yılmaz, H. Lead-free potassium sodium niobate piezoceramics for high-power ultrasonic cutting application: Modelling and prototyping. *Process. Appl. Ceram.* **2019**, *13*, 65–78. [CrossRef]
5. Saito, Y.; Takao, H.; Tani, T.; Nonoyama, T.; Takatori, K.; Homma, T.; Nagaya, T.; Nakamura, M. Lead-free piezoceramics. *Nature* **2004**, *432*, 84. [CrossRef]
6. Shams, S. Circular Economy Policy Barriers: An Analysis of Legislative Challenges in White Goods and Automotive Industry within the EU. Master's Thesis, The Royal Institute of Technology (KTH), Stockholm, Sweden, June 2020.
7. Lima, A.C.; Pereira, N.; Martins, P.L.A.; Lanceros-Mendez, S. Magnetic materials for magnetoelectric coupling: An unexpected journey. *Handb. Magn. Mater.* **2020**, *29*, 57–110.
8. Smolenskii, G.; Isupov, V. New ferroeleetries of com-plex composition. *Soy. Phys. Solid State* **1961**, *2*, 2.
9. Singha, A.; Praharaj, S.; Rout, D. Effect of sintering time on microstructure and electrical properties of lead-free sodium bismuth titanate perovskite. *Mater. Today Proc.* **2020**. [CrossRef]
10. Singh, P.; Pandey, R.; Singh, P. Tailoring the electrical and structural properties of sodium bismuth titanate with sintering temperature. *Mater. Today Proc.* **2020**, *44*, 166–169. [CrossRef]
11. Hiruma, Y.; Nagata, H.; Takenaka, T. Thermal depoling process and piezoelectric properties of bismuth sodium titanate ceramics. *J. Appl. Phys.* **2009**, *105*, 084112. [CrossRef]
12. Kimura, T.; Fukuchi, E.; Tani, T. Fabrication of textured bismuth sodium titanate using excess bismuth oxide. *Jpn. J. Appl. Phys.* **2005**, *44*, 8055. [CrossRef]
13. Halim, N.; Majid, W.A.; Velayutham, T. Ferroelectric, pyroelectric and piezoelectric properties of CeO_2-doped $Na_{0.5}Bi_{0.5}TiO_3$ ceramics. *SN Appl. Sci.* **2019**, *1*, 582. [CrossRef]
14. Zhou, C.; Liu, X. Dielectric and piezoelectric properties of bismuth-containing complex perovskite solid solution of $Bi_{1/2}Na_{1/2}TiO_3-$ $Bi(Mg_{2/3}$ $Nb_{1/3})O_3$. *J. Mater. Sci.* **2008**, *43*, 1016–1019. [CrossRef]
15. Liu, X.; Li, F.; Li, P.; Zhai, J.; Shen, B.; Liu, B. Tuning the ferroelectric-relaxor transition temperature in NBT-based lead-free ceramics by Bi nonstoichiometry. *J. Eur. Ceram. Soc.* **2017**, *37*, 4585–4595. [CrossRef]
16. Dong, G.; Fan, H.; Shi, J.; Li, Q. Large strain response with low driving field in $Bi_{1/2}Na_{1/2}TiO_3$–$Bi_{1/}2K_{1/2}TiO_3$–$Bi(Mg_{2/3}Nb_{1/3})O_3$ ceramics. *J. Am. Ceram. Soc.* **2018**, *101*, 3947–3955. [CrossRef]
17. Dong, G.; Fan, H.; Jia, Y. Effect of the element ratio in the doping component on the properties of 0.975 (0.8 $Bi_{1/2}Na_{1/2}TiO_3$–0.2 $Bi_{1/2}K_{1/2}TiO_3$)–$0.025Bi_{x/3}Mg_{y/3}Nb_{z/3}O_3$ ceramics. *J. Mater. Res.* **2020**. [CrossRef]
18. Wang, T.; Liu, J.; Kong, L.; Yang, H.; Wang, F.; Li, C. Evolution of the structure, dielectric and ferroelectric properties of $Na_{0.5}Bi_{0.5}TiO_3$-added $BaTiO_3$–Bi $(Mg_{2/3}Nb1_{/3})$ O_3 ceramics. *Ceram. Int.* **2020**, *46*, 25392–25398. [CrossRef]
19. Mishra, P.; Kumar, P. Effect of sintering temperature on dielectric, piezoelectric and ferroelectric properties of BZT–BCT 50/50 ceramics. *J. Alloys Compd.* **2012**, *545*, 210–215. [CrossRef]
20. Yoon, M.-S.; Khansur, N.H.; Ur, S.-C. The effect of pre-milling/pre-synthesis process and excess Ba on the microstructure and dielectric/piezoelectric properties of nano-sized 0.94 $[(Bi_{0.5}Na_{0.5})TiO_3]$–0.06 $[Ba(1 + x)TiO_3]$. *Ceram. Int.* **2010**, *36*, 1265–1275. [CrossRef]
21. Chaouchi, A.; Kennour, S.; d'Astorg, S.; Rguiti, M.; Courtois, C.; Marinel, S.; Aliouat, M. Characterization of sol–gel synthesised lead-free (1− x) $Na_{0.5}Bi_{0.5}TiO_3$–$xBaTiO_3$-based ceramics. *J. Alloys Compd.* **2011**, *509*, 9138–9143. [CrossRef]
22. Badapanda, T.; Venkatesan, S.; Panigrahi, S.; Kumar, P. Structure and dielectric properties of bismuth sodium titanate ceramic prepared by auto-combustion technique. *Process. Appl. Ceram* **2013**, *7*, 135–141. [CrossRef]
23. Pan, Z.; Wang, Q.; Chen, J.; Liu, C.; Fan, L.; Liu, L.; Fang, L.; Xing, X. Enhanced piezoelectric properties of tetragonal $(Bi_{1/2}K_{1/2})TiO_3$ lead-free ceramics by substitution of pure Bi-based Bi $(Mg_{2/3}Nb_{1/3})$ O_3. *J. Am. Ceram. Soc.* **2015**, *98*, 104–108. [CrossRef]
24. Bhandari, S.; Sinha, N.; Ray, G.; Kumar, B. Processing and properties of ferroelectric $Bi_{0.5}$ $(Na_{0.65}K_{0.35})_{0.5}$ TiO_3 ceramics under the effect of different sintering temperature. *Scr. Mater.* **2014**, *89*, 61–64. [CrossRef]

25. Yoshida, K.; Fujimori, H. Morphotropic Phase Boundary on K-substituted $Na_{0.5}Bi_{0.5}TiO_3$ synthesized with suppressing evaporation of bismuth and sodium. *Trans. Mater. Res. Soc. Jpn.* **2020**, *45*, 207–210. [CrossRef]
26. Amini, R.; Ghazanfari, M.R.; Alizadeh, M.; Ardakani, H.A.; Ghaffari, M. Structural, microstructural and thermal properties of lead-free bismuth–sodium–barium–titanate piezoceramics synthesized by mechanical alloying. *Mater. Res. Bull.* **2013**, *48*, 482–486. [CrossRef]
27. Naceur, H.; Megriche, A.; Maaoui, M.E. Frequency-dependant dielectric characteristics and conductivity behavior of Sr1−x ($Na_{0.5}Bi_{0.5}$)x$Bi_2Nb_2O_9$ (x = 0.0, 0.2, 0.5, 0.8 and 1.0) ceramics. *Orient. J. Chem.* **2013**, *29*, 937–944. [CrossRef]
28. Singh, P.; Pandey, R.; Singh, P. Polyol-mediated synthesis of Bi-deficient Mg^{2+}-doped sodium bismuth titanate and study of oxide ion migration behaviour with functional properties. *J. Alloys Compd.* **2020**, *860*, 158492.
29. Tsurumi, T.; Harigai, T. *Dielectric and Optical Properties of Perovskite Artificial Superlattices, Handbook of Advanced Dielectric, Piezoelectric and Ferroelectric Materials*; Elsevier: Amsterdam, The Netherlands, 2008; pp. 971–1005.
30. Bacha, E.; Renoud, R.; Terrisse, H.; Borderon, C.; Richard-Plouet, M.; Gundel, H.; Brohan, L. Electrophoretic deposition of $BaTiO_3$ thin films from stable colloidal aqueous solutions. *J. Eur. Ceram. Soc.* **2014**, *34*, 2239–2247. [CrossRef]

coatings

MDPI

Article

Effects of Polysilane Addition to Chlorobenzene and High Temperature Annealing on $CH_3NH_3PbI_3$ Perovskite Photovoltaic Devices

Takeo Oku [1,*], Masaya Taguchi [1], Atsushi Suzuki [1], Kaede Kitagawa [1], Yugo Asakawa [1], Satoshi Yoshida [1], Masanobu Okita [2], Satoshi Minami [2], Sakiko Fukunishi [2] and Tomoharu Tachikawa [2]

1 Department of Materials Science, The University of Shiga Prefecture, 2500 Hassaka, Hikone, Shiga 522-8533, Japan; of21mtaguchi@ec.usp.ac.jp (M.T.); suzuki@mat.usp.ac.jp (A.S.); ov21kkitagawa@ec.usp.ac.jp (K.K.); ov21yasakawa@ec.usp.ac.jp (Y.A.); satoshi.glass.jp@gmail.com (S.Y.)
2 Osaka Gas Chemicals Co., Ltd., 5-11-61 Torishima, Konohana-ku, Osaka 554-0051, Japan; okita@ogc.co.jp (M.O.); s-minami@ogc.co.jp (S.M.); fukunishi@ogc.co.jp (S.F.); t-tachikawa@ogc.co.jp (T.T.)
* Correspondence: oku@mat.usp.ac.jp; Tel.: +81-749-28-8368

Abstract: $CH_3NH_3PbI_3$ perovskite photovoltaic devices treated with a polysilane layer were fabricated and characterized. Decaphenylcyclopentasilane (DPPS) in chlorobenzene solution was deposited at the surface of the perovskite layer, and the resulting device was annealed at 140–260 °C. The photoconversion efficiencies of the DPPS-treated device remained high even after 255 days in ambient air. Raman scattering spectroscopy and ab initio molecular orbital calculations of DPPS suggested that it increased hole transport efficiency in the treated devices, which was confirmed from the high shunt resistances of the DPPS-treated devices.

Keywords: perovskite; $CH_3NH_3PbI_3$; solar cells; polysilane; decaphenylcyclopentasilane; stability; chlorobenzene; calculation; Raman scattering

Citation: Oku, T.; Taguchi, M.; Suzuki, A.; Kitagawa, K.; Asakawa, Y.; Yoshida, S.; Okita, M.; Minami, S.; Fukunishi, S.; Tachikawa, T. Effects of Polysilane Addition to Chlorobenzene and High Temperature Annealing on $CH_3NH_3PbI_3$ Perovskite Photovoltaic Devices. *Coatings* **2021**, *11*, 665. https://doi.org/10.3390/coatings11060665

Academic Editors: Tongtong Xuan and Alessandro Latini

Received: 13 April 2021
Accepted: 26 May 2021
Published: 31 May 2021

1. Introduction

Si-based photovoltaic cells are a widely used energy technology. However, the manufacture of Si-based devices is complex, and the band structure of silicon involves an indirect transition type. Conversely, $CH_3NH_3PbI_3$ ($MAPbI_3$) has the advantages of a direct bandgap, a high conversion efficiency, and an ability to be solution processed [1–5]. Hence, perovskite compounds are regarded as candidates for new generation photovoltaic materials. However, perovskites are normally unstable in air. The instability of $MAPbI_3$ has been attributed to migration and desorption of CH_3NH_3 (MA) and reactions with moisture in air [6,7]. Therefore, the stability of perovskite solar cells must be improved to enable their practical application in modules [8,9].

Various doped perovskite crystals have been widely studied with the aim of improving their stability and photovoltaic properties [10–13]. Although perovskite solar cells doped with $CH_3(NH_2)_2$ [14–16], $CH_3CH_2NH_3$ [17,18], or $C(NH_2)_3$ [19,20] have been developed and studied, these organic molecules may still induce instabilities. Substitutions of CH_3NH_3 by doping sodium [21], potassium [22–24], rubidium, or cesium are expected to be effective for suppressing desorption of CH_3NH_3 sites in the $MAPbI_3$. Doping Rb^+ or Cs^+ to the $MAPbI_3$ also reduced defect densities and increased grain sizes [25–28]. First-principles calculation also indicated that co-doping of alkali metals and copper at the MA and Pb sites, respectively, lowered the distortion and energies of the crystal structures [29,30]. In fact, photovoltaic properties of the co-doped perovskite solar cells were improved and stable, even after one year [31].

Another approach to improving the stability of perovskite solar cells is incorporating polymeric materials into the perovskite devices [32–37]. Polymers have been shown to promote device stability in perovskite cells [38]. For example, coating a thin layer of poly(methyl methacrylate) on top of the perovskite layer forms a cross-linked network structure, which protects the cell from moisture and oxygen [38–40]. Poly(propylene carbonate) is similarly effective for improving stability, owing to the formation of large crystals of crosslinked perovskite particles with few defects [34].

The stability of the $MAPbI_3$ compounds is affected by moisture and oxygen in the air and is also influenced by the hole transport layer (HTL). A common HTL in perovskite solar cells is 2,2′,7,7′-tetrakis-[N,N-di(p-methoxyphenyl)amine]-9,9′-spirobifluorene (spiro-OMeTAD); however, this HTL is expensive and has poor stability. Alternative low cost and more stable HTLs have been developed [36,37], and other HTL materials, such as polysilanes, have also been utilized with organic photovoltaic devices [41,42].

In contrast to organic polymers, polysilanes have two important features. First, polysilanes are *p*-type semiconductors, which promote hole transfer. Second, polysilanes are more stable at elevated temperatures above 300 °C than ordinary organic materials. Polysilanes may function as a protective layer when deposited on perovskite compounds. Hence, polysilanes, such as decaphenylcyclopentasilane (DPPS), have been applied as HTLs [43,44] and as additives in the photoactive layer [45] of $MAPbI_3$ perovskite devices. DPPS has been found to promote a uniform perovskite morphology, which increases device power conversion efficiencies. However, chlorobenzene is typically used to dissolve and deposit DPPS by solution processing. Chlorobenzene can also have effects on device performance, which have not been investigated separately from its use as a solvent for DPPS [46–48]. Thus, there is a need to separately investigate the effects of DPPS and chlorobenzene in detail.

The purpose of the present work was to investigate the photovoltaic properties and stabilities of perovskite photovoltaic devices treated with a DPPS layer. The cells were treated by high temperature annealing in ambient air. The effects of treating devices with DPPS in chlorobenzene on the photovoltaic properties and microstructures were investigated. When only the DPPS is used as the HTL, the obtained conversion efficiencies are not enough; consequently, the DPPS/spiro-OMeTAD bilayer was applied in the present work. The chlorobenzene-treated devices were also compared to investigate the effect of the chlorobenzene. To increase the conversion efficiency by raising the fabrication temperatures of the devices, the device preparation time was shortened.

2. Experimental and Calculation Procedures

Figure 1a shows the fabrication process of the devices. Detailed conditions for the fabrication process have been described in previous reports [10,48–51]. All fabrication processes were performed under atmospheric conditions in ambient air, and the temperature and humidity were ~20 °C and ~30%, respectively. A compact TiO_2 layer and a mesoporous TiO_2 layers were formed on a fluorine-doped tin oxide (FTO) substrate by annealing at 550 °C. To prepare the perovskite compounds, solutions of $PbCl_2$ (Sigma Aldrich, Tokyo, Japan, 111.2 mg) and CH_3NH_3I (Tokyo Chemical Industry, 190.7 mg) with the desired molar ratio were mixed in *N*,*N*-dimethylformamide (Sigma Aldrich, 0.5 mL) at 60 °C for 24 h. These perovskite precursor solutions were normally spin-coated during the first coating. During the second and third spin-coating steps, an air-blowing method was employed [50], as illustrated in Figure 1b. The cells were maintained at 90 °C during the air-blowing. DPPS (Osaka Gas Chemicals, OGSOL SI-30-15, Osaka, Japan, 10 mg) solutions were prepared in chlorobenzene (0.5 mL) and dropped onto the perovskite layer during the last 15 s of the third spin-coating of the perovskite precursor solutions, as shown in Figure 1b. A suitable temperature to initiate the reaction of the starting materials ($3CH_3NH_3I$ and $PbCl_2$) for forming $MAPbI_3$ and $2CH_3NH_3Cl$ (as a gas) is 140 °C [10,50]. The devices with DPPS layers were annealed at temperatures in the range of 140 to 260 °C for 1 to 30 min. Then, a spiro-OMeTAD layer was formed as an HTL by spin-coating, and the spiro-OMeTAD layer was formed below the gold electrodes for all the fabricated devices in the present work. Finally, gold (Au) electrodes were formed by evaporation.

All the fabricated cells in the present work were put into dark storage at a temperature of 22 °C and ~30% humidity in ambient air.

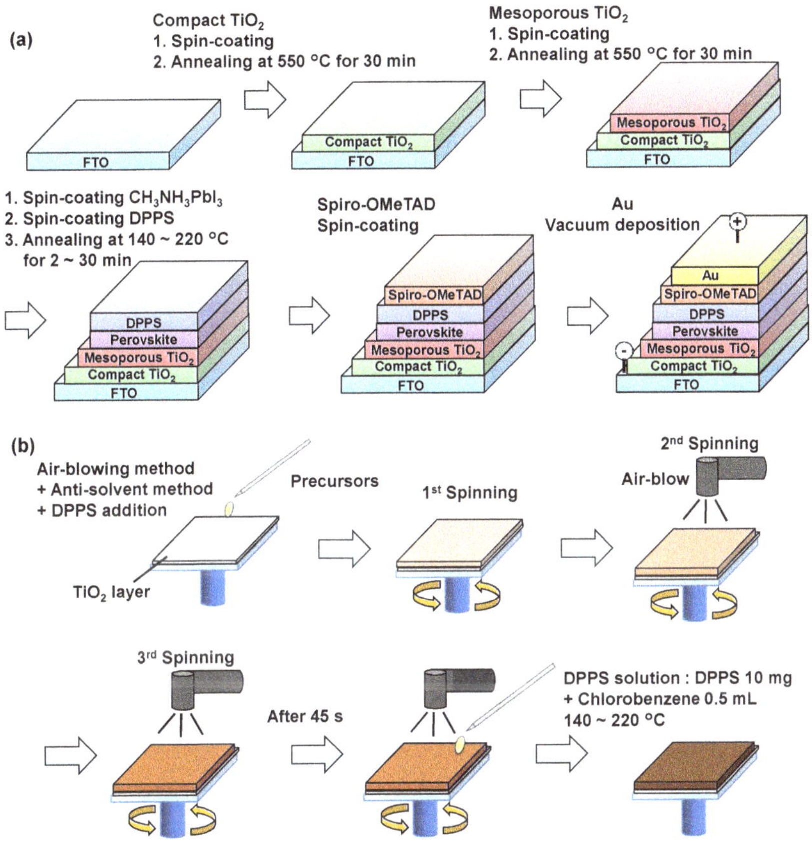

Figure 1. (**a**) Schematic illustration of the fabrication process of the perovskite photovoltaic devices. (**b**) Detailed illustration of DPPS insertion process.

Detailed conditions for the characterization of the devices have been described in previous reports [51–53]. Microstructural analysis was conducted by an X-ray diffractometer (Bruker, Billerica, MA, USA, D2 PHASER). Raman scattering spectra were measured with a Raman microspectrometer (JASCO, Tokyo, Japan, NRS-5100). The surface morphologies of the perovskite layers were examined using an optical microscope (Nikon, Tokyo, Japan, Eclipse E600). The current density voltage characteristics of the fabricated devices were measured (Keysight, Santa Rosa, CA, USA, B2901A) under a solar simulator (San-ei Electric, Osaka, Japan, XES-301S) with irradiation at 100 mW cm^{-2}. Geometry optimization and energy calculation of the DPPS molecule were performed by the ab initio calculation using the restricted open-shell Hartree–Fock (RHF) method as the approximated wavefunctions with STO-3G* basis set (Gaussian 09) [21,54,55]. The electron density distributions around the lowest unoccupied molecular orbital (LUMO) and the highest occupied molecular orbital (HOMO) were calculated from total self-consistent-field density. The isovalue for the MO on the surface were adjusted to be 0.02 Hartree. The hybrid orbital was widely mapped on the atoms, yielding delocalization. The detailed electron density and Mulliken charge was described in the log file. The electrostatic potential was calculated from the Mulliken population analysis. As the Mulliken population analysis, the charge distribution of atoms was calculated while considering the atomic orbital overlapping and molecular orbital coefficient. The electrostatic potential was displayed on the surface of the isoelectron density and was mapped with electron density from total self-consistent-field density. The density of electrostatic potential was adjusted to be 0.03. Maximum and minimum energy values with the Hartree unit are shown by scale bars. The positive charge was distributed as blue electrostatic potential around the atom. Raman scattering spectra and the vibration modes were calculated by RHF with STO-3G* using frequency mode.

3. Results and Discussion

Figure 2a shows current density voltage (*J*-*V*) curves of the fabricated solar cells. The performance of the cells is summarized in Table 1. The measured parameters were as follows: V_{OC}: open-circuit voltage, J_{SC}: short-circuit current density, FF: fill factor, η: conversion efficiency, η_{ave}: averaged efficiency of four cells, R_S: series resistance, and R_{Sh}: shunt resistance. Devices prepared with only chlorobenzene had an η of 3.87% after annealing at 140 °C. To enable a comparison of the chlorobenzene-treated devices with the DPPS-treated devices, a pair of these devices were annealed at 190 °C. The device prepared with DPPS in chlorobenzene had η of 9.40%. The same device fabricated at 220 °C had an initially higher η value of 10.04%. All cell parameters were improved for the devices treated with DPPS in chlorobenzene. The J_{SC} and FF markedly increased compared with those values of the chlorobenzene-treated devices.

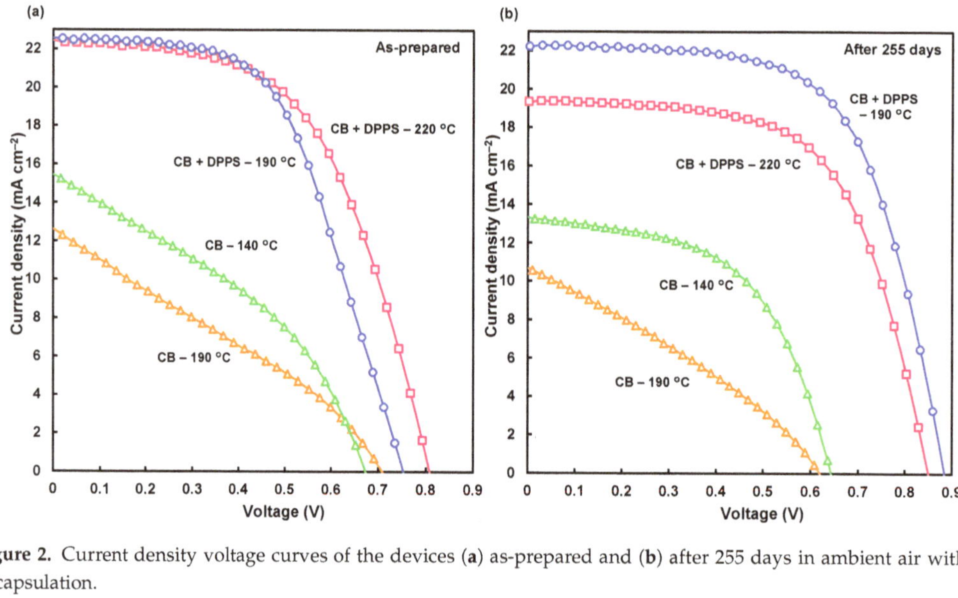

Figure 2. Current density voltage curves of the devices (**a**) as-prepared and (**b**) after 255 days in ambient air without encapsulation.

Stabilities of the photovoltaic parameters after preparation in ambient air were measured for the cells over 255 days, and *J*-*V* characteristics are shown in Figure 2b. After 255 days, the highest photoconversion efficiency of 12.4% was obtained for the DPPS device prepared at 190 °C (Table 1). The DPPS device prepared at 220 °C had good stability over the extended time, as shown in Figure 3. Whereas the photoconversion efficiency of the chlorobenzene-treated devices prepared at 190 °C decreased over time, η for the devices prepared with DPPS in chlorobenzene at 190 °C increased; hence, DPPS effectively increased the photovoltaic properties when subjected to high temperature annealing. Although the FF values increased for all the devices, V_{OC} decreased for the chlorobenzene-treated device after 255 days. Conversely, V_{OC} for the DPPS-treated devices increased. The hysteresis index (HI) is also calculated and listed in Table 1. The HI values were estimated from the next equation [56]: HI = $(J_{RH} - J_{FH})/J_{RH}$, where J_{RH} is the current density at the half open-circuit voltage for the reverse scan, and J_{FH} is the current density at the half open-circuit voltage for the forward scan. When there is no hysteresis, the HI is equal to 0. Although the HI values of CB devices were lower than those of CB + DPPS devices for the as-prepared cells, HI values of CB + DPPS and CB devices decreased and increased after 255 days, respectively. The parameters of the DPPS devices either remained stable or increased over time, which is attributed to the reduced influence of moisture, oxygen, and spiro-OMeTAD in the DPPS protected perovskite layers. Shunt resistances were high for the DPPS-added devices, which is likely because of the effects of DPPS on hole transport and electron blocking [45,48].

Table 1. Photovoltaic parameters of perovskite photovoltaic devices. * Prepared from $CH_3NH_3I{:}PbI_2$ = 1:1 and without air blowing.

Solution	Annealing (°C, min)	J_{SC} (mA cm^{-2})	V_{OC} (V)	FF	R_S (Ω cm^2)	R_{Sh} (Ω cm^2)	η (%)	η_{ave} (%)	HI
*	100, 15	15.8	0.800	0.550	8.89	525	6.94	6.76	0.029
CB	140, 12	15.4	0.674	0.373	12.35	72	3.87	3.65	−0.082
CB	190, 6	12.6	0.708	0.298	18.87	71	2.66	2.30	−0.034
CB + DPPS	140, 60	12.6	0.903	0.660	8.13	916	7.51	6.06	0.076
CB + DPPS	190, 30	22.5	0.753	0.554	11.92	5670	9.40	8.99	0.334
CB + DPPS	220, 2	22.4	0.808	0.554	9.33	831	10.04	9.40	0.125
CB + DPPS	250, 1.5	20.5	0.875	0.524	9.90	268	9.40	8.71	0.043
CB + DPPS	260, 1	13.5	0.884	0.483	24.37	709	5.75	5.42	0.327
After 255 days	-	-	-	-	-	-	-	-	-
CB	140, 12	13.3	0.643	0.543	8.35	330	4.64	4.54	0.102
CB	190, 6	10.6	0.620	0.311	23.71	86	2.05	1.70	−0.087
CB + DPPS	190, 30	22.2	0.884	0.634	7.35	5830	12.44	11.84	0.086
CB + DPPS	220, 2	19.3	0.849	0.618	8.03	2700	10.15	9.39	0.044

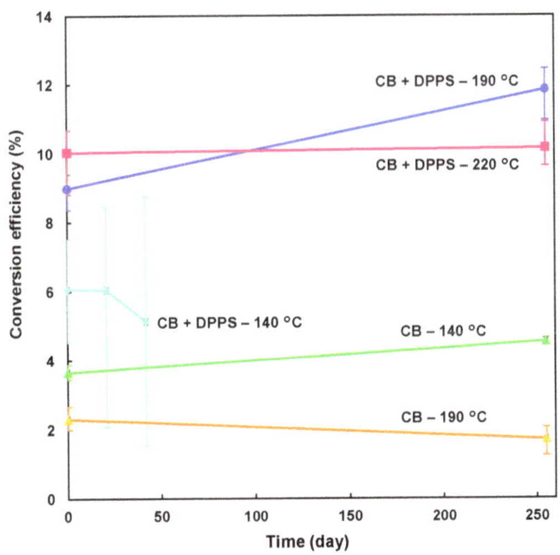

Figure 3. Stabilities of the present devices.

Optical microscope images of the perovskites in the present devices measured after 255 days are shown in Figure 4. The perovskite grains were dispersed and divided by space for the chlorobenzene-treated devices, as observed in Figure 4a,b. On the other hand, the morphologies of the perovskite changed drastically by adding DPPS, and the perovskite grains seems to form smoother surface. Hence, the DPPS-treated devices had fewer grain boundaries and a greater surface coverage of grains. This morphology should suppress carrier recombination and reduce carrier losses. These effects were confirmed by the low R_S and high R_{Sh} values for the DPPS-treated devices.

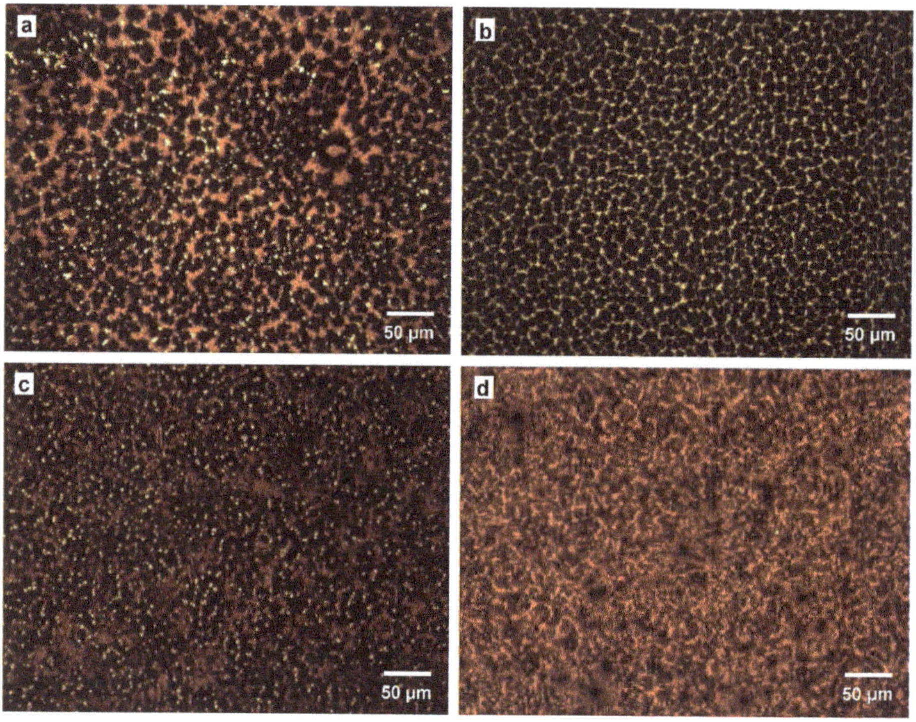

Figure 4. Optical microscope images of cells prepared with chlorobenzene and annealed at (**a**) 140 °C and (**b**) 190 °C; and cells prepared with DPPS in chlorobenzene annealed at (**c**) 190 °C and (**d**) 220 °C.

Figure 5a shows X-ray diffraction (XRD) results of the devices after 255 days. The diffraction patterns were indexed to a cubic perovskite structure [10,49]. The (100) diffraction peaks were higher for both the devices treated with chlorobenzene and DPPS in chlorobenzene prepared at 190 °C, as observed in the enlarged XRD patterns in Figure 5b.

The lattice constant of the perovskite compound decreased slightly at 220 °C (Table 2), which indicated desorption of MA. The lattice constant was smallest for the chlorobenzene device prepared at 190 °C, indicating that the MA desorption was associated with a decrease in conversion efficiency. The device treated with DPPS in chlorobenzene at 190 °C had the largest lattice constant, indicating that the MA desorption was suppressed, contributing to the high efficiency.

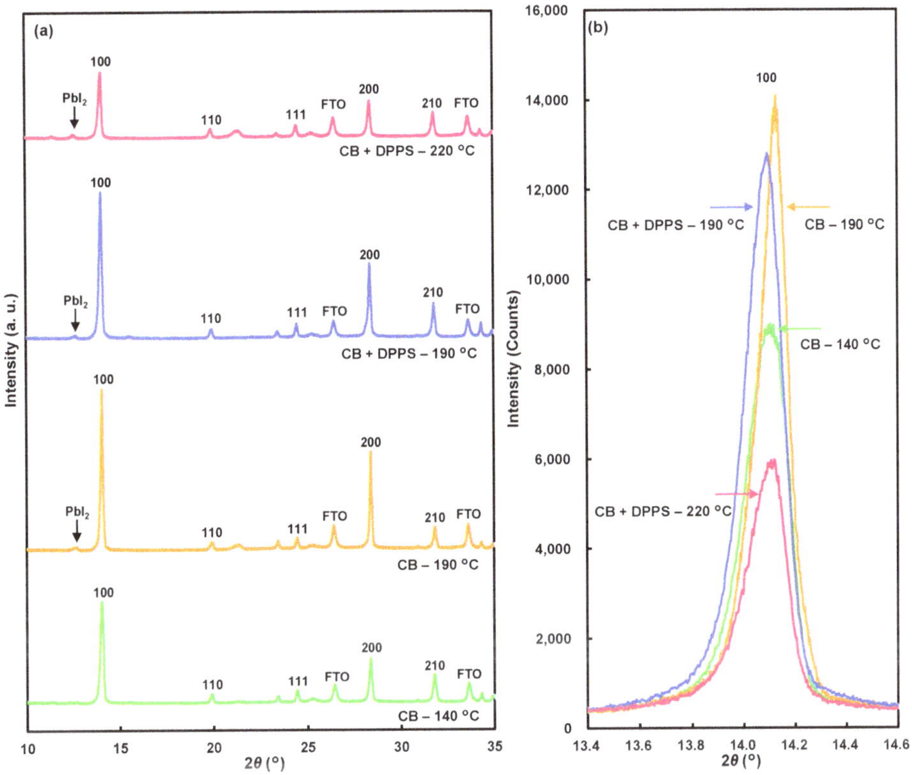

Figure 5. (**a**) X-ray diffraction patterns of the devices. (**b**) Enlarged patterns of 100 reflections.

Table 2. Crystallographic data of perovskite films.

Solution	Annealing (°C, min)	Lattice Constant a (Å)	Orientation I_{100}/I_{210}
CB	140, 12	6.270(0)	3.6
CB	190, 6	6.265(1)	7.5
CB + DPPS	190, 30	6.276(0)	4.4
CB + DPPS	220, 2	6.270(1)	2.7

The crystal orientation of the perovskite grains was estimated from the ratios of the 100 intensity (I_{100}) to the 210 intensity (I_{210}) in the XRD patterns, as summarized in Table 2. When the crystal planes in the perovskite crystallites were randomly aligned, the intensity ratio of I_{100}/I_{210} was 2.08 [10]. For the device prepared with chlorobenzene at 190 °C, I_{100}/I_{210} was 7.5, which indicates that the (100) planes were comparatively well aligned with the FTO substrate. For the devices treated with DPPS in chlorobenzene, I_{100}/I_{210} decreased to 4.4 after annealing at 190 °C to 2.7 for devices annealed at 220 °C. Hence, the DPPS treatment of the perovskite promoted more randomly aligned structures. Small PbI_2 peaks were observed for both devices treated with chlorobenzene and DPPS in chlorobenzene after annealing at 190 °C. Further formation of PbI_2 was suppressed during annealing at 220 °C, which indicates that the DPPS protected the $MAPbI_3$ crystals against decomposition to PbI_2.

Raman spectroscopy measurements of DPPS and the present as-prepared photovoltaic devices are shown in Figure 6, together with calculated data for the DPPS. The Raman scattering peaks at ~600, ~1100, ~1540, and ~3040 cm^{-1} are respectively assigned to Si-Si, phenyl-Si, phenyl group, and phenyl C-H groups of the DPPS. Several peaks were assigned to MA, CH_3, and NH_3 for the devices treated with chlorobenzene and DPPS in chlorobenzene. These peaks derive from internal vibrations of MA relating to its local symmetry in the crystal symmetry [57,58]. Because the amount of DPPS was quite small, no clear differences were apparent in the Raman spectra of the devices treated with chlorobenzene and DPPS in chlorobenzene.

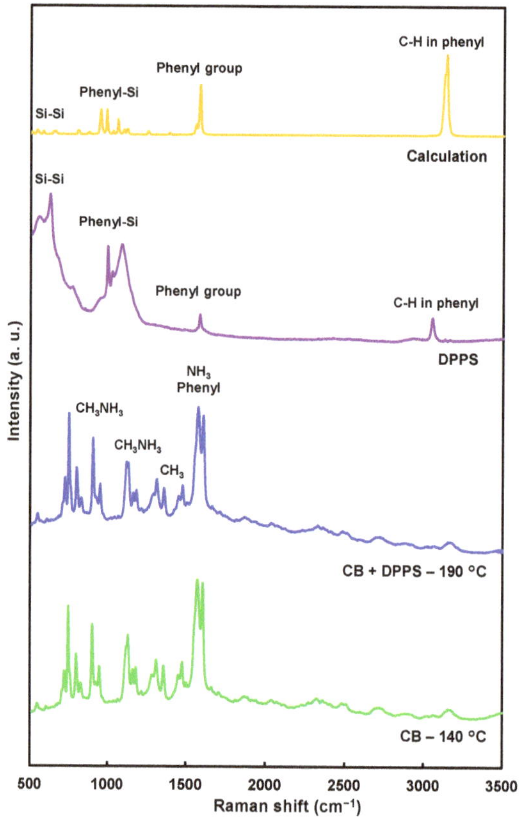

Figure 6. Raman spectra of DPPS and devices.

In this study, the DPPS was dissolved in chlorobenzene, which is often used as an antisolvent to promote grain growth and form smooth surface structures on perovskite films, resulting increased current densities [59–61]. Although temperatures around ~100 °C are commonly used to fabricate perovskite devices, high temperatures above ~180 °C are required to improve the efficiencies of DPPS-treated cells. Thus, DPPS affects the morphology and photoelectronic properties by a different mechanism from that of chlorobenzene. The DPPS layer suppresses MA desorption and DPPS is also a *p*-type semiconductor, which has hole transporting properties that inhibit hole and electron recombination.

The *J*-*V* characteristics of the champion DPPS-treated device with the highest conversion efficiency in the present work are also shown in Figure 7, and the measured photovoltaic parameters are listed in Table 3. The device was annealed at 190 °C for 5 min. Although the conversion efficiency of this as-prepared device was lower than that prepared at 190 °C for 30 min, its efficiency increased to ~15% after 66 days. Changes of the (100) XRD reflections for the champion device in the present work are shown in Figure 7b, and the crystallographic data are summarized in Table 4. The perovskite crystallites were randomly aligned after 10 days, and the intensity ratio of I_{100}/I_{210} increased from 1.9 to 2.6

after 66 days, which indicates that the (100) planes were comparatively well aligned. In addition, the crystallite size increased from 486 to 617 Å after 66 days. This indicates that the increase of the conversion efficiencies would be caused by the crystal growth of the perovskite compounds during room temperature aging. This crystallization mechanism even after the annealing at the high temperature of 190 °C would be explained by the DPPS treatment, which might slow the diffusion of ions and crystal growth during annealing. Then, the non-crystalized phase that remained might contribute to the crystal growth during the aging.

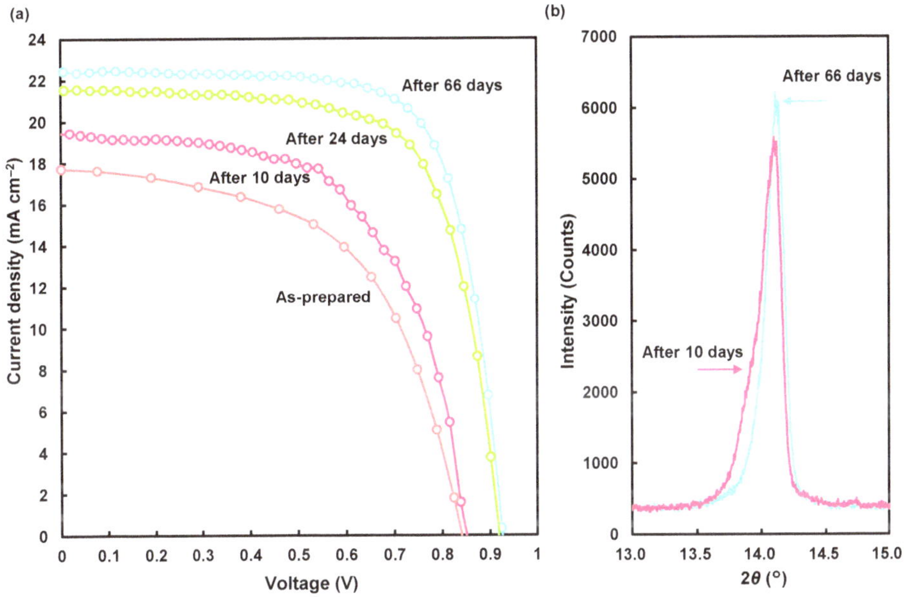

Figure 7. (**a**) Changes of the *J-V* curves for the champion device in the present work. (**b**) X-ray diffraction patterns of 100 reflections.

Table 3. Photovoltaic parameters of champion device, treated with DPPS in chlorobenzene and annealed at 190 °C for 5 min.

Time (Day)	J_{SC} (mA cm^{-2})	V_{OC} (V)	FF	R_S (Ω cm^2)	R_{Sh} (Ω cm^2)	η (%)	η_{ave} (%)	HI
0	17.7	0.839	0.558	4.41	908	8.30	6.58	0.268
10	19.4	0.850	0.593	3.49	1440	9.80	8.81	0.034
24	21.6	0.918	0.699	3.96	1610	13.82	12.48	0.046
58	22.3	0.930	0.724	3.66	2120	15.03	13.31	0.073
66	22.4	0.923	0.727	3.72	3070	15.10	13.38	0.085

Table 4. Changes of crystallographic data of perovskite films.

Time (Day)	Lattice Constant *a* (Å)	Orientation I_{100}/I_{210}	Crystallite Size D_{100} (Å)
10	6.272(1)	1.9	486
66	6.268(1)	2.6	617

Schematic and optimized structural models of DPPS are shown in Figure 8a,b, respectively. Pentagonal Si bonding is present in both models. Figure 8c,d shows an electrostatic potential map of DPPS and its cutaway view, respectively, as calculated by ab initio methods based on the HF. The electrostatic potential was positive (blue) around the cyclopentasilane and proton, as shown in Figure 8c,d. Calculated electronic structures of the DPPS at the HOMO and the LUMO energy levels are shown in Figure 8e,f, respectively. The phases of electron densities in the Si-3p and C-2p orbitals were inverted, as indicated by the green and red coloration. The electronic charge of the HOMO was broadly distributed over the phenyl rings and Si-Si chains, which contributed to the carrier transport and electronic properties. The length of the main Si-Si chain also affects the localization of σ electrons, which determines the LUMO level.

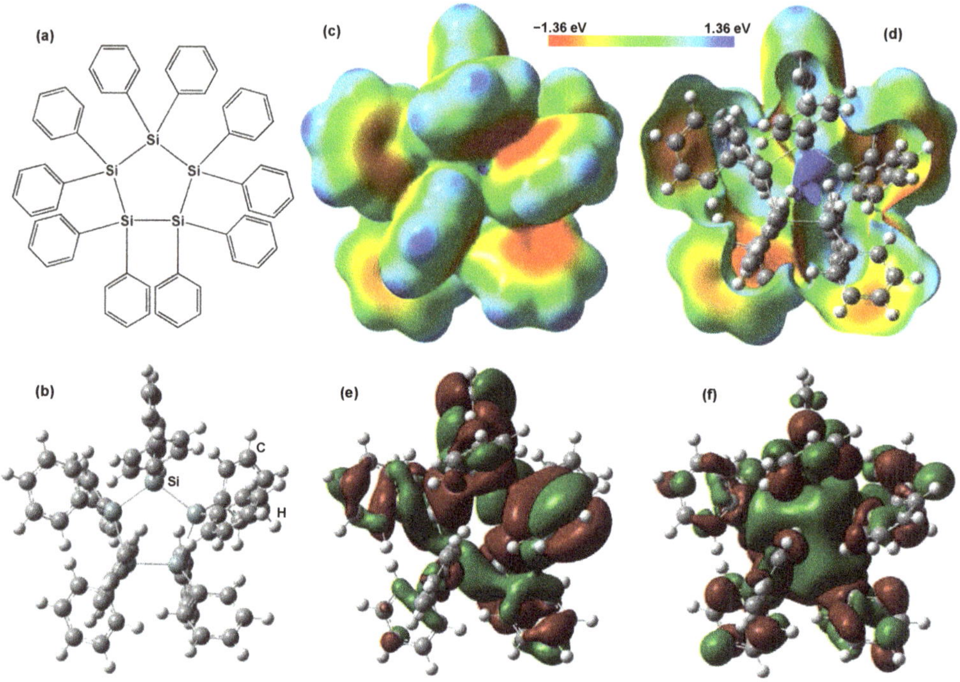

Figure 8. (**a**) Schematic and (**b**) optimized models of DPPS. (**c**) Electrostatic potential, (**d**) cutaway view of (**c**), (**e**) HOMO, and (**f**) LUMO.

An energy level diagram of the present DPPS-treated perovskite cells is shown in Figure 9. The energy levels of the valence band maximum, conduction band minimum, HOMO, and LUMO are indicated in the diagram [62,63]. When the device was irradiated from the FTO glass side, carriers (holes and electrons) separate at the interfaces. Holes separated in the perovskite layer are carried through the PbI_2, DPPS, and spiro-OMeTAD to the gold electrode. Conversely, electrons are transported through titanium dioxide to the FTO. By inserting a DPPS layer between the photoactive layer and the HTL, holes are effectively transported from the valence band maximum of the $MAPbI_3$ to the Fermi level of Au. High shunt resistances were obtained for the DPPS-treated devices, which are attributed to the hole transporting and formation of smoother surface morphology by DPPS. Efficient carrier transport is likely caused by the specific arrangement of the phenyl group around the cyclopentasilane in the DPPS [41].

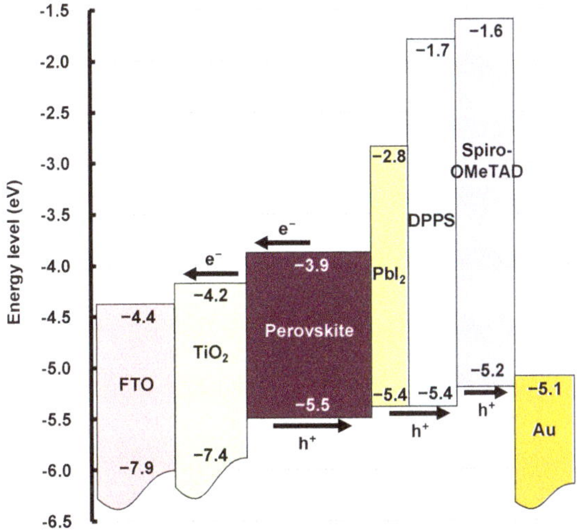

Figure 9. Energy level diagram of the present cell.

A small PbI_2 layer might be formed by MA desorption at the perovskite/DPPS interface at high temperatures. If this thin PbI_2 layer forms at the perovskite/DPPS interface during or after annealing, PbI_2 may act as a p-type semiconductor and an HTL [64,65]. Activation energies of ion migration of MA^+, I^- and Pb^{2+} in the $MAPbI_3$ were reported to be 0.84, 0.58, and 2.31 eV, respectively [66]. Since the activation energy of Pb^{2+} migration is higher than those of other ions, the formed PbI_2 layer may remain around the surface of the perovskite. The increased efficiency of the DPPS-treated devices might also be related to crystallization of amorphous grains. During the spin-coating of DPPS, a composite layer of DPPS and amorphous pre-perovskite compounds forms, which provides a solid interface for room temperature aging. Because DPPS can also function as a hole transport material [41], holes are efficiently transported at the interface, to improve the R_{sh} and V_{OC}. Since all the processes in the present work were performed in the ambient air, further improvement of photovoltaic properties is expected by controlling the environmental conditions.

4. Conclusions

In summary, the effects of a DPPS treatment on perovskite solar cells were investigated. The DPPS layer was inserted at the perovskite/spiro-OMeTAD interface. Conversion efficiencies improved by inserting the DPPS layer during spin-coating of $MAPbI_3$ and annealing above 190 °C. A cell fabricated at 220 °C had the highest photoconversion efficiency among the as-fabricated cells, and the conversion efficiencies of all devices remained stable over more than 8 months in air. In addition, a device fabricated at 190 °C had the highest efficiency following room temperature aging. The DPPS layer acts as both a protective layer for the perovskite and as an HTL. Although a small amount of PbI_2 was detected by XRD, the PbI_2 layer likely also functioned as an HTL. The perovskite grains grew more densely, and their surface coverage increased compared with that resulting from the ordinary chlorobenzene anti-solvent method. The DPPS treatment promoted fewer lattice defects and grain boundaries, which suppressed the leakage current and increased the JSC. The effectiveness of the DPPS on hole transport was also confirmed by ab initio molecular orbital calculations. These findings indicate that high temperature annealing of devices treated with DPPS in chlorobenzene is an effective and easy method for improving the photoconversion efficiencies and stability of $MAPbI_3$ solar cells.

Author Contributions: Conceptualization, T.O. and M.T.; methodology, T.O., M.T., A.S., K.K., Y.A. and S.Y.; formal analysis, T.O., M.T, A.S., K.K., Y.A. and S.Y.; investigation, M.T., Y.A., K.K., A.S. and S.Y.; resources, M.O., S.M., S.F. and T.T.; data curation, T.O., M.T., K.K. and Y.A.; writing—original draft preparation, T.O.; writing—review and editing, M.T., A.S., K.K., Y.A., S.Y., M.O., S.M., S.F. and T.T.; project administration, T.O.; funding acquisition, T.O. All authors have read and agreed to the published version of the manuscript.

Funding: This research was partly funded by the Super Cluster Program of the Japan Science and Technology Agency (JST) and by a Grant-in-Aid for Scientific Research (C) 21K04809.

Institutional Review Board Statement: Not Applicable.

Informed Consent Statement: Not Applicable.

Data Availability Statement: Data is contained within the article.

Conflicts of Interest: The authors declare no conflict of interest.

References

1. Gedamu, D.; Asuo, I.M.; Benetti, D.; Basti, M.; Ka, I.; Cloutier, S.G.; Rosei, F.; Nechache, R. Solvent-antisolvent ambient processed large grain size perovskite thin films for high-performance solar cells. *Sci. Rep.* **2018**, *8*, 12885. [CrossRef] [PubMed]
2. Mingyu, J.; Choi, I.W.; Go, E.M.; Cho, Y.; Kim, M.; Byongkyu, L.; Seonghun, J.; Yimhyun, J.; Choi, H.W.; Lee, J.; et al. Stable perovskite solar cells with efficiency exceeding 24.8% and 0.3-V voltage loss. *Science* **2020**, *369*, 1615–1620. [CrossRef]
3. Miyasaka, T.; Kulkarni, A.; Kim, G.M.; Öz, S.; Jena, A.K. Perovskite solar cells: Can we go organic-free, lead-free, and dopant-free? *Adv. Energy Mater.* **2020**, *10*, 1902500. [CrossRef]
4. Tong, J.; Song, Z.; Kim, D.M.; Chen, X.; Chen, C.; Palmstrom, A.F.; Ndione, P.F.; Reese, M.O.; Dunfield, S.P.; Reid, O.G.; et al. Carrier lifetimes of >1 μs in Sn-Pb perovskites enable efficient all-perovskite tandem solar cells. *Science* **2019**, *364*, 475–479. [CrossRef] [PubMed]
5. Wang, F.; Yang, M.; Yang, S.; Qu, X.; Yang, L.; Fan, L.; Yang, J.; Rosei, F. Iodine-assisted antisolvent engineering for stable perovskite solar cells with efficiency >21.3%. *Nano Energy* **2020**, *67*, 104224. [CrossRef]
6. Dunfield, S.P.; Bliss, L.; Zhang, F.; Luther, J.M.; Zhu, K.; van Hest, M.F.A.M.; Reese, M.O.; Berry, J.J. From defects to degradation: A mechanistic understanding of degradation in perovskite solar cell devices and modules. *Adv. Energy Mater.* **2020**, *10*, 1904054. [CrossRef]
7. Lee, J.W.; Kim, S.G.; Yang, J.M.; Yang, Y.; Park, N.G. Verification and mitigation of ion migration in perovskite solar cells. *APL Mater.* **2019**, *7*, 041111. [CrossRef]
8. Zhang, X.; Yin, J.; Nie, Z.; Zhang, Q.; Sui, N.; Chen, B.; Zhang, Y.; Qu, K.; Zhao, J.; Zhou, H. Lead-free and amorphous organic–inorganic hybrid materials for photovoltaic applications: Mesoscopic $CH_3NH_3MnI_3/TiO_2$ heterojunction. *RSC Adv.* **2017**, *7*, 37419–37425. [CrossRef]
9. Dong, H.; Wu, Z.; Xi, J.; Xu, X.; Zuo, L.; Lei, T.; Zhao, X.; Zhang, L.; Hou, X. Pseudohalide-induced recrystallization engineering for $CH_3NH_3PbI_3$ film and its application in highly efficient inverted planar heterojunction perovskite solar cells. *Adv. Funct. Mater.* **2017**, *28*, 1704836. [CrossRef]
10. Oku, T. Crystal structures of perovskite halide compounds used for solar cells. *Rev. Adv. Mater. Sci.* **2020**, *59*, 264–305. [CrossRef]
11. Travis, W.; Glover, E.N.K.; Bronstein, H.; Scanlon, D.O.; Palgrave, R.G. On the application of the tolerance factor to inorganic and hybrid halide perovskites: A revised system. *Chem. Sci.* **2016**, *7*, 4548–4556. [CrossRef]
12. Hoefler, S.F.; Trimmel, G.; Rath, T. Progress on lead-free metal halide perovskites for photovoltaic applications: A review. *Monatsh. Chem.* **2017**, *148*, 795–826. [CrossRef] [PubMed]
13. Tanaka, H.; Oku, T.; Ueoka, N. Structural stabilities of organic–inorganic perovskite crystals. *Jpn. J. Appl. Phys.* **2018**, *57*, 08RE12. [CrossRef]
14. Zhou, Y.; Yang, M.; Pang, S.; Zhu, K.; Padture, N.P. Exceptional morphology-preserving evolution of formamidinium lead triiodide perovskite thin films via organic-cation displacement. *J. Am. Chem. Soc.* **2016**, *138*, 5535–5538. [CrossRef]
15. Hu, M.; Liu, L.; Mei, A.; Yang, Y.; Liu, T.; Han, H. Efficient hole conductor-free, fully printable mesoscopic perovskite solar cells with a broad light harvester $NH_2CH{=}NH_2PbI_3$. *J. Mater. Chem. A* **2014**, *2*, 17115–17121. [CrossRef]
16. Suzuki, A.; Kato, M.; Ueoka, N.; Oku, T. Additive effect of formamidinium chloride in methylammonium lead halide compound-based perovskite solar cells. *J. Electron. Mater.* **2019**, *48*, 3900–3907. [CrossRef]
17. Wang, Y.; Zhang, T.; Li, G.; Xu, F.; Li, Y.; Yang, Y.; Zhao, Y. A mixed-cation lead iodide $MA_{1-x}EA_xPbI_3$ absorber for perovskite solar cells. *J. Energy Chem.* **2018**, *27*, 215–218. [CrossRef]
18. Nishi, K.; Oku, T.; Kishimoto, T.; Ueoka, N.; Suzuki, A. Photovoltaic characteristics of $CH_3NH_3PbI_3$ perovskite solar cells added with ethylammonium bromide and formamidinium iodide. *Coatings* **2020**, *10*, 410. [CrossRef]
19. Jodlowski, A.D.; Roldán-Carmona, C.; Grancini, G.; Salado, M.; Ralaiarisoa, M.; Ahmad, S.; Koch, N.; Camacho, L.; Miguel, G.; Nazeeruddin, M. Large guanidinium cation mixed with methylammonium in lead iodide perovskites for 19% efficient solar cells. *Nat. Energy* **2017**, *2*, 972–979. [CrossRef]

20. Kishimoto, T.; Suzuki, A.; Ueoka, N.; Oku, T. Effects of guanidinium addition to $CH_3NH_3PbI_{3-x}Cl_x$ perovskite photovoltaic devices. *J. Ceram. Soc. Jpn.* **2019**, *127*, 491–497. [CrossRef]
21. Suzuki, A.; Miyamoto, Y.; Oku, T. Electronic structures, spectroscopic properties, and thermodynamic characterization of sodium or potassium-incorporated $CH_3NH_3PbI_3$ by first principles calculation. *J. Mater. Sci.* **2020**, *55*, 9728–9738. [CrossRef]
22. Zheng, F.; Chen, W.; Bu, T.; Ghiggino, K.P.; Huang, F.; Cheng, Y.; Tapping, P.; Kee, T.W.; Jia, B.; Wen, X. Triggering the passivation effect of potassium doping in mixed-cation mixed-halide perovskite by light illumination. *Adv. Energy Mater.* **2019**, *9*, 1901016. [CrossRef]
23. Machiba, H.; Oku, T.; Kishimoto, T.; Ueoka, N.; Suzuki, A. Fabrication and evaluation of K-doped $MA_{0.8}FA_{0.1}K_{0.1}PbI_3(Cl)$ perovskite solar cells. *Chem. Phys. Lett.* **2019**, *730*, 117–123. [CrossRef]
24. Kandori, S.; Oku, T.; Nishi, K.; Kishimoto, T.; Ueoka, N.; Suzuki, A. Fabrication and characterization of potassium- and formamidinium-added perovskite solar cells. *J. Ceram. Soc. Jpn.* **2020**, *128*, 805–811. [CrossRef]
25. Bush, K.A.; Frohna, K.; Prasanna, R.; Beal, R.E.; Leijtens, T.; Swifter, S.A.; McGehee, M.D. Compositional engineering for efficient wide band gap perovskites with improved stability to photoinduced phase segregation. *ACS Energy Lett.* **2018**, 3, 428–435. [CrossRef]
26. Liu, C.; Kong, W.; Li, W.; Chen, H.; Li, D.; Wang, W.; Xu, B.; Cheng, C.; Jen, A.K.Y. Enhanced stability and photovoltage for inverted perovskite solar cells via precursor engineering. *J. Mater. Chem. A* **2019**, *7*, 15880–15886. [CrossRef]
27. Zhang, M.; Yun, J.S.; Ma, Q.; Zheng, J.; Lau, C.F.J.; Deng, X.; Kim, J.; Kim, D.; Seidel, J.; Green, M.A.; et al. High-efficiency rubidium-incorporated perovskite solar cells by gas quenching. *ACS Energy Lett.* **2017**, *2*, 438–444. [CrossRef]
28. Turren-Cruz, S.H.; Saliba, M.; Mayer, M.T.; Juárez-Santiesteban, H.; Mathew, X.; Nienhaus, L.; Tress, W.; Erodici, M.P.; Sher, M.J.; Bawendi, M.G.; et al. Enhanced charge carrier mobility and lifetime suppress hysteresis and improve efficiency in planar perovskite solar cells. *Energy Environ. Sci.* **2018**, *11*, 78–86. [CrossRef]
29. Ueoka, N.; Oku, T.; Suzuki, A. Additive effects of alkali metals on Cu-modified $CH_3NH_3PbI_{3-\delta}Cl_\delta$ photovoltaic devices. *RSC Adv.* **2019**, *9*, 24231–24240. [CrossRef]
30. Ueoka, N.; Oku, T. Effects of co-addition of sodium chloride and copper(II) bromide to mixed-cation mixed-halide perovskite photovoltaic devices. *ACS Appl. Energy Mater.* **2020**, *3*, 7272–7283. [CrossRef]
31. Ueoka, N.; Oku, T.; Suzuki, A. Effects of doping with Na, K, Rb, and formamidinium cations on $(CH_3NH_3)_{0.99}Rb_{0.01}Pb_{0.99}Cu_{0.01}$-$I_{3-x}(Cl, Br)_x$ perovskite photovoltaic cells. *AIP Adv.* **2020**, *10*, 125023. [CrossRef]
32. Chen, Z.; Dong, Q.; Liu, Y.; Bao, C.; Fang, Y.; Lin, Y.; Tang, S.; Wang, Q.; Xiao, X.; Bai, Y.; et al. Thin single crystal perovskite solar cells to harvest below-bandgap light absorption. *Nat. Commun.* **2017**, *8*, 1–7. [CrossRef] [PubMed]
33. Wang, F.; Shimazaki, A.; Yang, F.; Kanahashi, K.; Matsuki, K.; Miyauchi, Y.; Takenobu, T.; Wakamiya, A.; Murata, Y.; Matsuda, K. Highly efficient and stable perovskite solar cells by interfacial engineering using solution-processed polymer layer. *J. Phys. Chem. C* **2017**, *121*, 1562–1568. [CrossRef]
34. Han, T.H.; Lee, J.W.; Choi, C.; Tan, S.; Lee, C.; Zhao, Y.; Dai, Z.; Marco, N.D.; Lee, S.J.; Bae, S.H.; et al. Perovskite-polymer composite cross-linker approach for highly-stable and efficient perovskite solar cells. *Nat. Commun.* **2019**, *10*, 520. [CrossRef] [PubMed]
35. Kim, G.W.; Choi, H.; Kim, M.; Lee, J.; Son, S.Y.; Park, T. Hole transport materials in conventional structural (*n–i–p*) perovskite solar cells: From past to the future. *Adv. Energy Mater.* **2020**, *10*, 1903403. [CrossRef]
36. Calió, L.; Kazim, S.; Grätzel, M.; Ahmad, S. Hole-transport materials for perovskite solar cells. *Angew. Chem. Int. Ed.* **2016**, *55*, 14522–14545. [CrossRef]
37. Singh, R.; Singh, P.K.; Bhattacharya, B.; Rhee, H.-W. Review of current progress in inorganic hole-transport materials for perovskite solar cells. *Appl. Mater. Today* **2019**, *14*, 175–200. [CrossRef]
38. Bi, D.; Yi, C.; Luo, J.; Decoppet, J.D.; Zhang, F.; Zakeeruddin, S.M.; Li, X.; Hagfeldt, A.; Gratzel, M. Polymer-templated nucleation and crystal growth of perovskite films for solar cells with efficiency greater than 21%. *Nat. Energy* **2016**, *1*, 16142. [CrossRef]
39. Taguchi, M.; Suzuki, A.; Tanaka, H.; Oku, T. Fabrication and characterization of perovskite solar cells added with $MnCl_2$, YCl_3 or poly(methyl methacrylate). *AIP Conf. Proc.* **2018**, *1929*, 020012. [CrossRef]
40. Taguchi, M.; Suzuki, A.; Ueoka, N.; Oku, T. Effects of poly(methyl methacrylate) addition to perovskite photovoltaic devices. *AIP Conf. Proc.* **2019**, *2067*, 020018. [CrossRef]
41. Oku, T.; Nakagawa, J.; Iwase, M.; Kawashima, A.; Yoshida, K.; Suzuki, A.; Akiyama, T.; Tokumitsu, K.; Yamada, M.; Nakamura, M. Microstructures and photovoltaic properties of polysilane-based solar cells. *Jpn. J. Appl. Phys.* **2013**, *52*, 04CR07. [CrossRef]
42. Nakagawa, J.; Oku, T.; Suzuki, A.; Akiyama, T.; Yamada, M.; Fukunishi, S.; Kohno, K. Effects of PBr_3 addition to polysilane thin films on structures and photovoltaic properties. *Green Sustain. Chem.* **2017**, *7*, 20–34. [CrossRef]
43. Shirahata, Y.; Yamamoto, Y.; Suzuki, A.; Oku, T.; Fukunishi, S.; Kohno, K. Effects of polysilane-doped spiro-OMeTAD hole transport layers on photovoltaic properties. *Phys. Status Solidi A* **2017**, *214*, 1600591. [CrossRef]
44. Shirahata, Y.A.; Oku, T.; Fukunishi, S.; Kohno, K. Fabrication of perovskite-type photovoltaic devices with polysilane hole transport layers. *Mater. Sci. Appl.* **2017**, *8*, 209–222. [CrossRef]
45. Oku, T.; Nomura, J.; Suzuki, A.; Tanaka, H.; Fukunishi, S.; Minami, S.; Tsukada, S. Fabrication and characterization of $CH_3NH_3PbI_3$ perovskite solar cells added with polysilanes. *Int. J. Photoenergy* **2018**, 8654963. [CrossRef]
46. Taguchi, M.; Suzuki, A.; Oku, T.; Fukunishi, S.; Minami, S.; Okita, M. Effects of decaphenylcyclopentasilane addition on photovoltaic properties of perovskite solar cells. *Coatings* **2018**, *8*, 461. [CrossRef]

47. Taguchi, M.; Suzuki, A.; Oku, T.; Ueoka, N.; Minami, S.; Okita, M. Effects of annealing temperature on decaphenylcyclopentasilane-inserted $CH_3NH_3PbI_3$ perovskite solar cells. *Chem. Phys. Lett.* **2019**, *737*, 136822. [CrossRef]
48. Oku, T.; Kandori, S.; Taguchi, M.; Suzuki, A.; Okita, M.; Minami, S.; Fukunishi, S.; Tachikawa, T. Polysilane-inserted methylammonium lead iodide perovskite solar cells doped with formamidinium and potassium. *Energies* **2020**, *13*, 4776. [CrossRef]
49. Oku, T.; Zushi, M.; Imanishi, Y.; Suzuki, A.; Suzuki, K. Microstructures and photovoltaic properties of perovskite-type $CH_3NH_3PbI_3$ compounds. *Appl. Phys. Express* **2014**, *7*, 121601. [CrossRef]
50. Oku, T.; Ohishi, Y.; Ueoka, N. Highly (100)-oriented $CH_3NH_3PbI_3$(Cl) perovskite solar cells prepared with NH_4Cl using an air blow method. *RSC Adv.* **2018**, *8*, 10389–10395. [CrossRef]
51. Oku, T.; Ohishi, Y.; Suzuki, A.; Miyazawa, Y. Effects of NH_4Cl addition to perovskite $CH_3NH_3PbI_3$ photovoltaic devices. *J. Ceram. Soc. Jpn.* **2017**, *125*, 303–307. [CrossRef]
52. Ueoka, N.; Oku, T.; Tanaka, H.; Suzuki, A.; Sakamoto, H.; Yamada, M.; Minami, S.; Miyauchi, S.; Tsukada, S. Effects of PbI_2 addition and TiO_2 electron transport layers for perovskite solar cells. *Jpn. J. Appl. Phys.* **2018**, *57*, 08RE05. [CrossRef]
53. Oku, T.; Ohishi, Y. Effects of annealing on $CH_3NH_3PbI_3$(Cl) perovskite photovoltaic devices. *J. Ceram. Soc. Jpn.* **2018**, *126*, 56–60. [CrossRef]
54. Suzuki, A.; Oku, T. Effects of transition metals incorporated into perovskite crystals on the electronic structures and magnetic properties by first-principles calculation. *Heliyon* **2018**, *4*, e00755. [CrossRef]
55. Suzuki, A.; Oku, T. Effects of mixed-valence states of Eu-doped $FAPbI_3$ perovskite crystals studied by first-principles calculation. *Mater. Adv.* **2021**, *2*. in press. [CrossRef]
56. Chen, H.-W.; Sakai, N.; Ikegami, M.; Miyasaka, T. Emergence of hysteresis and transient ferroelectric response in organo-lead halide perovskite solar cells. *J. Phys. Chem. Lett.* **2014**, *6*, 164–169. [CrossRef]
57. Pérez-Osorio, M.A.; Lin, Q.; Phillips, R.T.; Milot, R.L.; Herz, L.M.; Johnston, M.B.; Giustino, F. Raman spectrum of the organic–inorganic halide perovskite $CH_3NH_3PbI_3$ from first principles and high-resolution low-temperature Raman measurements. *J. Phys. Chem. C* **2018**, *122*, 21703–21717. [CrossRef]
58. Brivio, F.; Frost, J.M.; Skelton, J.M.; Jackson, A.J.; Weber, O.J.; Weller, M.T.; Goñi, A.R.; Leguy, A.M.A.; Barnes, P.R.F.; Walsh, A. Lattice dynamics and vibrational spectra of the orthorhombic, tetragonal, and cubic phases of methylammonium lead iodide. *Phys. Rev. B* **2015**, *92*, 144308. [CrossRef]
59. Jeon, N.J.; Noh, J.H.; Kim, Y.C.; Yang, W.S.; Ryu, S.; Seok, S. Solvent engineering for high-performance inorganic–organic hybrid perovskite solar cells. *Nat. Mater.* **2014**, *13*, 897–903. [CrossRef] [PubMed]
60. Xiao, M.; Huang, F.; Huang, W.; Dkhissi, Y.; Zhu, Y.; Etheridge, J.; Weale, A.G.; Bach, U.; Cheng, Y.B.; Spiccia, L. A fast deposition-crystallization procedure for highly efficient lead iodide perovskite thin-film solar cells. *Angew. Chem. Int. Ed.* **2014**, *53*, 9898–9903. [CrossRef] [PubMed]
61. Tavakoli, M.M.; Yadav, P.; Prochowicz, D.; Sponseller, M.; Osherov, A.; Bulovic, V.; Kong, J. Controllable perovskite crystallization via antisolvent technique using chloride additives for highly efficient planar perovskite solar cells. *Adv. Energy Mater.* **2019**, *9*, 1803587. [CrossRef]
62. Noh, J.H.; Im, S.H.; Heo, J.H.; Mandal, T.N.; Seok, S.I. Chemical management for colorful, efficient, and stable inorganic–organic hybrid nanostructured solar cells. *Nano Lett.* **2013**, *13*, 1764–1769. [CrossRef]
63. Haga, Y.; Harada, Y. Photovoltaic characteristics of phthalocyanine-polysilane composite films. *Jpn. J. Appl. Phys.* **2001**, *40*, 855–861. [CrossRef]
64. Chen, Q.; Zhou, H.; Song, T.B.; Luo, S.; Hong, Z.; Duan, H.S.; Dou, L.; Liu, Y.; Yang, Y. Controllable self-induced passivation of hybrid lead iodide perovskites toward high performance solar cells. *Nano Lett.* **2014**, *14*, 4158–4163. [CrossRef] [PubMed]
65. Ueoka, N.; Oku, T. Stability characterization of PbI_2-added $CH_3NH_3PbI_{3-x}Cl_x$ photovoltaic devices. *ACS Appl. Mater. Interfaces* **2018**, *10*, 44443–44451. [CrossRef] [PubMed]
66. Eames, C.; Frost, J.M.; Barnes, P.R.F.; O'Regan, B.C.; Walsh, A.; Islam, M.S. Ionic transport in hybrid lead iodide perovskite solar cells. *Nat. Commun.* **2015**, *6*, 7497. [CrossRef] [PubMed]

coatings

MDPI

Article

Synthesis and Characterization of CeO_2/CuO Nanocomposites for Photocatalytic Degradation of Methylene Blue in Visible Light

Alia Raees [1], Muhammad Asghar Jamal [1,*], Ikram Ahmed [2,*], Mika Silanpaa [3] and Tahani Saad Algarni [4]

1 Department of Chemistry, Government College University, Faisalabad 38000, Pakistan; aliaraees@gmail.com
2 Department of Chemistry, University of Sahiwal, Sahiwal 54000, Pakistan
3 Department of Civil and Environmental Engineering, Florida International University, Miami, FL 33301, USA; mikaesillanpaa@gmail.com
4 Chemistry Department, College of Science, King Saud University, Riyadh 11451, Saudi Arabia; tahanis@ksu.edu.sa
* Correspondence: asgharjamal@gcuf.edu.pk (M.A.J.); drikramahmad@uosahiwal.edu.pk (I.A.)

Abstract: Removal of hazardous organic dyes from polluted water bodies requires the introduction of strong adsorbents and photocatalysts to industrial wastewaters. Herein, photocatalytic CeO_2 nanoparticles and CeO_2/CuO nanocomposite were synthesized following a co-precipitation method for low cost elution of methylene blue (MB) from water. The crystallinity and surface structure of the as-prepared materials have been analyzed using characterization techniques including X-ray powder diffraction (XRPD), field emission scanning electron microscopy (FE-SEM), energy-dispersive spectroscopy (EDS), ultra-violet visible spectroscopy (UV–Vis), and Fourier-transform infrared spectroscopy (FTIR). The average particle size of both the nano scaled samples were approximately 20–30 nm. The photocatalytic properties of CeO_2/CuO were investigated under visible light against methylene blue (MB). The results showed 91% photodegradation of MB organic pollutant in 3 h as monitored by UV–Vis spectroscopy. Absorbance peaks appeared at around 670 nm corresponding to degradation of MB. Such output displayed the effectiveness of Ce nanocomposites for environmental benefits. Hence, CeO_2/CuO nanocomposite could be useful for treatment of industrial wastewaters by removing hazardous MB dye.

Keywords: nanocomposites; Ce–Cu oxide; Co- precipitation; photocatalyst; dye degradation

check for updates

Citation: Raees, A.; Jamal, M.A.; Ahmed, I.; Silanpaa, M.; Saad Algarni, T. Synthesis and Characterization of CeO_2/CuO Nanocomposites for Photocatalytic Degradation of Methylene Blue in Visible Light. *Coatings* **2021**, *11*, 305. https://doi.org/10.3390/coatings11030305

Academic Editor: Emerson Coy

Received: 24 January 2021
Accepted: 2 March 2021
Published: 8 March 2021

Publisher's Note: MDPI stays neutral with regard to jurisdictional claims in published maps and institutional affiliations.

1. Introduction

Currently, water pollution is considered one of the important factors affecting the environment [1]. Industrial effluents, especially from printing, dyeing and textile industries, have synthetic dyestuff which is lethal to aquatic life and also results in water contamination due to their frequent elusion into water bodies [2–6]. A considerable quantity of many dyes is released in effluents from dyeing processes and dye production units. For better living conditions, this issue must be solved [7,8]. Scientists have overcome numerous struggles to introduce and establish a substantial process for elution of hazardous dyes from the industrial effluents. However, adsorption has emerged as a preferred technique for this purpose owing to its cost effectiveness, easy operations, design simplicity, and environment friendliness [9]. Such adsorption phenomenon can be made more efficient by means of involving some photocatalytic materials that enhance the photo-degradation of hazardous dyes.

Additionally, researchers are now focusing on semiconducting material acting as photocatalysts which utilize photoelectrons to oxidize pollutants and is considered a green and cost-effective method to fix these issues [10–12]. These photocatalytic remediation processes have also attracted huge attention owing to their goodness in degradation of hazardous organic dyes [2,3,13]. In case of large and dense aggregates of particles, the inner particles become inactive in contrast to surface particles resulting in diminished photocatalytic activity [14] so introduction of nanosized photocatalysts results in improved photocatalytic activity, also the agglomeration of nanoparticles could be fixed by making such nanocomposites [6].

Among the lanthanides, cerium is selected due to its unique Ce^{+3}/Ce^{+4} redox couple, which is capable of shifting between Ce_2O_3 and CeO_2 in different redox conditions [6,15,16]. In lattice structure, the presence of Ce^{+3} is responsible for the oxygen defects in the nanoparticles by producing the oxygen vacancy to fulfill the charge deficiency [17,18]. Further Ce^{+4} ions have been reported to behave as electron-trapping sites to diminish the charge-pair recombination, and so enhance photocatalytic activity [19]. Ce^{+3} and Ce^{+4} have different optical properties due to different electronic structures, thus are more active for the oxidation process due to generation of more oxygen vacancies [20,21].

Moreover, the semiconductor's photocatalytic activity could be enhanced by doping oxygen vacancies in it in a certain amount [22,23]. Doping of different semiconductors having similar band potentials could effectively construct heterojunctions which improve their charge separation efficiency. These heterojunction interfaces could behave as a transportation channel to enhance separation of electron hole pairs [24]. Only few studies have focused on physically improved CeO_2 nanostructures on doping for environmental remediation [25]. Photocatalytic properties of CeO_2 nanoparticles fabricated with diverse transition metal moieties like Mn, Ti, Fe, and Co were compared and it was reported that this doping affects the CeO_2 nanoparticles morphology, enhances their surface area as well as absorption properties, and also leads to a decrease in rate of recombination of electron-hole pairs [26–28].

In addition, methylene blue is an aniline dye used in coloring of multiple manufactured items, especially wool, silk, and cotton. It is greatly detected in textile wastewater effluents. It is highly risky to human, animal, and even plant life. Therefore, removal of MB remains a prime priority of industrial societies.

In the present work, we report the impact of composite formation on the structural, photosensitive, and photocatalytic bearings of CeO_2 nanostructures. The Ce–Cu nanocomposites are prepared by co-precipitation method. These as-prepared materials have been analyzed employing multiple characterization technologies and the results are described in detail. Moreover, their photocatalytic output was examined in visible light irradiation purposely for degradation of methylene blue.

2. Materials and Method

2.1. Materials

The chemical reagents utilized in the current experimental scheme were purely of analytical grade and were used without further processing. Cerium nitrate hexahydrate used as cerium precursor and cobalt nitrate hexahydrate used as cobalt precursor were purchased from Uni Chem. Chemical reagents Co., Ltd. MB was taken from Sigma Aldrich (Berlin, Germany). Double distilled water was preferably used for preparation of aqueous solutions.

2.2. Synthesis of CeO_2 Nanoparticles

CeO_2-NPs (nanoparticles) were synthesized following the co-precipitation technique. Acting as precursor, 10.85 g of Ce $(NO_3)_3{\cdot}6H_2O$ was dissolved in double distilled water (250 mL). This mixture was heated along with stirring and the pH of solution was changed to 11 by continuous pouring of 0.1 M NaOH drop-wise. Dispersion was thus obtained. This growth solution was heated with stirring at 70 °C for 6–7 h. Subsequently, the precipitates were collected, filtered, and washed with water and ethanol, dried in oven at 85 °C, and then calcined at 600 °C for 7 h.

2.3. Preparation of CeO_2-CuO Nanocomposite

Cerium copper nanocomposite was synthesized using the same method. A 0.1 M solution of Ce $(NO_3)_3 \cdot 6H_2O$ and 5% of 0.1 M copper nitrate hexahydrate solution were mixed, and the pH of the solution raised to 11 by adding 0.1 M NaOH drop-wise with continuous stirring and heating at 70 °C. After 6–7 h of stirring, filtrating and washing of precipitates with double distilled water and ethanol was performed. The oven-dried residue at 85 °C was grinded and calcined at 600 °C for 7 h. Moreover, the physical picture of synthesis of nanocomposite can be understood by Figure 1.

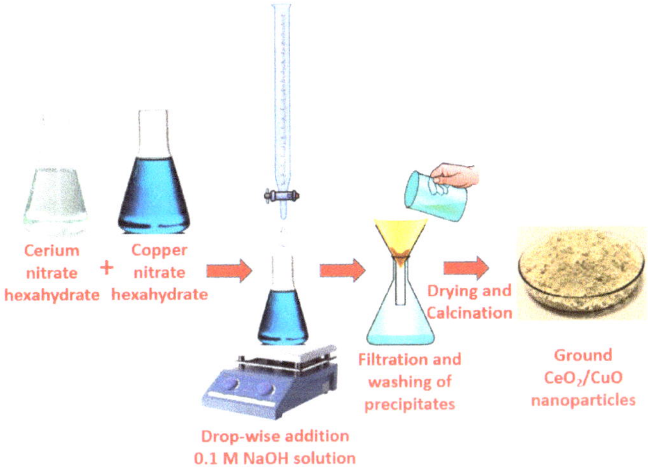

Figure 1. Synthesis scheme of CeO_2/CuO nanocomposite.

2.4. Photocatalytic Activity

Aqueous solution of MB was used to estimate the catalytic activity of the photocatalysts. Extent of photodegradation by the photocatalysts under sunlight irradiation was estimated using the UV/Vis spectrophotometer. Reaction solutions were made by adding the required amount of the as-prepared nanocomposites into 500 mL of 0.03 mM MB solution exhibiting initial pH 5. This suspended mixture was then followed by stirring in the dark for approximately 30 min until an equilibrium was achieved.

The suspension was stirred in the dark for 30 min for adsorption–desorption equilibrium. The MB-containing aqueous solution and the added photocatalyst were placed in direct sunlight with constant stirring. The analytical samples from the suspension were collected at regular intervals of time, i.e. 30 min, centrifuged and filtered to remove the photocatalyst. Moreover, UV–visible spectrophotometer was used to analyze the MB concentration in analytical samples. Figure 2 represents a proposed mechanism of MB degradation using synthesized catalysts.

A conceivable mechanism involves the electron quenching/injection from photo-excited molecules of methylene blue to CeO_2/CuO. Molecular oxygen in the solution mixture was then reduced followed by oxidative decomposition of MB, termed as photosensitization. The degradation mechanism fundamentally depends upon electron-hole separation, i.e., *e–h* charge. Herein, visible light radiation energy corresponds to band gap energy of catalytic material which excites the electrons to conduction band from valence bond. In this way, the hole created in the valence band provides a platform to degradation (oxidation/reduction) of methylene blue aided by the creation of free radicals. Excited electrons strike the nearest oxygen and form superoxide anion radical (O_2^-), which further reacts with hydroxyl radicals and protons from water to form H_2O_2. In this system, this electron-hole separation is generated by visible light by water splitting into radicals and these radicals are means of dye degradation.

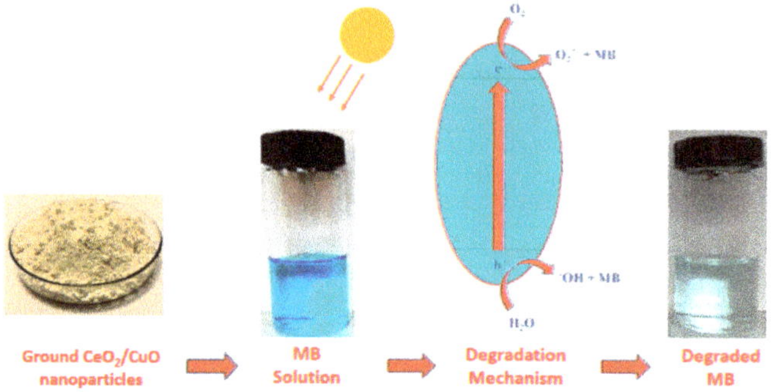

Figure 2. Photocatalytic degradation of methylene blue at of CeO_2/CuO nanoparticles.

Percentage degradation was estimated with help of Equation (1) [29].

$$Percentage\ of\ degradation = C_0 - C_t \ / \ C_0 \times 100 \quad (1)$$

Here, C_0 corresponds to initial absorbance of the MB solution and C_t represents the absorbance of the solution at the aforementioned time.

2.5. Characterization of Nanocomposites

The crystalline nature of the as-prepared photocatalysts was observed by using diffractometer with a scan rate 0.4° per minute in the 2θ range from 10° to 70°. The average crystallite size is calculated by applying the Scherrer equation [30].

$$D = 0.\ 89\ \lambda / \beta \cdot \cos\theta \quad (2)$$

D represents the average crystallite size of the samples, λ stands for X-ray wavelength, β refers to the full-width-half-maximum (FWHM) in radians, while θ denotes the Bragg's angle.

Fourier-transform infrared (FT-IR) spectral analysis is done on a FT-IR spectrophotometer (Perkin Elmer, Überlingen England) with the range between 500 and 4000 cm^{-1} for functional group determination. Surface morphology of the prepared samples was determined using (cube compact model, Emcraft Seoul, South Korea) scanning electron microscope (SEM). With this instrument, the elemental analysis and sample purity were also analyzed by energy dispersive spectroscopy (EDS). The UV–vis absorbance spectra were acquired for the nanocomposites using a UV–Vis spectrophotometer (Tensor II BRUKER, Massachusetts, USA). The spectra were recorded at room temperature in air at the wavelength in range of 200 to 800 nm. The dye concentration was analyzed by UV/Vis professional double beam spectrophotometer (C-7200S, Peak Instruments, Buxton, UK).

3. Results and Discussion

The phase structure of the as-proposed catalytic materials and their respective average crystallite size were determined using X-ray powder diffraction (XRPD) analysis. Figure 3 shows XRD spectra of both nanocomposites CeO_2/ CuO. The diffraction peaks were observed at 2θ = 33.2 (200), 28.2 (111), and 56.2 (311) showing the face centered cubic phase of CeO_2 in the synthesized catalysts. This pattern matches with JCPDS file no. 65-2975 [4]. The XRD peaks correlates closely with XRD patterns of a reported ceria [31]. The peak at 47.6 (220) corresponds to characteristic of CuO as reported earlier [32,33]. The sharp peaks observed in the XRD patterns indicated that the prepared nanocomposites were highly crystallized. The calculated average crystalline sizes of the cerium and cerium copper nanocomposites comes to be 20.73 and 23.22 nm respectively.

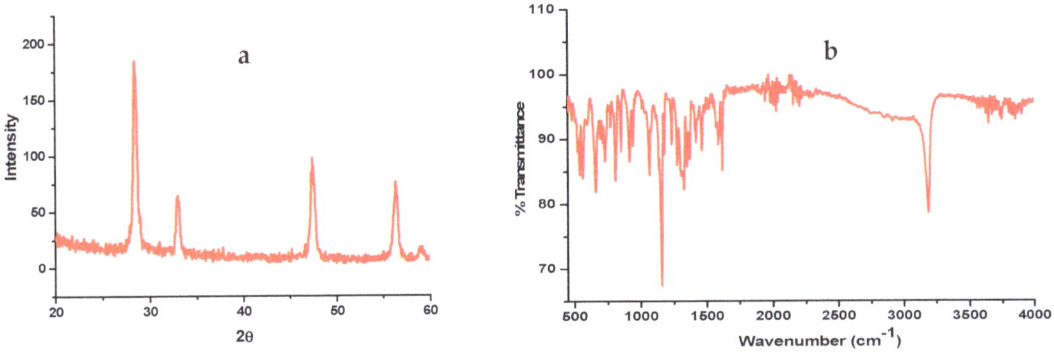

Figure 3. (**a**) X-ray power diffraction (XRPD) pattern and (**b**) FTIR spectrum of CeO_2-CuO nanocomposite.

The FTIR patterns of catalysts is displayed in Figure 3b. In spectrum, three main FTIR regions are observed, first between 3500 and 3000 cm^{-1}, second in the range of 1300 and 1800 cm^{-1}, and third in the range of 500 cm^{-1} [28]. The peaks at 3700 cm^{-1} are related to the O–H stretching mode of OH^- of the adsorbed water on the surface of the catalyst. The wider absorption peak appearing at 3446 cm^{-1} is associated to O–H stretching vibrations of the OH^- group. The band at 1541 cm^{-1} is due to H–O–H bending vibration mode of water. The third low wave number region absorption below 500 cm^{-1} could be assigned to Ce–O and Cu–O stretching vibrational mode as oxides form bonds in this region [34].

Figure 4 reveals the surface morphologies of the cerium nanocomposites. SEM micrographs of the Ce–Cu nanocomposite is shown in Figure 4a. In the micrograph, large aggregates consisting of fine particles of CeO_2 are seen. The average particle size cis approximately 25–30 nm [2,8]. Although, the images revealed some agglomerations, it is obvious that these nanocomposites form a heterogeneous surface structure that assist in catalysis.

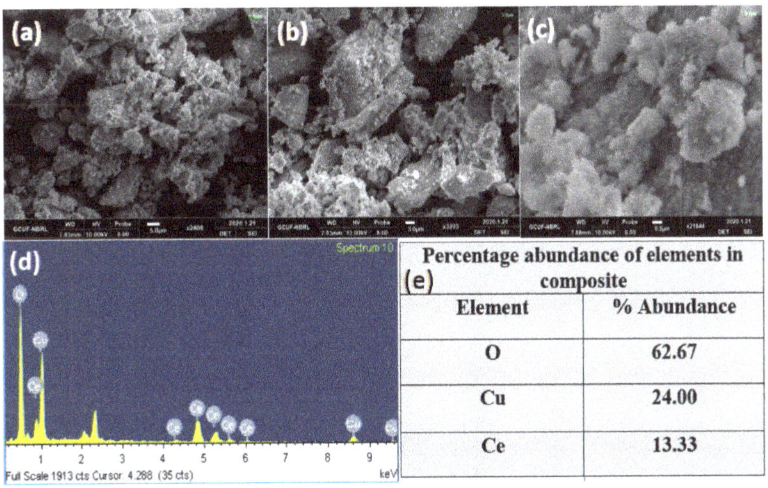

Figure 4. SEM images of CeO_2-CuO nanocomposite (**a**–**c**) at different magnification of (**a**) 5.0 µm, (**b**) 3.0 µm, and (**c**) 0.5 µm. (**d**) EDS Pattern of CeO_2-CuO nanocomposite (**e**) Elemental analysis of EDS pattern of CeO_2-CuO nanocomposite

Energy dispersive spectroscopy (EDS) was performed in addition to SEM at the same instrument and it was assessed that the as-proposed photocatalytic materials are quite pure, and almost appropriate and predictable percentage abundance appeared. The EDS spectrum is displayed in Figure 4d which confirms the sample purity of as-synthesized composites. However, Figure 4e presents the % abundance of elements in composite material and it validates the presence of the anticipated amount of every element put in the catalysts' synthesis.

The UV–vis spectrophotometer was used to estimate the absorption wavelength of the prepared nanocomposites. This is indispensable to estimate the energy of incident light radiation, which is either corresponding to or larger than the photocatalytic band-gap energy, so that sufficient electrons can be excited the conduction band of the photocatalysts [35]. Consequently, the absorption wavelength of the created CeO_2/CuO composites is measured using a spectrophotometer. The output expressed in Figure 5a presents that the pure semiconductor oxides have a very sharp band edge [36], whereas the nanocomposite showed absorbance over a wider range, which led to a red shift in of the spectrum. This wider band edge for amorphous Ce_2O_3 has been speculated to arise from the formation of Ce^{3+} ions that have induced some localized mid-gap states in the band gap [37,38]. Furthermore, an increase in delocalization in organic dye molecules leads to small energy gap between ground and excited states, therefore a red shift was observed. The red shifts experienced in absorbance reflect an increase in π-electrons delocalization in the MB molecule [39]. As the UV–vis response proves, the nanocomposite can absorb the visible light and produce a large number of photo-generated charge pairs under sunlight irradiation. These results are also in accordance with the photocatalytic reactions results, that the holes and electrons participate commendably in oxidation and reduction reactions to degrade organic compounds [40].

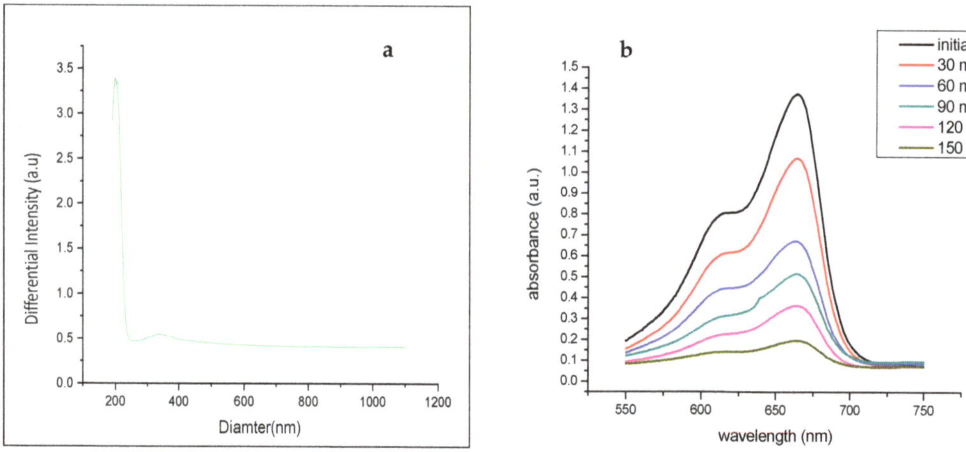

Figure 5. (**a**) UV/Vis spectrum of Cerium nanocomposites (CeO_2-CuO) and (**b**) time-dependent optical absorbance spectra of methylene blue (MB) showing the gradual decrease in concentration with time of CeO_2-CuO used as catalyst.

4. Measurement of Photocatalytic Activity

There are many dye degradation studies using metal oxides as photocatalysts [35,41–44]. These research articles revealed the usefulness of CeO_2 in photocatalytic reactions. Chaudhary et al., 2020 [45] studied synthesis of ceria nanoparticles which give photocatalytical activity against methylene blue and dye degraded in 9 h. Rao et al., 2015 [46] synthesized CuO nanoparticles and used them for photocatalytical activity of methylene blue and in 6 h. This study focuses on synthesis of the CeO/CuO nanocomposite which enhances photocatalytical activity and degrade it into 150 min (Table 1).

Table 1. Comparing catalysts performance of CeO, CuO, and CeO/CuO nanocomposite.

Sr#	Composite Name	Dye Degraded	Dose of Catalyst	Dose of Dye	Time of Degradation	Condition	Reference
1	CeO nanoparticles	MB	20 mg/L	15 ml	9 h	Visible light irradiation	[45]
2	CuO nanoparticles	MB	20 mg	10 mg/L	6 h	Visible light irradiation	[46]
3	CeO/CuO Nanocomposite	MB	1 g/L	0.03 mM	150 min	Visible light irradiation	This work

The combination of metal oxides incorporated in a composite may result in enrichment of surface oxygen defects. The increased concentration of surface oxygen defects can hold more photo generated electrons and holes individually and confirm their availability in decomposition organic pollutants. It accelerated the degradation of dyes and thus enhanced photo catalysis rate [35,45]. The photocatalytic response of the CeO_2/CuO composite were evaluated using degradation of methylene blue under direct sunlight exposure. Figure 2 shows the discoloration of MB by CeO_2/CuO catalyst under visible light irradiation which indicates the degradation of MB accompanied with CeO_2/CuO composites.

In this experiment, 0.03 mM solution of MB was degraded using 1 g/L of CeO_2/CuO nanocomposites. It is seen that without catalyst there was no degradation of the dye solution observed in bright sunlight which exposed no decolorization. This shows that light itself plays no part in the discoloration of MB. Initially, the reaction mixture was retained in the dark for 30 min to discern the extent of adsorption. After that, the mixture was shifted to visible light followed by constant stirring. With the procession of reaction, 4 mL of the aliquot solutions were separated through a pipette after 30 min sequel, and measured the absorption in UV–Vis spectrophotometers. Moreover, the reaction mixture was kept in the dark for 30 min, which experienced a little drop in the dye concentration. Hence, it is obvious that the catalyst is not so efficient in dark. It is obvious from Figure 6a that complete degradation of MB was achieved after 2.5 h, so CeO_2/CuO are observed to degrade the dye effectively. Moreover, Figure 6b shows that photocatalyzed degradation of MB is administered by first order kinetics equation because the plot between $-\ln (A/A_o)$ vs. time demonstrated linearity in the presence of as-proposed CeO_2/CuO nanocatalysts [46].

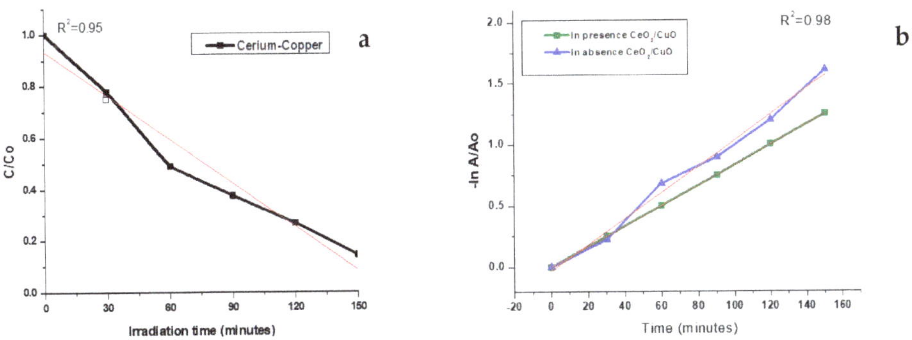

Figure 6. *Cont.*

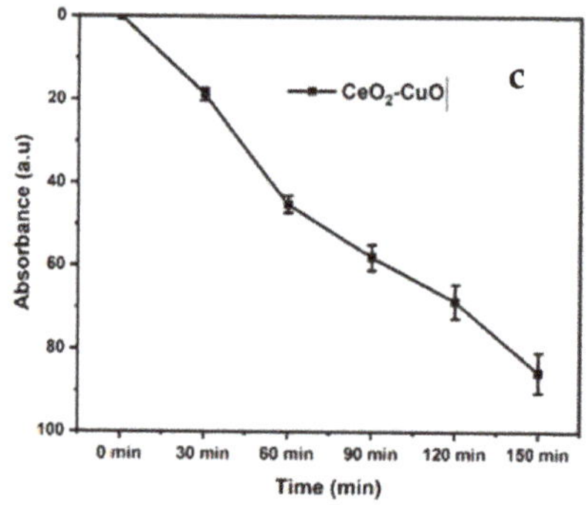

Figure 6. (**a**) Decrease in concentration and % degradation of MB dye with CeO_2/CuO and (**b**) natural logarithms of absorbance of MB plotted as a function of irradiation time (**c**) standard deviation error bar of CeO_2/CuO.

A few of the most efficient photocatalysts for dye degradation are compared with as-proposed photocatalysts in Table 2. It can be observed that the catalysts with maximum efficiency took prolonged time, however more fast catalysts revealed lower efficiency. Hence, the as-synthesized photocatalyst could be more efficient as well as faster degradation catalysts towards MB degradation.

Table 2. Summary of some reported bimetallic oxides for photocatalytic dye degradation.

Photocatalyst	% Efficiency	Degradation Time (min)	References
Y_2O_3/CeO_2	95.5	240	[47]
La_2O_3/CeO_2	70	120	[48]
$Urea/CeO_2$	70	21	[49]
MnO_2/CeO_2	90.4	–	[50]
$Mn\text{-}TiO_2$	89	24	[51]
$ZnO\text{-}TiO_2$	80	120	[52]
CeO_2/CuO	85.66	150	–

5. Conclusions

Well crystallized CeO_2/CuO nanoparticles were successfully prepared at low temperature via simple co-precipitation routes. Multiple characterization technologies were adapted to investigate the physio chemical and optical nature of the as-synthesized photocatalysts. The XRD results show that well-crystallized nanocomposites have been obtained and the average crystallite size estimated was 20–25 nm. With respect to catalytic significance, the as-synthesized CeO_2/CuO nanoparticles demonstrated many selective and commendable photocatalytic possessions toward MB degradation. CeO_2/CuO nanomaterials degraded the methylene blue up to approximately to 85.66% in 150 min, hence proving the strong catalytic performance of catalysts. Therefore, CeO_2/CuO nanocomposites could act as a promising candidate for MB degradation in a multitude of industrial wastewater treatment plants.

Author Contributions: Conceptualization, A.R and M.A.J.; methodology, M.A.J.; software, T.S.A.; validation, A.R., M.A.J. and I.A.; formal analysis, I.A.; investigation, I.A.; resources, M.S.; data curation, T.S.A.; writing—original draft preparation, A.R.; writing—review and editing, M.A.J.; visualization, M.A.J.; supervision, M.A.J.; project administration, I.A.; funding acquisition, T.S.A. All authors have read and agreed to the published version of the manuscript.

Funding: This work was funded by the Department of Chemistry, Government College University Faisalabad, Pakistan, and Researchers Supporting Project Number (RSP-2020/254) King Saud University, Riyadh, Saudi Arabia.

Institutional Review Board Statement: Not applicable.

Informed Consent Statement: Not applicable.

Data Availability Statement: No new data were created or analyzed in this study. Data sharing is not applicable to this article.

Conflicts of Interest: The authors declare no conflict of interest.

References

1. Zhang, Q.; Zhao, X.; Duan, L.; Shen, H.; Liu, R. Controlling oxygen vacancies and enhanced visible light photocatalysis of CeO_2/ZnO nanocomposites. *J. Photochem. Photobiol. A Chem.* **2020**, *392*, 112156. [CrossRef]
2. Rajendran, S.; Khan, M.M.; Gracia, F.; Qin, J.; Gupta, V.K.; Arumainathan, S. Ce^{3+}-ion-induced visible-light photocatalytic degradation and electrochemical activity of ZnO/CeO_2 nanocomposite. *Sci. Rep.* **2016**, *6*, 31641. [CrossRef]
3. Ameen, S.; Akhtar, M.S.; Seo, H.-K.; Shin, H.-S. Solution-processed CeO_2/TiO_2 nanocomposite as potent visible light photocatalyst for the degradation of bromophenol dye. *Chem. Eng. J.* **2014**, *247*, 193–198. [CrossRef]
4. Saravanan, R.; Karthikeyan, N.; Govindan, S.; Narayanan, V.; Stephen, A. Photocatalytic degradation of organic dyes using ZnO/CeO_2 nanocomposite material under visible light. *Adv. Mater. Res.* **2012**, *584*, 381–385. [CrossRef]
5. Aravind, M.; Aravind, M.; Ahmad, A.; Ahmad, I.; Amalanathan, M.; Naseem, K.; Mary, S.M.M.; Parvathiraja, C.; Hussain, S.; Algarni, T.S.; et al. Critical green routing synthesis of silver NPs using jasmine flower extract for biological activities and photocatalytical degradation of methylene blue. *J. Environ. Chem. Eng.* **2020**, *9*, 104877.
6. Ahmad, A.; Ahmad, A.; Jini, D.; Aravind, M.; Parvathiraja, C.; Ali, R.; Kiyani, M.Z.; Alothman, A. A novel study on synthesis of egg shell based activated carbon for degradation of methylene blue via photocatalysis. *Arabian J. Chem.* **2020**, *13*, 8717–8722. [CrossRef]
7. Khataee, A.; Gholami, P.; Kalderis, D.; Pachatouridou, E.; Konsolakis, M. Preparation of novel CeO_2-biochar nanocomposite for sonocatalytic degradation of a textile dye. *Ultrason. Sonochem.* **2018**, *41*, 503–513. [CrossRef]
8. Arshad, T.; Khan, S.A.; Faisal, M.; Shah, Z.; Akhtar, K.; Asiri, A.M.; Ismail, A.A.; Alhogbi, B.G.; Khan, S.B. Cerium based photocatalysts for the degradation of acridine orange in visible light. *J. Mol. Liq.* **2017**, *241*, 20–26. [CrossRef]
9. Rafatullah, M.; Sulaiman, O.; Hashim, R.; Ahmad, A. Adsorption of methylene blue on low-cost adsorbents: A review. *J. Hazard. Mater.* **2010**, *177*, 70–80. [CrossRef]
10. Hussain, S.; Hussain, S.; Khan, A.J.; Arshad, M.; Javed, M.S.; Ahmad, A.; Shah, S.S.A.; Khan, M.R.; Akram, S.; Ali, S.; et al. Charge storage in binder-free 2D-hexagonal CoMoO4 nanosheets as a redox active material for pseudocapacitors. *Ceram. Int.* **2020**, *47*.
11. Naseem, K.; Zia Ur Rehman, M.; Ahmad, A.; Dubal, D.; AlGarni, T.S. Plant Extract Induced Biogenic Preparation of Silver Nanoparticles and Their Potential as Catalyst for Degradation of Toxic Dyes. *Coatings* **2020**, *10*, 1235. [CrossRef]
12. Pirhashemi, M.; Habibi-Yangjeh, A.; Pouran, S.R. Review on the criteria anticipated for the fabrication of highly efficient ZnO-based visible-light-driven photocatalysts. *J. Ind. Eng. Chem.* **2018**, *62*, 1–25. [CrossRef]
13. Gröttrup, J.; Schütt, F.; Smazna, D.; Lupan, O.; Adelung, R.; Mishra, Y.G. Porous ceramics based on hybrid inorganic tetrapodal networks for efficient photocatalysis and water purification. *Ceram. Int.* **2017**, *43*, 14915–14922. [CrossRef]
14. Zhang, X.Z.; Xu, P.H.; Liu, G.W.; Ahmad, A.; Chen, X.H.; Zhu, Y.L.; Qiao, G.J. Synthesis, characterization and wettability of Cu-Sn alloy on the Si-implanted 6H-SiC. *Coatings* **2020**, *10*, 906. [CrossRef]
15. Kashif, M.; Jaafar, E.; Bhadja, P.; Low, F.W.; Sahari, S.K.; Hussain, S.; Al-Tamrah, S.A. Effect of Potassium Permanganate on Morphological, Structural and Electro-Optical Properties of Graphene Oxide Thin Films. *Arabian J. Chem.* **2020**, *14*, 102953.
16. Saleem, M.; Irfan, M.; Tabassum, S.; Alothman, Z.; Javed, M.S.; Hussain, S.; Zubair, M. Experimental and Theoretical Study of Highly Porous Lignocellulose Assisted Metal Oxide Photoelectrodes for Dye-sensitized Solar Cells. *Arabian J. Chem.* **2020**, *14*, 102937.
17. Miri, A.; Sarani, M. Biosynthesis, characterization and cytotoxic activity of CeO_2 nanoparticles. *Ceram. Int.* **2018**, *44*, 12642–12647. [CrossRef]
18. Choudhury, B.; Chetri, P.; Choudhury, A. Oxygen defects and formation of Ce^{3+} affecting the photocatalytic performance of CeO_2 nanoparticles. *RSC Adv.* **2014**, *4*, 4663–4671. [CrossRef]
19. Li, Z.; Zhang, J.; Lv, J.; Lu, L.; Liang, C.; Dai, K. Sustainable synthesis of CeO_2/CdS-diethylenetriamine composites for enhanced photocatalytic hydrogen evolution under visible light. *J. Alloy. Compd.* **2018**, *758*, 162–170. [CrossRef]
20. Li, Z.; Yu, T.; Zou, Z. Degradation in photocatalytic activity induced by hydrogen-related defects in nano-$LiNbO_3$ material. *Appl. Phys. Lett.* **2006**, *88*, 071917. [CrossRef]
21. Li, T.B.; Chen, G.; Zhou, C.; Shen, Z.Y.; Jin, R.C.; Sun, J.X. New photocatalyst BiOCl/BiOI composites with highly enhanced visible light photocatalytic performances. *Dalton Trans.* **2011**, *40*, 6751–6758. [CrossRef]
22. Wang, J.; Xia, Y.; Dong, Y.; Chen, R.; Xiang, L.; Komarneni, S. Defect-rich ZnO nanosheets of high surface area as an efficient visible-light photocatalyst. *Appl. Catal. B Environ.* **2016**, *192*, 8–16. [CrossRef]
23. Li, X.; Song, J.; Liu, Y.; Zeng, H. Controlling oxygen vacancies and properties of ZnO. *Curr. Appl. Phys.* **2014**, *14*, 521–527. [CrossRef]
24. Habibi-Yangjeh, A.; Shekofteh-Gohari, M. Novel magnetic Fe_3O_4/ZnO/$NiWO_4$ nanocomposites: Enhanced visible-light photocatalytic performance through pn heterojunctions. *Sep. Purif. Technol.* **2017**, *184*, 334–346. [CrossRef]

25. Saranya, J.; Ranjith, K.S.; Saravanan, P.; Mangalaraj, D.; Kumar, R.T.R. Cobalt-doped cerium oxide nanoparticles: Enhanced photocatalytic activity under UV and visible light irradiation. *Mater. Sci. Semicond. Process.* **2014**, *26*, 218–224. [CrossRef]
26. Yue, L.; Zhang, X.-M. Structural characterization and photocatalytic behaviors of doped CeO_2 nanoparticles. *J. Alloy. Compd.* **2009**, *475*, 702–705. [CrossRef]
27. Arul, N.S.; Mangalaraj, D.; Chen, P.C.; Ponpandian, N.; Meena, P.; Masuda, Y. Enhanced photocatalytic activity of cobalt-doped CeO_2 nanorods. *J. Sol-Gel Sci. Technol.* **2012**, *64*, 515–523. [CrossRef]
28. Nasir, M.; Bagwasi, S.; Jiao, Y.; Chen, F.; Tian, B.; Zhang, J. Characterization and activity of the Ce and N co-doped TiO_2 prepared through hydrothermal method. *Chem. Eng. J.* **2014**, *236*, 388–397. [CrossRef]
29. Abdollahi, Y.; Abdullah, A.H.; Zainal, Z.; Yusof, N.A. Photocatalytic degradation of p-Cresol by zinc oxide under UV irradiation. *Int. J. Mol. Sci.* **2012**, *13*, 302–315. [CrossRef] [PubMed]
30. Holzwarth, U.; Gibson, N. The Scherrer equation versus the 'Debye-Scherrer equation'. *Nat. Nanotechnol.* **2011**, *6*, 534. [CrossRef]
31. Zhou, G.; Shah, P.R.; Montini, T.; Fornasiero, P.; Gorte, R.J. Oxidation enthalpies for reduction of ceria surfaces. *Surf. Sci.* **2007**, *601*, 2512–2519. [CrossRef]
32. Lanje, A.S.; Sharma, S.J.; Pode, R.B.; Ningthoujam, R.S. Synthesis and optical characterization of copper oxide nanoparticles. *Adv. Appl. Sci. Res.* **2010**, *1*, 36–40.
33. Wongpisutpaisan, N.; Charoonsuk, P.; Vittayakorn, N.; Pecharapa, W. Sonochemical synthesis and characterization of copper oxide nanoparticles. *Energy Procedia* **2011**, *9*, 404–409. [CrossRef]
34. Magdalane, C.M.; Kaviyarasu, K.; Judith Vijaya, J.; Jayakumar, C.; Maaza, M.; Jeyaraj, B. Photocatalytic degradation effect of malachite green and catalytic hydrogenation by UV–illuminated CeO_2/CdO multilayered nanoplatelet arrays: Investigation of antifungal and antimicrobial activities. *J. Photochem. Photobiol. B Biol.* **2017**, *169*, 110–123. [CrossRef]
35. Saravanan, R.; Gupta, V.K.; Narayanan, V.; Stephen, A. Comparative study on photocatalytic activity of ZnO prepared by different methods. *J. Mol. Liq.* **2013**, *181*, 133–141. [CrossRef]
36. Kang, Y.; Yang, Y.; Yin, L.-C.; Liu, G.; Cheng, H.-M. An amorphous carbon nitride photocatalyst with greatly extended visible-light-responsive range for photocatalytic hydrogen generation. *Adv. Mater.* **2015**, *27*, 4572–4577. [CrossRef] [PubMed]
37. Khan, M.M.; Ansari, S.A.; Pradhan, D.; Han, D.H.; Lee, J.; Cho, M.H. Defect-induced band gap narrowed CeO_2 nanostructures for visible light activities. *Ind. Eng. Chem. Res.* **2014**, *53*, 9754–9763. [CrossRef]
38. Ansari, S.; Khan, M.M.; Ansari, M.O.; Kalathil, S.; Lee, J.; Cho, M.H. Band gap engineering of CeO_2 Nanostructure by electrochemically active biofilm for visible light applications. *RSC Adv.* **2014**, *4*, 16782–16791, Erratum *New J. Chem.* **2016**, *40*, 3000–3009. [CrossRef]
39. Nathans, J. Determinants of visual pigment absorbance: Identification of the retinylidene Schiff's base counterion in bovine rhodopsin. *Biochemistry* **1990**, *29*, 9746–9752. [CrossRef] [PubMed]
40. Saravanan, R.; Gupta, V.K.; Narayanan, V.; Stephen, A. Visible light degradation of textile effluent using novel catalyst ZnO/γ-Mn_2O_3. *J. Taiwan Inst. Chem. Eng.* **2014**, *45*, 1910–1917. [CrossRef]
41. Chae, B.W.; Amna, T.; Shamshi Hassan, M.; Al-Deyab, S.S.; Khil, M.-S. CeO_2-Cu_2O composite nanofibers: Synthesis, characterization photocatalytic and electrochemical application. *Adv. Powder Technol.* **2017**, *28*, 230–235. [CrossRef]
42. Chen, Q.; He, Q.; Lv, M.; Liu, X.; Wang, J.; Lv, J. The vital role of PANI for the enhanced photocatalytic activity of magnetically recyclable N–$K_2Ti_4O_9$/$MnFe_2o4$/PANI composites. *Appl. Surf. Sci.* **2014**, *311*, 230–238. [CrossRef]
43. Wen, X.-J.; Niu, C.-G.; Ruan, M.; Zhang, L.; Zeng, G.-M. AgI nanoparticles-decorated CeO_2 microsheets photocatalyst for the degradation of organic dye and tetracycline under visible-light irradiation. *J. Colloid Interface Sci.* **2017**, *497*, 368–377. [CrossRef] [PubMed]
44. Yusoff, N.; Huang, N.M.; Muhamad, M.R.; Kumar, S.V.; Lim, H.N.; Harrison, I. Hydrothermal synthesis of CuO/functionalized graphene nanocomposites for dye degradation. *Mater. Lett.* **2013**, *93*, 393–396. [CrossRef]
45. Subhan, M.A.; Ahmed, T.; Uddin, N.; Azad, A.K.; Begum, K. Synthesis, characterization, PL properties, photocatalytic and antibacterial activities of nano multi-metal oxide NiO· CeO_2· ZnO. *Spectrochim. Acta Part A Mol. Biomol. Spectrosc.* **2015**, *136*, 824–831. [CrossRef]
46. Ahmad, I.; Khan, S.B.; Kamal, T.; Asiri, A.M. Visible light activated degradation of organic pollutants using zinc–iron selenide. *J. Mol. Liq.* **2017**, *229*, 429–435. [CrossRef]
47. Magdalane, C.M.; Kaviyarasu, K.; Priyadharsini, G.M.A.; Bashir, A.K.H.; Mayedwa, N.; Matinise, N.; Isaev, A.B.; Al-Dhabi, N.A.; Arasu, M.V.; Arokiyaraj, S.; et al. Improved photocatalytic decomposition of aqueous Rhodamine-B by solar light illuminated hierarchical yttria nanosphere decorated ceria nanorods. *J. Mater. Res. Technol.* **2019**, *8*, 2898–2909. [CrossRef]
48. Magdalane, C.M.; Kaviyarasu, K.; Matinise, N.; Mayedwa, N.; Mongwaketsi, N.; Letsholathebe, D.; Mola, G.T.; Al-Dhabi, N.A.; Arasu, M.V.; Henini, M.; et al. Evaluation on La_2O_3 garlanded ceria heterostructured binary metal oxide nanoplates for UV/visible light induced removal of organic dye from urban wastewater. *South Afr. J. Chem. Eng.* **2018**, *26*, 49–60. [CrossRef]
49. Sreekanth, T.V.M.; Nagajyothi, P.C.; Reddy, G.R.; Shim, J.; Yoo, K. Urea assisted ceria nanocubes for efficient removal of malachite green organic dye from aqueous system. *Sci. Rep.* **2019**, *9*, 1–9. [CrossRef] [PubMed]
50. Ojha, G.P.; Pant, B.; Park, S.-J.; Park, M.; Kim, H.-Y. Synthesis and characterization of reduced graphene oxide decorated with CeO_2-doped MnO_2 nanorods for supercapacitor applications. *J. Colloid Interface Sci.* **2017**, *494*, 338–344. [CrossRef]

51. Poudel, M.B.; Yu, C.; Kim, H.J. Synthesis of conducting bifunctional polyaniline@ Mn-TiO_2 nanocomposites for supercapacitor electrode and visible light driven photocatalysis. *Catalysts* **2020**, *10*, 546. [CrossRef]
52. Pant, B.; Ojha, G.P.; Kuk, Y.-S.; Kwon, O.H.; Park, Y.W.; Park, M. Synthesis and Characterization of ZnO-TiO_2/Carbon Fiber Composite with Enhanced Photocatalytic Properties. *Nanomaterials* **2020**, *10*, 1960. [CrossRef] [PubMed]

coatings

MDPI

Article

Designing Optically & Utilization of Thermopile Chip with Resonant Cavity Absorber Structure as IR Absorber

Haigang Hou [1], Jian Yang [1], Guiwu Liu [1], Junlin Liu [1], Mudassar Abbas [2], Shahid Hussain [1,*], Haicheng Shao [1], Guanjun Qiao [1,*], Ayman A. Ghfar [3], Mohamed Ouladsmane [3], Muhammad Tariq Nazir [4,*] and Murefah Mana AL-Anazy [5]

1 School of Materials Science and Engineering, Jiangsu University, Zhenjiang 212013, China; neil_kanth@yahoo.com (H.H.); xzzhang2018@ujs.edu.cn (J.Y.); gwliu76@ujs.edu.cn (G.L.); liujunlin@ujs.edu.cn (J.L.); shaohaicheng311@163.com (H.S.)
2 School of Textile and Design, University of Management and Technology, Lahore 54770, Pakistan; mudassar.abbas@umt.edu.pk
3 Advanced Materials Research Chair, Chemistry Department P. O. Box 2455, College of Science, King Saud University, Riyadh 11451, Saudi Arabia; Aghafr@ksu.edu.sa (A.G.); mouladsmane@ksu.edu.sa (M.O.)
4 School of Mechanical and Manufacturing Engineering, University of New South Wales, Sydney, NSW 2052, Australia
5 Department of Chemistry P. O. Box 84428, College of Science, Princess Nourah bint Abdulrahman University, Riyadh 11671, Saudi Arabia; mmalanazy@pnu.edu.sa
* Correspondence: shahid@ujs.edu.cn (S.H.); gjqiao@ujs.edu.cn (G.Q.); tariq.nazir@unsw.edu.au (M.T.N.)

Abstract: This paper presents a novel thermopile chip in which the resonant cavity structure was fully utilized as an absorber by an optical design. The resonant cavity absorber structure was designed using Al as anthe bottom reflective metal layer, air as the intermediate dielectric layer, and $SiO_2/TiN/Si_3N_4$ sandwich layers as the top absorption layer, while the bottom reflective metal (Al) was deposited on the cold junctions of the thermopile. The simulation and calculation results show that the thermopile chip with resonant cavity absorber structure not only has great infrared absorption in the wide infrared absorption range but also can effectively prevent the cold junctions from absorbing infrared radiation and inhibit the rise of temperature. As a result, the temperature difference between the hot junctions and the cold junctions is increased, and the responsivity of the thermopile chip is further improved. Moreover, the duty cycle of the thermopile chip is greatly improved due to the double-layer suspension structure. Compared with the traditional thermopile chip structure, the sizes of the thermopile chip with the resonant cavity absorber structure can be further reduced while maintaining responsivity and specific detectivity.

Keywords: infrared detector; resonant cavity; energy applications; absorptance

Citation: Hou, H.; Yang, J.; Liu, G.; Liu, J.; Abbas, M.; Hussain, S.; Shao, H.; Qiao, G.; Ghfar, A.; Ouladsmane, M.; et al. Designing Optically & Utilization of Thermopile Chip with Resonant Cavity Absorber Structure as IR Absorber. *Coatings* **2021**, *11*, 302. https://doi.org/10.3390/coatings11030302

Received: 5 February 2021
Accepted: 3 March 2021
Published: 6 March 2021

1. Introduction

The thermopile IR detector is applied in many fields, because it has the advantages of no need for cooling and chopping, broad spectral response, low cost and simple output circuit [1–4]. The thermopile chip is the core component of thermopile IR detector and has been greatly optimized and has opened the door to mass production, with the help of conventional complementary metal oxide semiconductor (CMOS) and micro electromechanical system (MEMS) technologies. The thermopile chip is generally a central suspended structure with a supporting layer, thermopile, absorber, and other layers deposited on a silicon substrate. The hot junctions and the cold junctions of the thermopile are respectively distributed on the edge and center of the silicon substrate hollowed out in the center, and the hot junctions are covered by the absorber, as shown in Figure 1. The thermopile infrared detector contains two energy conversion processes of light-heat and heat-electricity, and its performance also depends on the superposition of the two conversions efficiency. Thus, the absorber can play a crucial role on the thermopile chip. The absorber is required for great absorption at both 3–5 μm and 8–14 μm, while a thermopile chip is used as the sensing element of an infrared thermometer [5,6], nondispersive infrared (NDIR) gas detector [7,8] or infrared imager [9–11]. The porous materials with dendritic and soft structure, like gold-black, can have great absorption at ultra-wide wavelength range and are typically used as the absorber of thermopile chip [12–17]. However, these porous materials are generally too fragile and not compatible with CMOS [18,19]. Owing to the interference of light, the resonant cavity absorber structures can also have great absorption at a specific band and are also often used as the absorber of thermopile chips. However, the resonant absorption structure is a stack of multilayer films, so its absorption characteristics largely depend on the matching relationship between the refractive index and thickness of the multilayer film, and often only absorb specific wavelength bands. In addition, the resonant cavity absorption structure is often sensitive to thickness errors, and the thickness of each layer of the film needs to be accurately controlled during the preparation process. In fact, an Si_3N_4 layer or $SiO_2/Si_3N_4/SiO_2$ sandwich structure is also usually used as the absorber of thermopile chips. However, the Si_3N_4 layer or $SiO_2/Si_3N_4/SiO_2$ sandwich structure has higher absorption only at 8.5–13.5 μm and lower absorption at 2.5–8 μm [20,21].

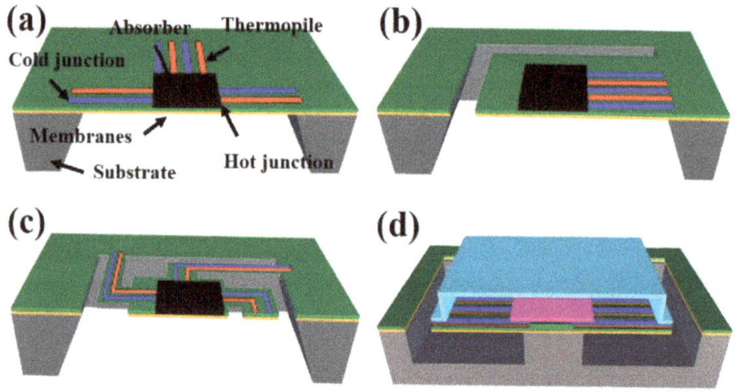

Figure 1. Typical thermopile chip structures: (**a**) close membrane structure, (**b**) cantilever beam structure (**c**) suspension bridge structure, and (**d**) the designed thermopile chip structure with resonant cavity.

The structures of thermopile chips are mainly divided into three types: close membrane structure, cantilever beam structure, and suspension bridge structure (Figure 1a–c) [22–25]. Compared to the cantilever beam and suspension bridge structures, the close membrane structure is the current mainstream structure of thermopile chip, which is provided with smaller thermal resistance, shorter response time, and a simpler manufacturing process. However, the above chip structures all have the problem of mutual restriction between the absorption zone and the thermocouple zone, and further reduction in thermopile chip size cannot maintain relatively higher responsivity and detectivity.

In this paper, we designed a novel thermopile chip in which the resonant cavity structure was fully utilized as an infrared absorber by an optical design (Figure 1d). In the resonant cavity absorber structure, the Al, air, and $SiO_2/TiN/Si_3N_4$ sandwich layers were used as the bottom reflective metal layer, intermediate dielectric layer and top absorption layer, respectively. The designed thermopile chip was proven to be provided with a high duty cycle, great infrared absorption, and a wide infrared absorption range.

2. Design of Resonant Cavity Absorber Structure

As shown in Figure 2a, the resonant cavity absorber structure generally consists of three parts: the top absorption metal layer, the middle dielectric layer, and the bottom reflective metal layer [26,27]. The absorption characteristics of the resonant cavity absorber structure are based on the theory of optical interference absorption. The incident light and reflected light will produce light interference effects at the top absorption metal layer when the optical thickness of middle dielectric layer is of a quarter-wavelength thickness, thereby achieving higher infrared absorption at specific band. Thus, it is difficult for the resonant cavity absorber structure to achieve ultra-wide spectral absorption like the porous materials and to realize a detector in various fields such as gas detection and non-contact human temperature measurement. Moreover, the matching of each thin film, especially the thickness, has a great influence on the absorption characteristics of the resonant cavity absorber structure. Thus, the resonant cavity absorber structure generally requires a precise design of the absorption structure and control of the manufacturing process.

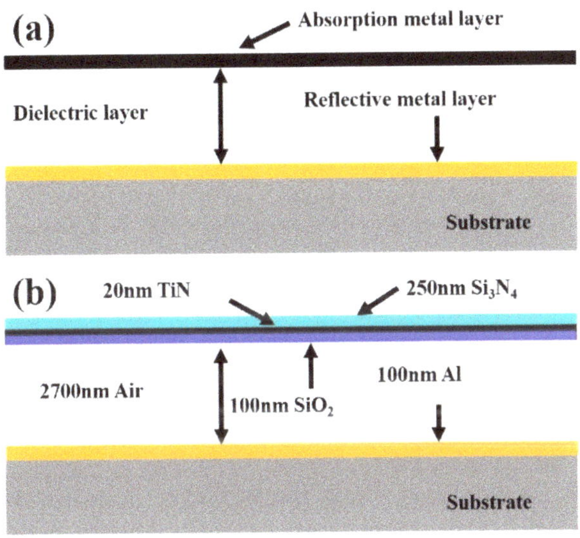

Figure 2. Schematics of (**a**) common and (**b**) the designed resonant cavity absorber structures.

The Essential Macleod is a simulation software for the transmission, reflection and absorption characteristics of optical multilayer films. In order to design a resonant cavity absorber structure with broad spectrum absorption and low film thickness error sensitivity, we first used the Essential Macleod to simulate and calculate the reflection and absorption characteristics of different materials based on the theory of optical interference absorption and then chose reasonable materials for each layer of the resonant cavity absorber structure, as shown in Figure 3. The simulation results show that the average reflectance of the Al layer could reach more than 97% at 2–14 µm. Thus, an Al layer was used as the bottom reflective metal layer of the resonant cavity absorber structure. According to the theory of optical interference absorption, It is easier to obtain a resonant cavity absorber structure with broad spectrum absorption with an intermediate dielectric layer with low refractive index [28]. The refractive index of air is 1, which is very suitable as the intermediate dielectric layer of the resonant cavity absorber structure. As seen in Figure 3, the Si_3N_4, SiO_2, and TiN have higher absorption at 3–14 µm, 8–10 µm, and 10–14 µm, respectively, which is why we chose the SiO_2/TiN/Si_3N_4 sandwich layers as the top absorption layer of the resonant cavity absorber structure. Moreover, the Si_3N_4 and SiO_2 layers are designed to be deposited on the upper and lower sides of TiN layer, respectively, where the Si_3N_4 layer can act as a passivation layer and have a certain protective effect on the TiN layer, and the SiO_2 layer can act as a supporting layer and play a positive role in supporting the overall structure. The SiO_2/TiN/Si_3N_4 sandwich layers as the top absorption layer can well reduce the difficulty of preparation and the sensitivity of the resonant cavity absorber structure to the film thickness error and improve the mechanical strength, stability and yield of the resonant cavity absorber structure.

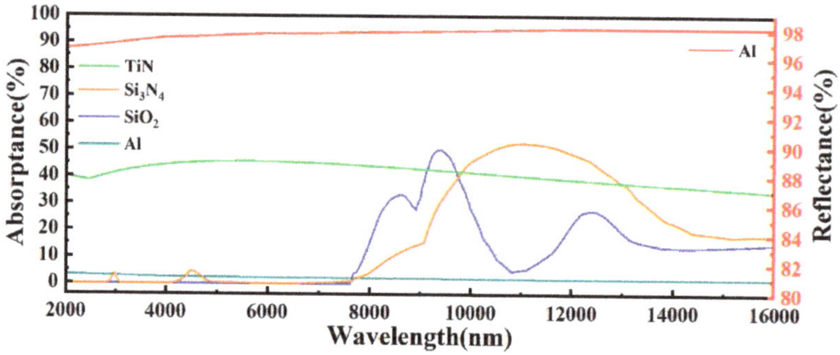

Figure 3. Calculated spectra of the TiN, Si_3N_4, SiO_2, and Al layers.

Based on the theory of optical interference absorption, we used the Essential Macleod to simulate and calculate the optical thicknesses of the intermediate dielectric layer and top absorption metal layer, thereby adjusting the absorption wavelength range of the structure, as shown in Figure 4. The simulation and calculation results show that the resonant cavity absorber structure has great absorption at 3–5 μm and 8–14 μm when the thicknesses of the air cavity, TiN layer and Si_3N_4 layer are 2700, 20, and 250 nm, respectively, as shown in Figure 2b. As shown in the red spectral curve in Figure 4a, the average absorptances of the designed resonant cavity absorber structure are ~89.56% and ~93.51% at 3–5 μm and 8–14 μm, respectively, which is comparable to those of porous materials [12–17]. However, as shown in the black spectral curve in Figure 4a, the conventional standard cavity absorption structure designed with Al as the bottom reflective metal layer, SiO_2 as the intermediate dielectric layer, and Ti as the top absorption layer only has an absorptance close to 91% in 3–5 μm [18]. Therefore, the thermopile chip with the designed resonant cavity absorber structure can not only be used as the detector elements of the non-contact infrared thermometer and thermal imager, but also as the NDIR gas detector element for detection of CO_2, CO, NO_2, and CH_4 gases. Generally, the absorption of the resonant cavity absorber structure is highly sensitive to the thickness error of each layer, and the thickness of each layer needs to be accurately controlled during the preparation process to achieve the ideal absorption effect. In fact, the deposition processes of MEMS and CMOS both have a certain thickness error, so the Essential Macleod software was used to simulate the absorption spectra of the designed resonant cavity absorber structure, under mean thickness error of 5%. As shown in Figure 4b, the absorption spectra of the designed resonant cavity absorber structure have no obvious deviation in absorptance compared with that of the theoretical design (Figure 4b). Thus, under the mean thickness error (5%) of the general MEMS and CMOS deposition processes, the designed resonant cavity absorber structure still has high absorption and stability. The structure design using air as the middle dielectric layer and SiO_2/TiN/Si_3N_4 sandwich layers as the top absorption layer can effectively broaden the absorption wavelength range and reduce the sensitivity to thickness error.

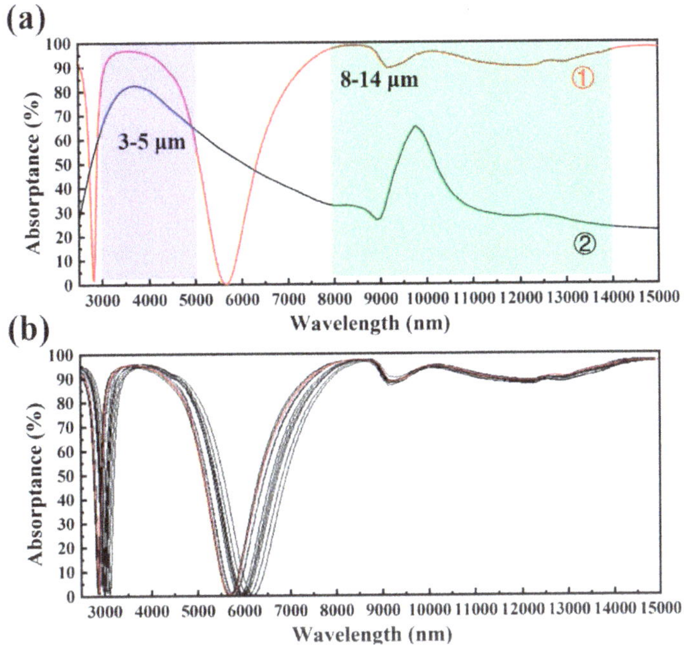

Figure 4. Theoretical absorption spectra of the designed resonant cavity absorber structures without (**a**) and (**b**) with mean thickness error of 5%.

3. Design of Thermopile Chip

Based on the designed resonant cavity absorber structure, we propose a new type of thermopile chip structure, as shown in Figure 5. In this paper, we adopted monocrystalline silicon as the substrate and N-/P-Poly-Si as the thermocouple. The SiO_2 layers were used as the support, insulating and passivation layers of the thermopile. The hot junctions of the thermopile were laid on the floating membrane, and the cold junctions were distributed on the center of substrate. Moreover, the N-Poly-Si and P-Poly-Si were connected using Al as interlayer at the hot junctions and the cold junctions of the thermopile, respectively.

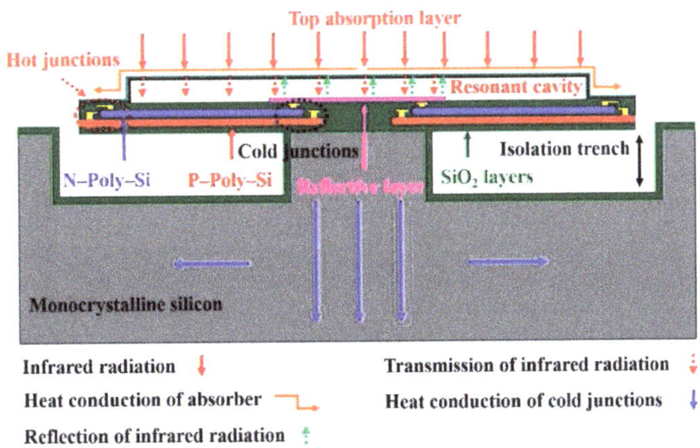

Figure 5. Schematics of structure and energy flow of thermopile chip with resonant cavity absorber structure.

The responsivity (R), noise equivalent power (NEP), specific detectivity (D *), and time constant (T) are the main three performance parameters for evaluating the thermopile chip, and R is the most obvious and direct way to characterize the thermopile chip by the ratio of ΔU to incident infrared radiation power (P_0) and can be calculated from Equation (1) [3]:

$$R = \frac{\Delta U}{P_0}$$

When infrared radiation is applied to the thermopile chip, there will be a temperature difference (T_{diff}) between the hot junctions and the cold junctions of the thermopile, and T_{diff} is converted into a measurable output voltage (ΔU) based on the Seebeck effect. The ΔU for a thermopile is then calculated from Equation (2) [3]:

$$\Delta U = NT_{diff}|\alpha_{\mathrm{A}} - \alpha_{\mathrm{B}}| = NT_{diff}\alpha_{\mathrm{AB}}$$

Therefore, increasing the T_{diff} between the hot junctions and the cold junctions of the thermopile is the most direct and effective way to improve the performance of the thermopile chip.

In order to better understand and design the structure of the thermopile chip and make the designed resonant cavity absorber structure be perfectly combined with the thermopile chip, we used the finite element analysis software to simulate the thermopile chip's thermal field. As shown in Figure 6, compared with other double-layer suspension thermopile chips, the thermopile chip we designed has a higher T_{diff} between the hot junctions and the cold junctions [29–31]. Since both the doped polysilicon and SiO_2 layers have certain absorption characteristic of infrared radiation, the cold junctions of the thermopile also have infrared absorption characteristic [20,21]. When the infrared radiation is irradiated on the surface of the thermopile chip, the cold junctions will also perform infrared absorption and light-to-heat conversion except the hot junctions and the infrared absorber. As shown in Figure 6a, the absorption of infrared radiation by the cold junctions will increase the temperature of the cold junction and thus weaken the responsivity of the thermopile chip. To solve this problem, we designed the cold junctions of the thermopile specifically, that is, the bottom reflective metal layer (Al) of the resonant cavity absorber structure was deposited on the cold junctions of the thermopile. Figure 3 shows that the Al layer has very low absorptance and very high reflectance (~97%) at 2–14 μm so that the Al layer can well reflect the infrared radiation irradiated on the cold junctions, prevent the cold junctions from absorbing infrared radiation, and inhibit the rise of temperature, resulting in further improvement of responsivity and specific detectivity (as shown in Figure 6b). Furthermore, the Al layer can also serve as the reflective metal layer of the resonant cavity absorber structure, forming the interference of light and strengthening the infrared absorption characteristic of the top absorption metal layer. The top absorption layer of the resonant cavity absorber structure can be connected to the hot junctions through the support column, which facilitates the transfer of the heat of the top absorption layer to the hot junctions as shown in Figure 6b. This can further increase the temperature difference between the hot junctions and the cold junctions, and reduce the response time of the thermopile chip. In particular, the intermediate dielectric layer (air) is located between the absorption layer and the thermopile, and as a result, the heat loss is reduced, and furthermore, the mutual restriction of area between the absorption zone and the thermocouple zone is also solved, leading to a very high duty cycle. Thus, the sizes of the designed thermopile chip can be further reduced while maintaining responsivity and specific detectivity compared with the three types of traditional thermopile chips.

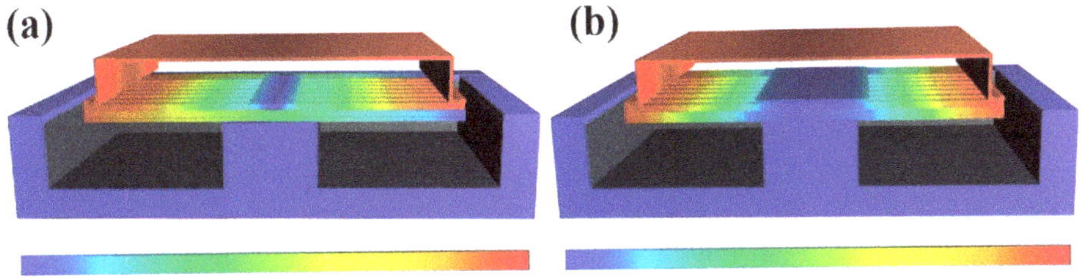

Figure 6. The (**a**) standard and (**b**) our design double suspension structures thermal field simulation.

4. Preparation Process of Thermopile Chip

On the basis of the designed structure of the thermopile chip, the preparation process was proposed, combining CMOS and MEMS processes. Firstly, an SiO_2 layer of ~200 nm thick was deposited on the surface of the silicon substrate as a masking layer using a thermal oxidation process, and then the photolithography and reactive ion etching (RIE) processes ware used to pattern and etch the SiO_2 layer (Figure 7a). Secondly, the isolation trench with depth of 6000 nm was obtained by the RIE process for the bare silicon substrate (Figure 7b). Thirdly, a thermal oxidation process was used again to generate an SiO_2 layer of ~200 nm thick on the etched silicon substrate as a release barrier (Figure 7c). Then, a polysilicon layer of ~3000 nm thick was grown by low pressure chemical vapor deposition (LPCVD) and then planarized, which served as a sacrificial release layer (Figure 7d). An SiO_2 layer of ~500 nm thick was deposited by LPCVD on the polysilicon layer, which served as a thermopile support layer (Figure 7e). A polysilicon layer of ~500 nm thick was deposited by LPCVD on the support layer as one layer of the thermocouple and doped with boron (B) by ion implantation to form P-type polysilicon, and then the P-type polysilicon was patterned by the RIE process (Figure 7f). Next, an SiO_2 layer of ~100 nm thick was deposited again by LPCVD as the insulating layer of the P-type polysilicon layer (Figure 7g). A polysilicon layer of ~500 nm thick was deposited by LPCVD on the insulating layer as another layer of the thermocouple and doped with phosphorus (P) by ion implantation to form N-type polysilicon, and then the N-type polysilicon was patterned by the RIE process (Figure 7h). An SiO_2 layer of ~100 nm thick was deposited by LPCVD as the insulating layer of the N-type polysilicon layer (Figure 7i). Subsequently, the two insulating layers were etched to form the connection holes at the hot junctions and the cold junctions by the RIE process (Figure 7j). Then, an Al layer of ~500 nm thick was deposited by magnetron sputtering technology (MST) and patterned and etched by the photolithography and RIE processes, which served as the electrical connection between the thermocouples and the electrode (Figure 7k). The withstanding temperature of the substrate was greater than 700 °C while employing LPCVD to deposit the SiO_2 or Si_3N_4 layer, which brought about serious damage to the deposited Al layer. Finally, the SiO_2 layer of ~100 nm thick, as the passivation layer of the thermopile, was deposited by plasma enhanced chemical vapor deposition (PECVD) (Figure 7l).

For the preparation of the resonant cavity absorber structure, an Al layer of ~100 nm thick was first deposited by MST as the bottom reflective metal layer, and then the photolithography and RIE processes were used to pattern and etch the Al layer (Figure 8a). The un-etched Al layer covered the cold junctions of the thermopile, which not only served as the reflective metal layer but also could reflect the infrared radiation irradiated at the cold junctions. Then, an amorphous silicon layer of ~2700 nm thick was deposited by PECVD as the sacrificial layer (Figure 8b). Subsequently, an SiO_2 layer of ~100 nm thick was deposited by PECVD as the support layer of the top absorption metal layer; next, a TiN layer of ~20 nm thick was deposited by MST as the top absorption metal layer, and then an Si_3N_4 layer of ~250 nm thick was deposited by PECVD as the passivation layer of the top absorption metal layer, thereby forming a complete SiO_2/TiN/Si_3N_4 resonant cavity absorber structure (Figure 8c). Finally, the sacrificial layer composed of amorphous silicon was dry-etched and released by the release hole etching and XeF_2 gas phase release processes to form the thermopile chip with the resonant cavity absorber structure (Figure 8d).

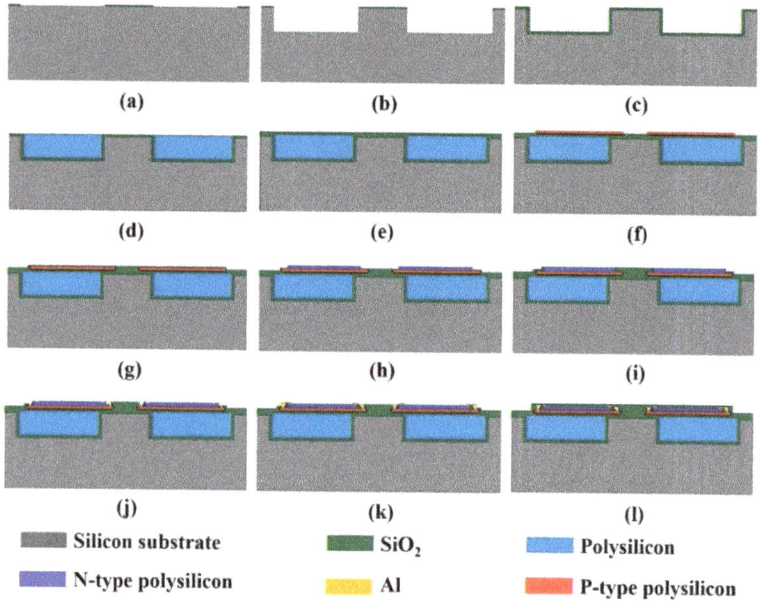

Figure 7. Schematic of preparation process of the designed thermopile. (**a**) deposition of SiO_2 layer of 200 nm thickness, (**b**) isolation trench with depth of 6000 nm, (**c**) thermal oxidation, (**d**) developing sacrificial release layer, (**e**) 500 nm thick thermopile support layer, (**f**) 500 nm thick polysilicon layer RIE process, (**g**) insulating layer, (**h**) polysilicon layer, (**i**) 100 nm insulating layer, (**j**) formation of connection holes, (**k**) 500 nm thick Al layer and (**l**) 100 nm thick SiO_2 layer.

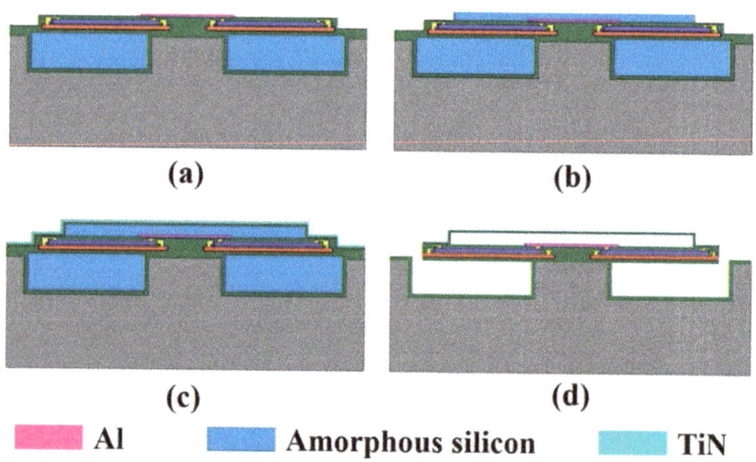

Figure 8. Schematic of preparation process of the designed resonant cavity absorber structure. (**a**) 100 nm thick bottom reflective metal layer, (**b**) 2700 nm thick amorphous silicon layer, (**c**) resonant cavity absorber structure, (**d**) final thermopile chip.

5. Conclusions

A novel thermopile chip with resonant cavity absorber structure was designed by an optical simulation in this work. In the resonant cavity absorber structure, the Al, air, and SiO_2/TiN/Si_3N_4 sandwich layers were innovatively used as the bottom reflective metal layer, intermediate dielectric layer, and top absorption layer, respectively. The designed resonant cavity absorber structure exhibited ~89.56% and ~93.51% average absorptances at 3–5 μm and 8–14 μm, respectively. In particular, the bottom reflective metal layer (Al) was deposited on the cold junctions of the thermopile, which made the Al layer not only act as the reflective metal layer to reflect the infrared radiation and to form the resonant cavity absorber structure, but also prevented the cold junctions from absorbing infrared radiation and inhibiting the rise of temperature. As a result, the temperature difference between the hot junctions and the cold junctions was increased, and the responsivity of the thermopile chip was improved. Moreover, the duty cycle of the thermopile chip was greatly improved due to the double-layer suspension structure. Compared with the traditional thermopile chips, the sizes of the designed thermopile chip can be further reduced while maintaining responsivity and specific detectivity.

Author Contributions: Conceptualization, H.H. and S.H.; methodology, J.Y., M.O., and G.L.; validation, J.L. and H.S.; formal analysis, S.H. and G.Q.; writing—original draft preparation, H.H. and J.Y.; writ-ing—review and editing, S.H. and M.T.N.; visualization, J.Y.; supervision, S.H. and M.T.N.; project administration, S.H., G.Q., A.G. and M.T.N.; funding acquisition, M.A. and M.M.A.-A. All authors have read and agreed to the published version of the manuscript.

Funding: This work was supported by the National Natural Science Foundation of China (51572111, 51950410596), the Key Research and Development Plan (BE2019094), The authors are grateful to the Deanship of Scientific Research, King Saud University for funding through Vice Deanship of Scientific Research Chairs and thankful to the financial support funded by the Deanship of Scientific Research at Princess Nourah bint Abdulrahman University through the Fast-track Research Funding Program; Research and Innovation Projects (KYCX19_1588), Six Talent Peaks Project (TD-XCL-004), 333 Talents project (BRA2017387) and Qing Lan Project ([2016]15) of Jiangsu Province.

Data Availability Statement: Data are available in the paper.

Conflicts of Interest: The authors declare no conflict of interest.

References

1. Graf, A.; Arndt, M.; Sauer, M.; Gerlach, G. Review of micromachined thermopiles for infrared detection. *Meas. Sci. Technol.* **2007**, *18*, R59–R75. [CrossRef]
2. Zhu, J.; Liu, X.; Shi, Q.; He, T.; Sun, Z.; Guo, X.; Liu, W.; Bin Sulaiman, O.; Dong, B.; Lee, C. Development Trends and Perspectives of Future Sensors and MEMS/NEMS. *Micromachines* **2019**, *11*, 7. [CrossRef]
3. Xu, D.; Wang, Y.; Xiong, B.; Li, T. MEMS-based thermoelectric infrared sensors: A review. *Front. Mech. Eng.* **2017**, *12*, 557–566. [CrossRef]
4. Stange, A.; Imboden, M.; Javor, J.; Barrett, L.; Bishop, D. Building a casimir metrology platform with a commercial MEMS sensor. *Microsyst. Nanoeng.* **2019**, *5*, 14. [CrossRef]
5. Chen, C.N. Temperature error analysis and parameter extraction of an 8–14 μm thermopile with a wavelength-independent absorber for tympanic thermometer. *IEEE Sens. J.* **2011**, *11*, 2310–2317.
6. Zhang, Z.; Zhang, H.; Liu, T. Study on body temperature detection of pig based on infrared technology: A review. *Artif. Intell. Agric.* **2019**, *1*, 14–26. [CrossRef]
7. Udina, S.; Carmona, M.; Carles, G.; Santander, J.; Fonseca, L.; Marco, S. A micromachined thermoelectric sensor for natural gas analysis: Thermal model and experimental results. *Sens. Actuators B Chem.* **2008**, *134*, 551–558. [CrossRef]
8. Vincent, T.; Gardner, J. A low cost MEMS based NDIR system for the monitoring of carbon dioxide in breath analysis at ppm levels. *Sens. Actuators B Chem.* **2016**, *236*, 954–964. [CrossRef]
9. Yin, Y.; Fang, Z.; Liu, Y.; Han, F. Temperature-insensitive structure design of micromachined resonant accelerome-ters. *Sensors* **2019**, *19*, 1544. [CrossRef] [PubMed]
10. Hirota, M.; Nakajima, Y.; Saito, M.; Uchiyama, M. 120×90 element thermoelectric infrared focal plane array with precisely patterned Au-black absorber. *Sens. Actuators A Phys.* **2007**, *135*, 146–151. [CrossRef]
11. Li, Y.L.; Zhou, H.; Li, T.; Wang, Y.; Liu, Y.X.; Wang, Y.L. CMOS-compatible 8 × 2 thermopile array. *Sens. Actuators A Phys.* **2010**, *161*, 120–126. [CrossRef]

12. Mao, H.; Chen, Y.; Ou, Y.; Ou, W.; Xiong, J.; You, C.; Tan, Q.; Chen, D. Fabrication of nanopillar forests with high infrared absorptance based on rough poly-Si and spacer technology. *J. Micromech. Microeng.* **2013**, *23*, 95033. [CrossRef]
13. De Luca, A.; Cole, M.T.; Hopper, R.H.; Boual, S.; Warner, J.H.; Robertson, A.R.; Ali, S.Z.; Udrea, F.; Gardner, J.W.; Milne, W.I. Enhanced spectroscopic gas sensors using in-situ grown carbon nanotubes. *Appl. Phys. Lett.* **2015**, *106*, 194101. [CrossRef]
14. Xia, Y.; Xue, C.Y.; Ou, W.; Yuqn, F.; Zhao, L. A design of thermopile infrared detector based on black silicon as absorption. *Instrum. Tech. Sens.* **2012**, *25*, 581–584.
15. Chen, C.-N. Fully quantitative characterization of CMOS–MEMS polysilicon/titanium thermopile infrared sensors. *Sens. Actuators B Chem.* **2012**, *161*, 892–900. [CrossRef]
16. Zhao, L.; He, X.; Li, J.; Gao, X.; Jia, J. Electrosprayed carbon-based black coatings for pyroelectric detectors. *Sens. Actuators A Phys.* **2013**, *196*, 16–21. [CrossRef]
17. He, Y.; Wang, Y.; Li, T. Performance Enhanced Thermopile with Rough Dielectric Film Black. *IEEE Electron Device Lett.* **2020**, *41*, 593–596. [CrossRef]
18. Roncaglia, A.; Mancarella, F.; Cardinali, G. CMOS-compatible fabrication of thermopiles with high sensitivity in the 3–5 μm atmospheric window. *Sens. Actuators B Chem.* **2007**, *125*, 214–223. [CrossRef]
19. Ashraf, S.; Mattsson, C.G.; Thungstrom, G.; Asharaf, S.; Asharf, S. Fabrication and Characterization of a SU-8 Epoxy Membrane-Based Thermopile Detector with an Integrated Multilayered Absorber Structure for the Mid-IR Region. *IEEE Sens. J.* **2019**, *19*, 4000–4007. [CrossRef]
20. Jutzi, F.; Wicaksono, D.H.B.; Pandraud, G.; de Rooij, N.; French, P.J. Far-infrared sensor with LPCVD-deposited low-stress Si-rich nitride absorber membrane: Part 2. Thermal property, and sensitivity. *Sens. Actuators A Phys.* **2009**, *152*, 119–125. [CrossRef]
21. Xu, D.; Xiong, B.; Wang, Y. Design, fabrication and characterization of a front-etched micromachined thermopile for IR detection. *J. Micromech. Microeng.* **2010**, *20*, 115004. [CrossRef]
22. Wang, K.; Xue, C.; Liang, T.; Jiao, B.; Zhang, W.; Chen, D.; Xiong, J. Thermopile Infrared Detector with Detectivity Greater Than 108 cmHz(1/2)/W. *J. Infrared Millim. Terahertz Waves* **2010**, *31*, 810–820. [CrossRef]
23. Lei, C.; Mao, H.; Yang, Y.; Ou, W.; Xue, C.; Yao, Z.; Ming, A.; Wang, W.; Wang, L.; Hu, J.; et al. A double-end-beam based infrared device fabricated using CMOS-MEMS process. *Sens. Rev.* **2016**, *36*, 240–248. [CrossRef]
24. Cao, L.M.; Zhang, Y.Z.; Sun, S.J.; Guo, H.; Zhang, Z. Structure design and test of MEMS thermocouple infrared detector. *Microsyst. Technol.* **2018**, *24*, 2463–2471.
25. Bao, A.; Lei, C.; Mao, H.; Li, R.; Guan, Y. Study on a High Performance MEMS Infrared Thermopile Detector. *Micromachines* **2019**, *10*, 877. [CrossRef] [PubMed]
26. Chang, K.-C.; Lee, Y.-C.; Sun, C.-M.; Fang, W. Novel absorber membrane and thermocouple designs for CMOS-MEMS thermoelectric infrared sensor. In Proceedings of the 2017 IEEE 30th International Conference on Micro Electro Mechanical Systems (MEMS), Las Vegas, NV, USA, 22–26 January 2017; IEEE: New York, NY, USA, 2017; pp. 1228–1231.
27. Shen, T.-W.; Chang, K.-C.; Sun, C.-M.; Fang, W. Performance enhance of CMOS-MEMS thermoelectric infrared sensor by using sensing material and structure design. *J. Micromech. Microeng.* **2018**, *29*, 025007. [CrossRef]
28. Hou, H.; Huang, Q.; Liu, G.; Qiao, G. Enhanced performances of CMOS-MEMS thermopile infrared detectors using novel thin film stacks. *Infrared Phys. Technol.* **2019**, *102*, 103058. [CrossRef]
29. Chou, B.C.S.; Mang, O. Thermopile Infrared Sensor, Thermopile Infrared Sensors Array, and Method of Manufacturing the Same. U.S. Patent 6335478 B1, 1 January 2002.
30. Xie, J.; Lee, C.; Feng, H. Design, Fabrication, and Characterization of CMOS MEMS-Based Thermoelectric Power Generators. *J. Microelectromech Syst.* **2010**, *19*, 317–324. [CrossRef]
31. Chen, Y.; Mao, H.; Tan, Q.; Xue, C.; Ou, W.; Chen, D.; Xiong, J. Design of thermopile-based infrared detectors with suspended absorber-thermopile Bi-layers. *J. Sens. Technol.* **2014**, *27*, 730–735.

coatings

Article

Electro-Oxidation of Ammonia at Novel Ag_2O-PrO_2/γ-Al_2O_3 Catalysts

Mariam Khan, Naveed Kausar Janjua, Safia Khan, Ibrahim Qazi, Shafaqat Ali and Tahani Saad Algarni

[1] School of Applied Sciences and Humanity, National University of Technology, Islamabad 44000, Pakistan; mariamkhan@nutech.edu.pk
[2] Department of Chemistry, Quaid-i-Azam University, Islamabad 45320, Pakistan; safiakhan@chem.qau.edu.pk
[3] Department of Materials Science and Engineering, Institute of Space Technology, Islamabad 44000, Pakistan; ibrahim.qazi@ist.edu.pk
[4] Department of Environmental Science and Engineering, Government College University, Faisalabad 38000, Pakistan; shafaqataligill@yahoo.com
[5] Department of Biological Sciences and Technology, China Medical University, Taichung 40402, Taiwan
[6] Chemistry Department, College of Science, King Saud University, Riyadh 11451, Saudi Arabia; tahanis@ksu.edu.sa
* Correspondence: nkjanjua@qau.edu.pk

Abstract: An $Ag_2O_{(x)}-PrO_{2(y)}/\gamma$-$Al_2O_3$ electrocatalyst series (X:Y is for Ag:Pr from 0 to 10) was synthesized, to use synthesized samples in electrochemical applications, a step in fuel cells advancements. $Ag_2O_{(x)}-PrO_{2(y)}/\gamma$-$Al_2O_3$/Glassy-Carbon was investigated for electrochemical oxidation of ammonia in alkaline medium and proved to be highly effective, having high potential utility, as compared to commonly used Pt-based electrocatalysts. In this study, gamma alumina as catalytic support was synthesized via precipitation method, and stoichiometric wt/wt.% compositions of Ag_2O-PrO_2 were loaded on γ-Al_2O_3 by co-impregnation method. The desired phase of γ-Al_2O_3 and supported nanocatalysts was obtained after heat treatment at 800 and 600 °C, respectively. The successful loadings of Ag_2O-PrO_2 nanocatalysts on surface of γ-Al_2O_3 was determined by X-rays diffraction (XRD), Fourier-transform Infrared Spectroscopy (FTIR), and energy dispersive analysis (EDX). The nano-sized domain of the sample powders sustained with particle sizes was calculated via XRD and scanning electron microscopy (SEM). The surface morphology and elemental compositions were examined by SEM, transmission electron microscopy (TEM) and EDX. The conductive and electron-transferring nature was investigated by cyclic voltammetry and electrochemical impedance (EIS). Cyclic voltammetric profiles were observed, and respective kinetic and thermodynamic parameters were calculated, which showed that these synthesized materials are potential catalysts for ammonia electro-oxidation. $Ag_2O_{(6)}-PrO_{2(4)}/\gamma$-$Al_2O_3$ proved to be the most proficient catalyst among all the members of the series, having greater diffusion coefficient, heterogeneous rate constant and lesser Gibbs free energy for this system. The catalytic activity of these electrocatalysts is revealed from electrochemical studies which reflected their potentiality as electrode material in direct ammonia fuel cell technology for energy production.

Keywords: ammonia electro-oxidation; cyclic voltammetry; electrochemical surface area (ECSA); electrocatalysts; nanocomposites

Citation: Khan, M.; Janjua, N.K.; Khan, S.; Qazi, I.; Ali, S.; Saad Algarni, T. Electro-Oxidation of Ammonia at Novel Ag_2O-PrO_2/γ-Al_2O_3 Catalysts. *Coatings* **2021**, *11*, 257. https://doi.org/10.3390/coatings11020257

Received: 25 January 2021
Accepted: 11 February 2021
Published: 22 February 2021

1. Introduction

Energy generation from hydrogen in a sustainable and continuous manner is the foremost concern of scientists and engineers across the globe [1,2]. Fuel cells are considered to be a great contributor of energy production which changes the chemical energy to electrical energy through an electrochemical reaction in the cell [3–5]. Among several fuel cells, the ammonia fuel cell is a better substitute of carbon-based conventional energy generating technologies [6,7]. Liquid ammonia is a promising hydrogen carrier, owing to its favorable properties, like high storage capacity, hydrogen density (17.8 wt.%, 10.7 kg H_2/100 L at 1 MPa and 298 K), carbon free fuel, high boiling point and odor; thus, the challenges associated with storage of hydrogen can be overcome [8]. Volumetric hydrogen density of ammonia is 1.5 times higher than pure hydrogen, because ammonia is effortlessly liquefiable at room temperature and 1 MPa [9]. Moreover, ammonia-derived electricity is under constant research for running gas turbines, steam turbines and alkaline fuel cells, which are being commercialized at lower-scale off-grid power plants [10,11]. However, the larger-scale implementation of ammonia fuel cells is currently hindered due to less efficient and much expensive electrodes [12]. A variety of electrode materials are being investigated for ammonia electro-oxidation in basic conditions [13–15]. Although Pt and Pt based materials are being employed as electrodes in ammonia fuel cells, due to sufficient catalytic significance, their industrial-scale applications still encounter critical challenges due to non-affordability [16,17]. Hence, development of novel Pt free cheaper electrodes is required as an alternative to accelerate ammonia fuel cells in energy generating technology [17]. Moreover, multiple factors affect the catalytic performance of electrodes towards ammonia electro-oxidation, including the nature of the support material and modifiers/promotor, as well as the synthesis conditions [18,19].

Gamma alumina (γ-Al_2O_3) is a highly used catalyst support, owing to its worthy mechanical properties, high thermal stability and ability to disperse the active oxide precursors [20,21]. Moreover, the introduction of metal oxides into γ-Al_2O_3 support is found to promote the catalyst's efficiency towards electro-oxidation processes [22–24]. With this intent, Ag_2O is used as a catalyst promotor in several oxidation processes [25,26]. Furthermore, lanthanide and lanthanide oxides seemed to be efficient promotors, exhibiting enhanced catalytic activity in many electro-oxidation processes [20]. Therefore, catalytic materials based on praseodymium oxide and praseodymium containing mixed metal oxides of high-surface area also provide efficient energy production [27]. Moreover, mixed metal oxides, including sieves supported metal oxides and bulk mixed metal oxides, have experienced an intense paradigm in electrocatalysis over the past few years [28–30].

Herein, a series of novel mixed metal oxides Ag_2O–PrO_2/γ-Al_2O_3 electrocatalysts were prepared by an incipient wet impregnation method. The aim of present research was to observe the effect of variation in percentage composition of PrO_2 and Ag_2O supported on γ-Al_2O_3 towards ammonia electro-oxidation. Physical and electrochemical properties of as-synthesized catalysts were investigated for NH_3 electro-oxidation via different techniques, i.e., XRD, FTIR, SEM, EDS, EIS and cyclic voltammetry. Figure 1 shows the schematic diagram for experimental procedure of electro-oxidation of ammonia over as synthesized electrocatalysts.

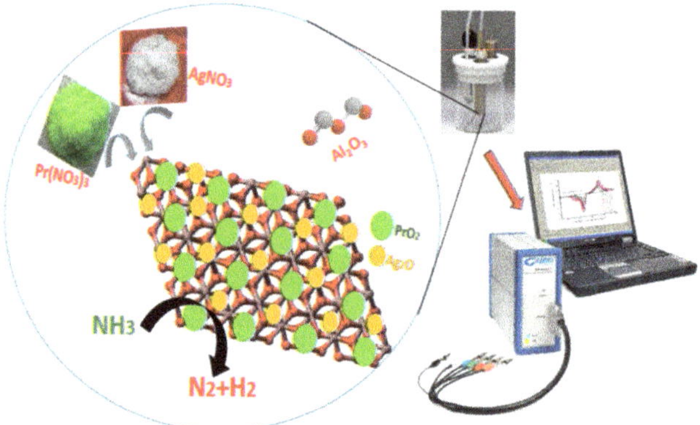

Figure 1. Graphical abstract elucidating the experimental, mechanistic and electrochemical insight.

2. Catalyst Preparation

2.1. Materials

Chemicals used for synthesis were bought from Sigma Aldrich (St. Louis, MO, USA). The chemicals used for the present work are aluminum nitrate ($Al(NO_3)_3{\cdot}9H_2O$), ammonia solution (NH_3), praseodymium nitrate hexahydrate ($Pr(NO_3)_3{\cdot}6H_2O$), silver nitrate ($AgNO_3$), potassium hydroxide (KOH), sulfur acid (H_2SO_4), potassium hexacyanoferrate (III) ($K_4[Fe(CN)_6]$), alumina powder and Nafion.

2.2. Synthesis of γ-Al_2O_3

In total, a 1 M solution of aluminum nitrate nonahydrate and a 3 M ammonia solution were prepared in deionized water. The ammonia solution was added dropwise into precursor solution, under constant stirring until formation of homogeneous precipitates of aluminum hydroxide. The temperature of solution was maintained at 60 °C. White gelatinous $Al(OH)_3$ precipitates were formed and were kept on heating, until semi-solid material was formed. The precipitates were dried at 200 °C overnight. The $Al(OH)_3$ precipitates were calcined at 800 °C, to form γ-alumina, for 2 h. The calcined sample was grinded with acetone, to form powdered γ-alumina.

2.3. Synthesis of Ag_2O–PrO_2/γ-Al_2O_3

The metal oxides composites catalysts were prepared via co-impregnation method. In this method, the solution of active element precursors was mixed before impregnating on support. We used an adapted method to synthesize the wt/wt.% alumina supported composites of silver and praseodymium oxides. Stoichiometric amounts of γ-alumina were taken and wetted with requisite volume of praseodymium nitrate hexahydrate 0.5 M and silver nitrate 0.5 M to load wt/wt.% Ag-Pr oxides. The mixtures were soaked overnight, evaporated and dried at 100 °C. The dried samples were calcined at 600 °C, for 2 h, and finely ground to Ag_2O–PrO_2/γ-Al_2O_3 catalysts.

3. Physical Characterization

3.1. X-Ray Diffraction Analysis

The size and phase of all synthesized materials ($Ag_2O_{(X)}$–$PrO_{2(Y)}$/γ-Al_2O_3) were examined by X-ray diffraction using (PANalytical X'PERT High Score's diffractometer, Malvern, UK), exhibiting Cu K_α radiation, works in the range of (10°–80°) Figure 2a,b. The peaks observed at 2θ of 37.0°, 40.20°, 45.78°, 61.30° and 66.99° were assigned to 311, 222, 400, 511 and 440 crystal planes. These are according to JCPDS (Joint Committee on Powder Diffraction Standards) card no. (29-0063) [31,32]. The gamma phase of synthesized alumina calcined at 800 °C is suggested by these peak positions. The XRD corroborate the characteristic peaks of Ag_2O (26.78°, 32.14°, 38.04° and 67.48°), PrO_2 (28.1°, 46.7°, 55.6° and 78.2°) and γ-Al_2O_3, which are related to JCPDS card no. (76-1393), (24-1006) and (29-0063), respectively [33,34]. The diffraction peaks at 67.48° are indexed to (222) planes of face-centered cubic silver oxide [35]. These peaks correspond to the successful loadings of promoting metal oxides on surface of support. Crystallinity of the catalysts increases with increasing the contents of silver oxide precursor on the γ-Al_2O_3 support as sharper peaks are observed with higher loadings of Ag_2O. No phase segregation is observed as obvious from XRD patterns.

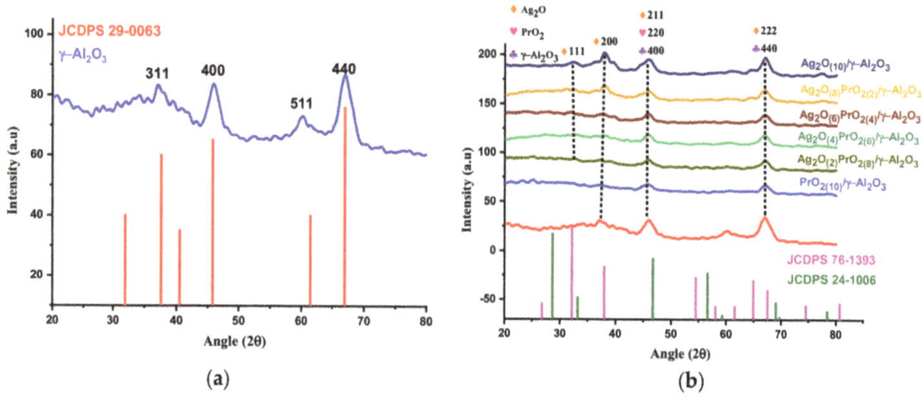

Figure 2. X-ray diffraction peaks of (**a**) γ-Al_2O_3 with Joint Committee on Powder Diffraction Standards (JCPDS) card numbers. (**b**) Overlay of XRD peaks of γ-Al_2O_3 and $Ag_2O_{(X)}$–$PrO_{2(Y)}$/γ-Al_2O_3, along with JCPDS card numbers.

The average crystallite sizes (D_{avg}), which are enlisted in Table 1, were calculated by the Scherrer equation [36]:

$$D_{avg}\ (\text{nm}) = 57.2k\lambda/\beta\cos\theta \tag{1}$$

where D_{avg} is the average crystallite size, constant k is shape factor (0.9), λ is the wavelength (0.154 nm), β corresponds to the peak width at half maximum intensity and θ is the peak position. D_{avg} for support (γ-Al_2O_3) is 18.3 nm, while for nanomaterials, it lies is in the range of 25–42 nm.

Table 1. Calculated values for average crystallite size and average particle size.

Electrocatalysts	$D_{avg(XRD)}$ (nm)	$D_{avg(SEM)}$ (nm)
$PrO_{2(10)}$/γ-Al_2O_3	41.3	26.02
$Ag_2O_{(2)}$–$PrO_{2(8)}$/γ-Al_2O_3	25.4	20.76
$Ag_2O_{(4)}$–$PrO_{2(6)}$/γ-Al_2O_3	28.9	29.41
$Ag_2O_{(6)}$–$PrO_{2(4)}$/γ-Al_2O_3	32.3	31.01
$Ag_2O_{(8)}$–$PrO_{2(2)}$/γ-Al_2O_3	20.3	23.02
$Ag_2O_{(10)}$/γ-Al_2O_3	25.0	31.04

3.2. *FTIR Analysis*

The Fourier-transform infrared spectroscopy (FTIR) spectra for γ-Al_2O_3 and all synthesized nanocomposites inserted in KBr powder system were investigated in wavelength range (400–4000) cm^{-1} by using Nicolet 5PC, Nicolet Analytical Instrument (Protea, Cambridgeshire, UK, Figure 3). The vibration bands observed in the range of (3500–4000) cm^{-1} correspond to O–H stretching vibrations. These bands appeared because of moisture absorbed by the samples [37]. C–O vibration bands due to adsorbed CO_2 also appeared in the range of (1500–2500) cm^{-1} [38]. The metal oxides vibration bands are seen between 300 and 700 cm^{-1} [39]. The bands in Figure 3b represent the stretching and bending vibration modes of Ag_2O [40] and stretching vibration bands of PrO_2 [41].

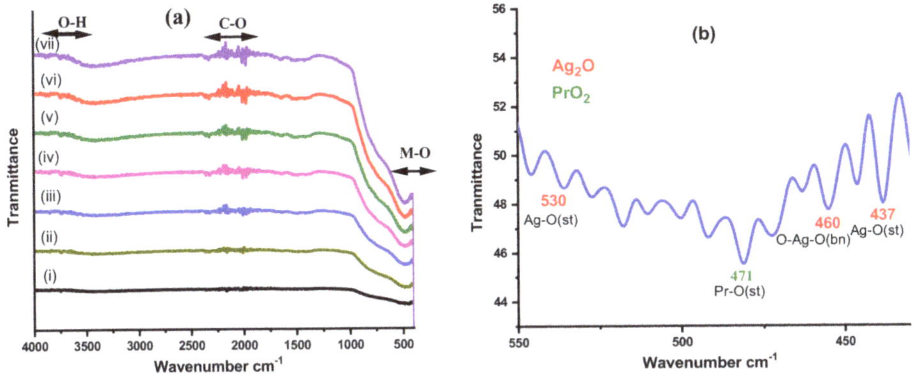

Figure 3. (**a**) FTIR spectra of γ-Al_2O_3 and $Ag_2O_{(X)}$–$PrO_{2(Y)}$/γ-Al_2O_3 nanomaterials. (**b**) FTIR spectrum representing vibration modes of Ag-O and Pr-O bonds.

3.3. SEM, TEM and EDX Analysis

Surface structure investigation of the nanomaterials was evaluated via scanning electron microscopy (SEM), armed with energy dispersive analysis (EDX), by using MIRA3 TESCAN (Brno, Czech Republic) microscope and transmission electron microscopy (TEM, Thermofisher, Manchester, UK). The SEM images for synthesized nanocomposites are given in figures on different scales. In Figure 4A, the first SEM micrograph corresponds to pure γ-Al_2O_3, which appeared to be like a plate-like interconnected network, which, upon impregnation, converted to globular structures. From Figure 4B, it is clear that there is no segregation of phase, which substantiates that metal oxide precursors are homogeneously dispersed on surface of support. The average particle size from these SEM micrographs was estimated by using the ImageJ software (Version: 1.52v) and is tabulated in Table 1. Figure 4D represents the distributions of diameters of nanoparticles ($Ag_2O_{(6)}$–$PrO_{2(4)}$/γ-Al_2O_3) via histograms presenting the D_{avg}. The estimated average particle sizes from SEM are also in dimensions and slightly in accordance with that calculated by XRD. Figure 4C,D represents the EDX spectra of $Ag_2O_{(4)}$–$PrO_{2(6)}$/γ-Al_2O_3 composition which confirms the presence of desired components in synthesized nanomaterial. These EDX analyses are in accordance with XRD, as we have not seen any extra peak in XRD diffraction pattern. The wt/wt.% elemental composition of all supported nanocomposites is given in Table 2.

The Transmission Electron Microscopy (TEM) analysis in Figure 4F showed that nanocomposites are in good contact with the surface of support and have nanosized morphology also shown by SEM micrographs. The TEM images supported the agreement that particles are closed spheres with a smooth surface and are uniform in size. There is a low contrast shell around them that is attributed to the layer of silver oxides [42]. The histogram in inset (b) of Figure 4F describes the distribution of nanoparticles presenting the synthesized materials that are nanosized, and inset (a) describes the area of TEM micrograph where distributions of particles are taken.

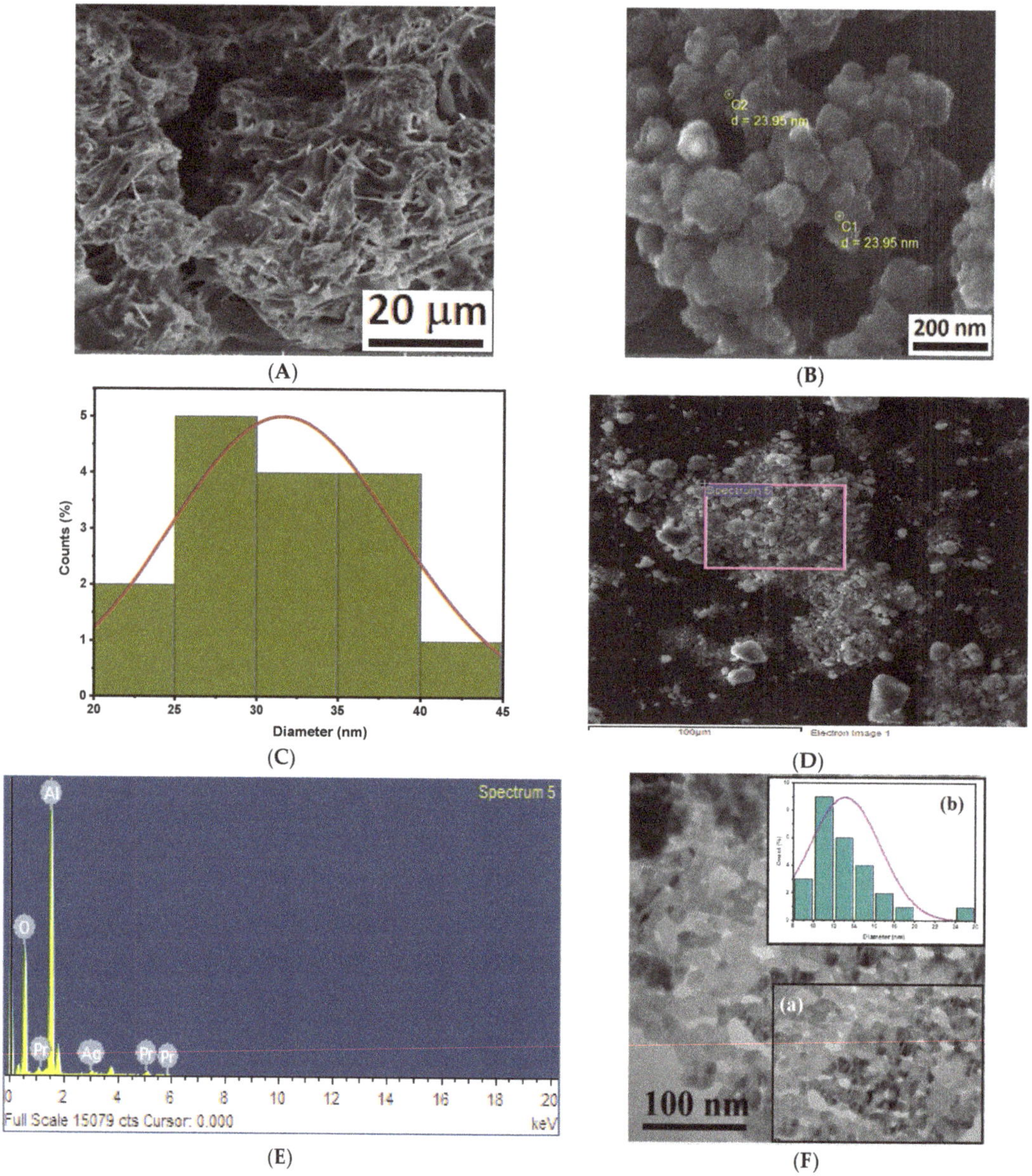

Figure 4. SEM micrographs for (**A**) Al_2O_3 and (**B**) $Ag_2O_{(10)}/\gamma\text{-}Al_2O_3$ nanocatalysts. (**C**) Particle size distribution of $Ag_2O_{(6)}\text{–}PrO_{2(4)}/\gamma\text{-}Al_2O_3$. (**D**,**E**) EDX spectrum of $Ag_2O_{(4)}\text{–}PrO_{2(6)}/\gamma\text{-}Al_2O_3$ composition. (**F**) TEM analysis of sample $Ag_2O_{(2)}\text{–}PrO_{2(8)}/\gamma\text{-}Al_2O_3$, the inset (a) describes the area where distribution of particles taken and inset (b) represents the distributions of particles.

Table 2. Elemental compositions (wt.%) from EDX analysis of samples.

Samples	wt.% Al	wt.% O	wt.% Pr	wt.% Ag
$PrO_{2(10)}/\gamma\text{-}Al_2O_3$	26.90	69.90	9.20	–
$Ag_2O_{(2)}–PrO_{2(8)}/\gamma\text{-}Al_2O_3$	42.40	48.86	8.19	1.56
$Ag_2O_{(4)}–PrO_{2(6)}/\gamma\text{-}Al_2O_3$	42.57	51.88	5.87	3.45
$Ag_2O_{(6)}–PrO_{2(4)}/\gamma\text{-}Al_2O_3$	38.70	57.76	3.54	6.30
$Ag_2O_{(8)}–PrO_{2(2)}/\gamma\text{-}Al_2O_3$	35.24	61.00	1.54	7.39
$Ag_2O_{(10)}/\gamma\text{-}Al_2O_3$	32.95	64.36	–	8.95

4. Electrochemical Characterization

An electrochemical study of the as-prepared electrocatalysts was performed, using Gamry potentiostat interface 1000 (Gamry, Warminster, PA, USA). A three-electrode system was established by modifying glassy carbon (GC) as working electrode, platinum wire as counter electrode and Ag/AgCl (3 M KCl) as reference electrode. The glassy carbon (GC) electrode was first polished by using alumina slurry and then cleaned with ethanol, prior to pouring the catalyst ink prepared in ethanol, and a solution of 5% Nafion (2.0 µL) was added to catalyst, to bind the powder catalyst at the GC surface. We used 5% Nafion as binders, as it has greater stability over others polymer. It also helps in the passivation of surface from by-products formed during reactions, to protect the catalyst from chemical attack [43].

4.1. Estimation of Electrochemical Surface Area (ECSA) of As-Synthesized Electrocatalysts

Electrochemical active surface area (ECSA) of $Ag_2O–PrO_2/\gamma\text{-}Al_2O_3$ samples was estimated by cyclic voltammetric profile, using a redox couple having 5 mM potassium ferrocyanide in 3 M KCl, as shown in Figure 5a. The peak current responses at various scan rates are plotted against $v^{1/2}$, applying the Randles–Ševčík equation [44]:

$$I_p = 2.69 \times 10^5 \cdot n^{3/2} \cdot A \cdot (D^o)^{1/2} \cdot v^{1/2} \cdot C \quad (2)$$

where, A is ECSA (cm^2); I_p is peak current (µA); n represents the number of electrons transferred (1 in this case); D^o is the diffusion co-efficient, which is $0.76 \times 10^{-5}\ cm^2 \cdot s^{-1}$ at standard temperature and pressure (S.T.P); and C is the concentration of $K_4[Fe(CN)_6]$. Corresponding ECSA values are calculated from slope, and the comparative ECSA for the whole series is presented in Figure 5b. The $Ag_2O_{(6)}–PrO_{2(4)}/\gamma\text{-}Al_2O_3$ modified electrode has a greater ECSA value, as compared to other modified electrodes, which means this composition provides greater surface area to catalyze the respective reaction on its surface and facilitates the reaction more efficiently.

4.2. Electrochemical Impedance Spectroscopy EIS

The electron transfer capacities of all modified electrodes were inspected via EIS with Fe^{2+}/Fe^{3+} system in 0.1 M KOH. The Nyquist plots observed for $Ag_2O_{(X)}–PrO_{2(Y)}/\gamma\text{-}Al_2O_3$ modified electrodes are presented in Figure 6, and respective EIS parameters are tabulated in Table 3. The electron transfer resistance systematically reduced from composition (Ag:Pr 0:10) to (Ag:Pr 6:4), and then it increased with increase of Ag content, which endorses that 6:4 wt.% ratio of Ag_2O and PrO_2 on alumina surface is optimum for electrochemical applications. The low value of R_{ct} reflects the greater conductivity and electrocatalytic activity of $Ag_2O_{(6)}–PrO_{2(4)}/\gamma\text{-}Al_2O_3$, as compared to other compositions. The variation in electrochemical behavior of all modified electrodes is due to relative ease in electron transferring, which reveals that nanocatalysts are well dispersed on the alumina surface, in the case of $Ag_2O_{(6)}–PrO_{2(4)}/\gamma\text{-}Al_2O_3$, as compared to the other compositions.

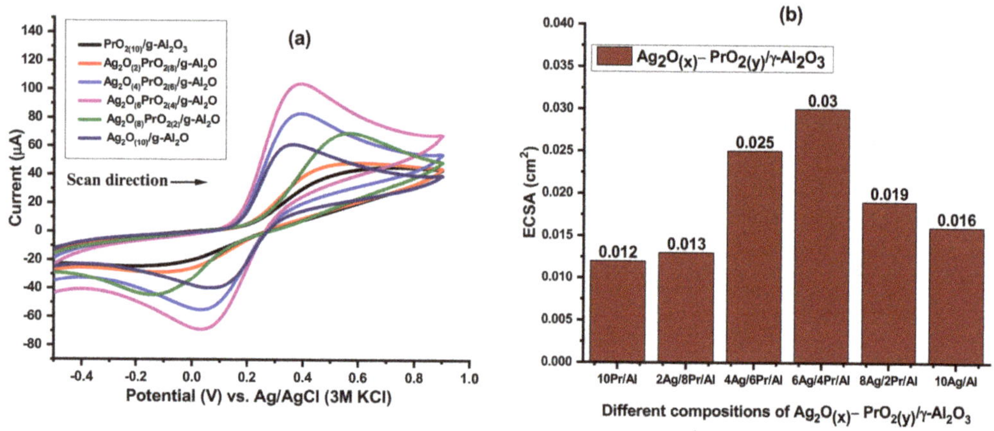

Figure 5. (**a**) Cyclic voltammograms for $Ag_2O_{(X)}$–$PrO_{2(Y)}$/γ-Al_2O_3 in $K_4[Fe(CN)_6]$ + 3 M KCl redox solution. (**b**) Bar graph presenting the variation of electrochemical surface area (ECSA) with composition of catalysts.

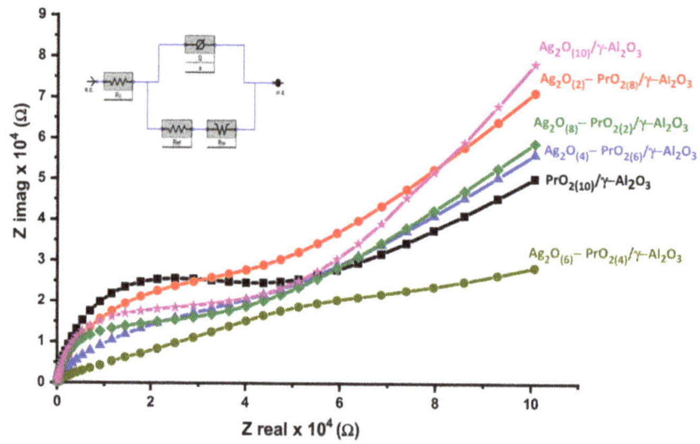

Figure 6. Nyquist plots of $Ag_2O_{(x)}$–$PrO_{2(y)}$/γ-Al_2O_3 modified electrodes recorded in 5 mM $K_4[Fe(CN)_6]$ + 3 M KCl. Inset represents the equivalent circuit.

Table 3. Parameters corresponding to electrochemical impedance (EIS) response of electrocatalysts.

Electrocatalysts	R_s (Ω)	R_{ct} (kΩ)	CPE (μF)	A	R_w (Ω)	$k_{app}/10^{-9}$ (cm·s^{-1})
$PrO_{2(10)}$/γ-Al_2O_3	390.0	50.0	3.70	0.90	24.0	1.10
$Ag_2O_{(2)}$–$PrO_{2(8)}$/γ-Al_2O_3	389.0	40.4	3.00	0.89	29.0	1.30
$Ag_2O_{(4)}$–$PrO_{2(6)}$/γ-Al_2O_3	388.0	27.0	1.40	0.90	24.4	1.97
$Ag_2O_{(6)}$–$PrO_{2(4)}$/γ-Al_2O_3	243.0	24.7	43.0	0.90	21.9	2.20
$Ag_2O_{(8)}$–$PrO_{2(2)}$/γ-Al_2O_3	388.2	29.7	2.00	0.85	25.6	1.80
$Ag_2O_{(10)}$/γ-Al_2O_3	388.3	36.5	2.20	0.92	29.7	1.50

The modification of electrodes has no effect on solution resistance (R_s) and Warburg resistance (R_w), as these are characteristics of electrolyte solution and diffusion of electroactive specie (which are the same in all measurements), while charge transfer resistance (R_{ct}) and constant phase element (CPE) are affected by modification of electrodes, because these are related to conductive properties of the active material. CPE constitutes two elements, and Q and α represent capacitance and surface roughness, respectively. The value of α varies from 0 to 1 [45]. The modified electrode systems have α value ranging from 0.85 to 0.92, revealing that catalysts show fair surface roughness, which agrees with the SEM results. The electron transfer rate constant for all modified system was calculated by Equation (3) [46].

$$k_{app} = RT/F^2 \cdot R_{ct} \cdot C \quad (3)$$

Where k_{app} is electron transfer rate constant, F is Faraday's constant and R is universal constant in SI units.

It can be seen from Table 2 that the value of electron transfer rate constant for $Ag_2O_{(6)}$–$PrO_{2(4)}/\gamma$-Al_2O_3 is higher than the other series materials, showing its higher capacity to facilitate the reaction.

5. Results and Discussion

5.1. Ammonia Electro-Oxidation at $Ag_2O_{(x)}$–$PrO_{2(y)}/\gamma$-Al_2O_3 Modified GC Electrodes

As-synthesized nanocomposites $Ag_2O_{(x)}$–$PrO_{2(y)}/\gamma$-Al_2O_3 were investigated to check their activities towards the oxidation of NH_3 in alkaline medium. The potential window for ammonia oxidation was from −0.3 to 0.6 mV, in forward and reverse scan. When ammonia oxidation was carried out on modified electrodes at room temperature, two anodic peaks were observed which are related to adsorption/oxidation of ammonia in positive going scan (I and II) Figure 7a. A strong cathodic peak (III) is observed at around 0.05 volts, in the absence (pink curve) and presence (green curve) of ammonia, which is the reduction peak of silver oxide converted into metallic silver [47,48]. An anodic shoulder peak (I) appearing at 0.23–0.27 V is due to the oxidation of pre-adsorbed hydrogen/nitrogen-containing intermediates species at the surface of modified electrode, which shows the structural sensitivity of nanocatalysts for these adsorbed species [49,50]. There is appearance of cathodic shoulder peak (IV) in ammonia oxidation catalyzed on the surface of higher members of synthesized series (Figure 8c,d), which is due to desorption of hydrogen on the surface [50]. With higher scan rates, the anodic and cathodic shoulder peaks diminished. The onset potential for ammonia oxidation is 0.03 V vs. Ag/AgCl (i.e., 0.023 V vs. NHE approximately) on the surface of modified electrodes and the oxidation peak appear in the range of 0.32–0.51 volts, which is a low input potential range, as compared to reported systems illustrated in Table 4, in the same medium.

The cyclic voltammetric response of bare glassy carbon and γ-Al_2O_3/GC with (red curve) and without (black curve) Ag_2O–PrO_2 loadings was studied in a 5 mM NH_3 + 0.1 M KOH system, as shown in Figure 7b,c. The electrochemical oxidation of ammonia in KOH is not observed on bare glassy carbon and γ-Al_2O_3/GC, while the peak appeared on $Ag_2O_{(x)}$–$PrO_{2(y)}/\gamma$-Al_2O_3/GC, confirming the electrocatalytic response of Ag_2O–PrO_2/γ-Al_2O_3 towards the electrochemical oxidation of ammonia, as elucidated by Figure 7b,c.

Table 4. Comparison of oxidation potentials for different systems.

Sr. No.	Modified Electrode Systems	Oxidation Potential Range (V)	Reference
1	Carbon-supported Pt/HOPG electrode	0.55–0.75	[51]
2	Pt film electrode/Si prism	0.45–0.85	[52]
3	Pt disk electrode and Pt/PBI/MWNT	0.45–0.90	[53]
4	Ag–Pr/Al/GC electrode	0.32–0.51	This work

GC, glassy carbon.

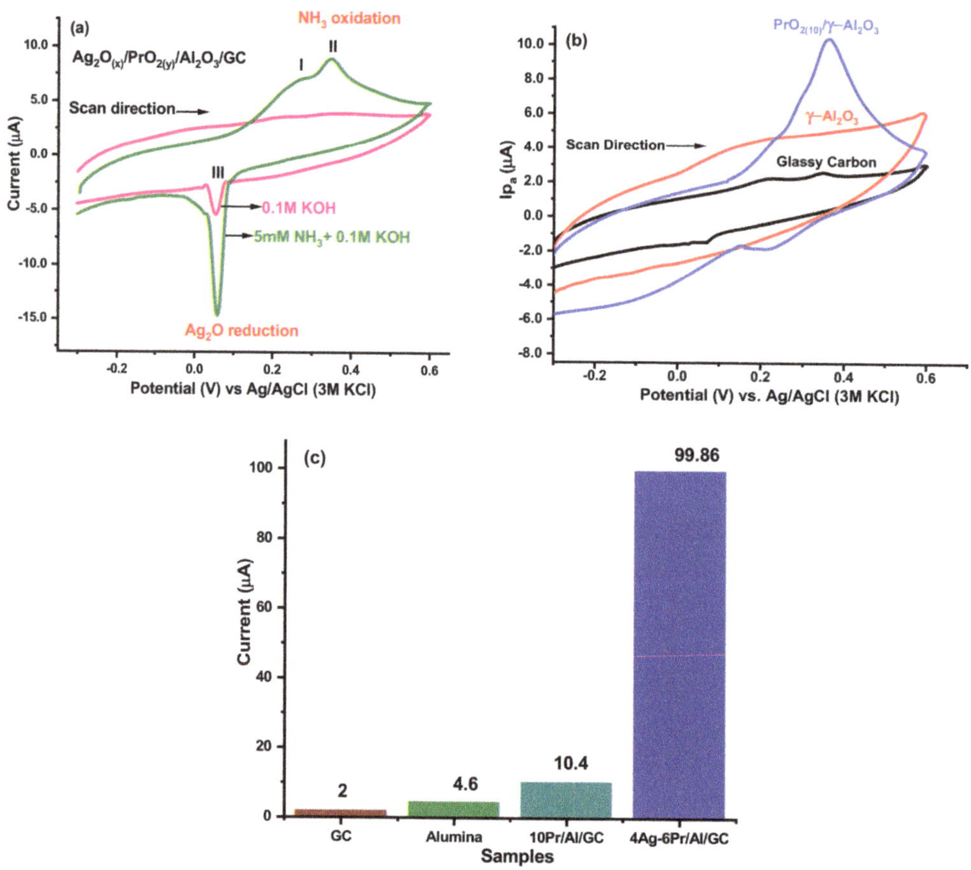

Figure 7. (**a**) Electrochemical responses and peak position in 0.1 M KOH and 5 mM NH_3. (**b**) Cyclic voltammograms towards ammonia oxidation over the surface of bare glassy carbon, γ-Al_2O_3 and $PrO_{2(10)}/\gamma$-Al_2O_3 modified electrodes 0.1 M KOH and 5 mM NH_3. (**c**) Comparative peak current response for bare GC, γ-Al_2O_3, and lowest and optimal composition, respectively.

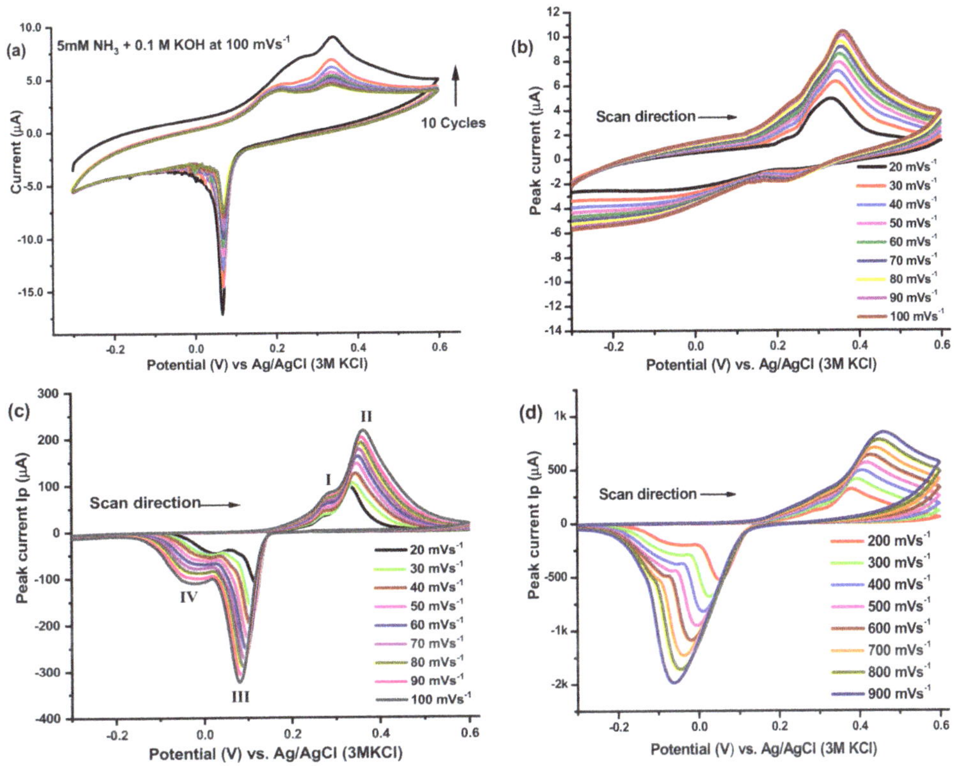

Figure 8. (**a**) Ten cycles in 5m mM NH_3 + 0.1M KOH at $Ag_2O_{(2)}$–$PrO_{2(8)}$/γ-Al_2O_3/GC. Scan rate variation for (**b**) $PrO_{2(10)}$/γ-Al_2O_3, (**c**) $Ag_2O_{(6)}$–$PrO_{2(4)}$/γ-Al_2O_3 at low scan rates and (**d**) $Ag_2O_{(6)}$–$PrO_{2(4)}$/γ-Al_2O_3, at high scan rates.

For commercialization of electrocatalysts, it is essential to check the electrochemical stability of catalysts; for this purpose, multiple scans have been observed at the surface of modified electrode in 5 m mM NH_3 + 0.1 M KOH. The current became constant, and no further decrease was observed, which shows the stability of modified electrode in the system (Figure 8a). The ammonia oxidation was observed at different scan rates on modified GC electrode and is given in Figure 8b–d. Figure 8b tells about the redox behavior of $PrO_{2(10)}$/γ-Al_2O_3 response towards ammonia electro-oxidation. It is revealed from the scan rate effect for all the electrochemical systems that the current increases with the increase of scan rate, and the peak potential is shifted towards more a positive potential region. A linear increase in peak current of ammonia oxidation is observed with sweep rate that specifies facilitation of electron transfer process of ammonia electro-oxidation; therefore, as-synthesized nanocatalysts behave like the adsorptive species on the surface of glassy carbon [54]. Figure 8c,d represents the voltammetric behavior of $Ag_2O_{(6)}$–$PrO_{2(4)}$/γ-Al_2O_3 at lower and higher scan rates, from which it is seen that separation in positive and negative peak potentials is smaller at low scan rate values, i.e., 20 to 100 mV^{-1}; hence, it is assumed that the facilitation of ammonia oxidation on surface of all catalysts is more manifested at the lower scan rates. Figure 8c,d shows the peak current dependence upon scan rate, which elaborates that the process of ammonia electro-oxidation over as-proposed catalysts is a diffusion-controlled reaction. Moreover, Figure 9a,b expresses the comparative peak current output of all compositions at lower and higher scan rates, which shows that $Ag_2O_{(6)}$–$PrO_{2(4)}$/γ-Al_2O_3 has a maximum current for NH_3 oxidation, as compared to the other members of the series.

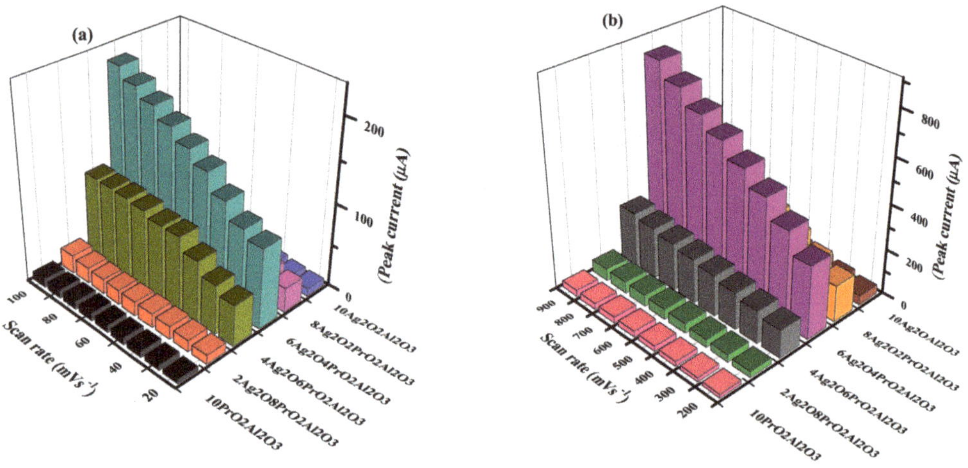

Figure 9. Comparative Ip_a values at modified electrodes surface on (**a**) low and (**b**) high scan.

The shift in peak potential with increasing scan rate is further confirmed by linear relationship of Ep_a and scan rate as shown in Figure 10b, which is one of the criteria to judge irreversibility of system [55,56]. The diffusion character of ammonia oxidation is substantiated by linear dependence of anodic peak current on scan rate, which is depicted by plots of ln Ip_a vs. ln ν [55,57], as shown in Figure 10c The slope values close to 0.5 represent diffusion-controlled electrode processes, whereas close to 1.0 will be for adsorption-controlled processes [58–60]. The behavior of current function with scan rate is also examined for oxidation of ammonia figures (Figure 10D). The current function decreases exponentially with increasing sweep rate, which is suitable for electrode process. The negative slopes from these exponential plots determine that electrochemical oxidation of ammonia on the surface of all modified electrodes is expedited and the process is electrocatalytic [61].

It is revealed from the scan rate effect for all the electrochemical systems that the current increases with the increase of scan rate, and the peak potential is shifted towards a more positive potential region.

5.2. *Kinetics of Ammonia Electro-Oxidation*

The diffusion character of ammonia oxidation is demonstrated by linear dependence of anodic peak current on scan rate, which is depicted by plots of ln Ip_a vs. ln v (Figure 10C). For the determination of diffusion coefficient, the Randles–Ševčík equation is used [62]:

$$I_p = (2.99 \times 10^5) \cdot n \cdot \{(1-\alpha) \cdot n_\alpha\} \cdot A \cdot D^{\circ 1/2} \cdot C \cdot \nu^{1/2} \quad (4)$$

where I_p is the anodic peak current; n is the no. of electrons involved in reaction; α is the transfer co-efficient whose value lies in the (0.3–0.7) range; n_α is the number of electrons in the rate-determining step, i.e., 3; A is area of the electrode, which is 0.07 cm^2; D° is the diffusion co-efficient in $cm^2 \cdot s^{-1}$; C is the bulk concentration in $mol \cdot cm^{-3}$; and ν is the scan rate in $mV \cdot s^{-1}$ The parameter α is calculated by using the following:

$$Ep_a - Ep_{a/2} = [0.048 / (\alpha n)] \quad (5)$$

where, Ep_a is anodic peak potential and $Ep_{a/2}$ is peak potential at $Ip_{a/2}$.

It is deduced that the reaction is diffusion controlled at low sweep rates, while it is kinetically controlled at higher scan rate (Figure 10C) [55,57].

The mass transport coefficient (m_T) is given by Equation (5) [63]:

$$m_T = [D^\circ/(RT/(F\cdot\nu))]^{1/2} \tag{6}$$

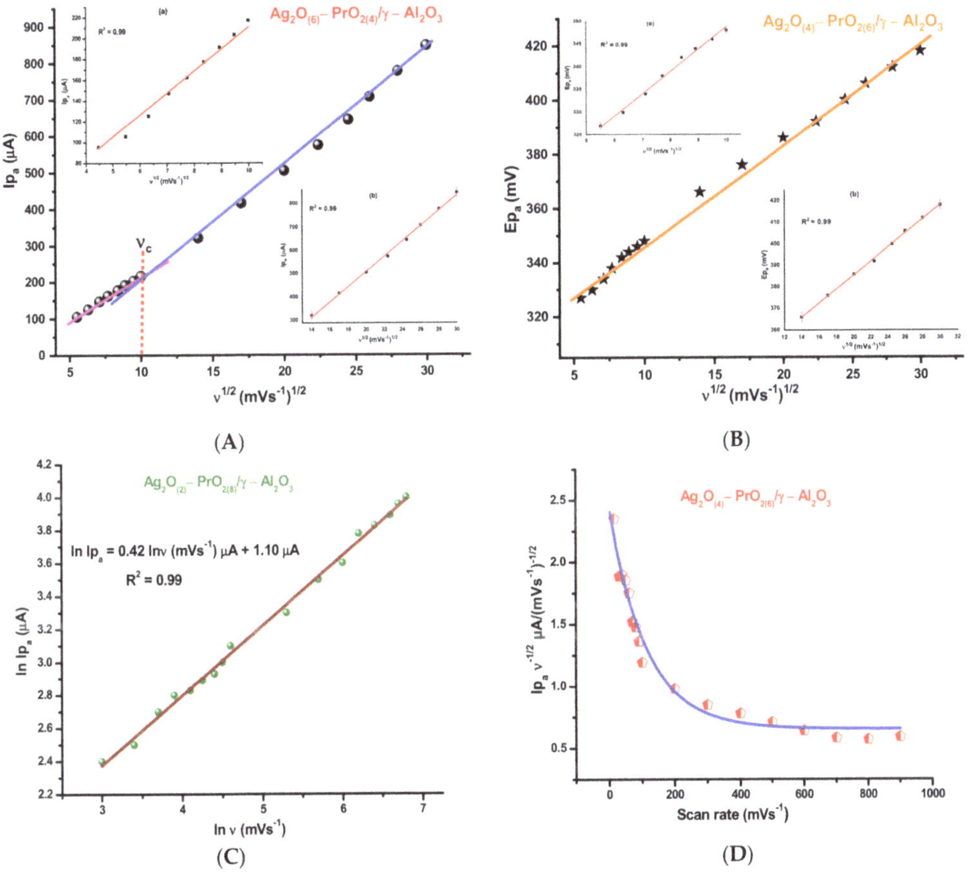

Figure 10. (**A**) Dependence of Ip_a on $\nu^{1/2}$; insets (a) and (b) show linear fitting for lower and higher scan rates for $Ag_2O_{(6)}$–$PrO_{2(4)}/\gamma$-Al_2O_3/GC. (**B**) Characteristic shift in peak potential with $\nu^{1/2}$ for $Ag_2O_{(4)}$–$PrO_{2(6)}/\gamma$-Al_2O_3/GC, insets (a) and (b) show linear fitting for lower and higher scan rates. (**C**) Dependence of Ip_a on potential scan rate in double logarithm coordinates for ammonia oxidation on $Ag_2O_{(2)}$–$PrO_{2(8)}/\gamma$-Al_2O_3/GC. (**D**) Variation of current function ($Ip_a\ \nu^{-1/2}$) with ν for $Ag_2O_{(4)}$–$PrO_{2(6)}/\gamma$-Al_2O_3/GC.

The values of the ammonia diffusion and mass transport coefficients for the entire series of the synthesized samples are given in Table 4. The diffusion coefficient and mass transport coefficient are higher for $Ag_2O_{(6)}$–$PrO_{2(4)}/\gamma$-Al_2O_3, revealing that it is a better electro-active material for NH_3 oxidation among all the samples. The value of diffusion coefficients for ammonia oxidation in alkaline medium (10^{-9}) is smaller than that of H_2O (10^{-5}), due to presence of hydroxyl groups of the supporting electrolyte present in bulk [64]. These hydroxyl groups from the KOH electrolyte hinder the diffusion process of ammonia; that could be the reason for the low diffusion coefficient. The diffusion coefficient value estimated for ammonia over $Ag_2O_{(6)}$-$PrO_{2(4)}/\gamma$-Al_2O_3 nanocomposite seems similar to ammonia diffusion over the Pt-Ni composite [65]. Moreover, these diffusion coefficient values are obtained for the current of major peak, i.e., around 0.4 V, which increases linearly with the scan rate, as seen from Figure 8. It suggests that oxidation of ammonia over $Ag_2O_{(6)}$-$PrO_{2(4)}/\gamma$-Al_2O_3 is governed by diffusion significantly.

The catalytic properties of synthesized materials can be determined by heterogeneous kinetics for ammonia oxidation. The value of heterogeneous rate constant was determined by taking cyclic voltammograms and varying the concentration of ammonia (Figure 11a). The Reinmuth equation is used to calculate heterogeneous rate constants, k°, for all supported mixed metal oxide electrocatalysts.

$$\text{Ip} = 0.227n \cdot F \cdot A \cdot C \cdot k^{\circ} \tag{7}$$

where k^{o} is the heterogeneous rate constant, and F is Faraday's constant. By plotting the anodic peak current vs. concentration of NH_3, the values of the rate constant, k^{o}, for oxidation are obtained from the slope, as illustrated in Table 5, which reveals the facilitated electron transfer process, and it gives valuable information about the nature of system, which is irreversible. The heterogeneous rate constant decreases in the following order:

$$k^{\circ}_{(6Ag\text{-}4Pr/Al)} > k^{\circ}_{(4Ag\text{-}6Pr/Al)} > k^{\circ}_{(8Ag\text{-}2Pr/Al)} > k^{\circ}_{(10Ag/Al)} > k^{\circ}_{(2Ag\text{-}8Pr/Al)} > k^{\circ}_{(10Pr/Al)}$$

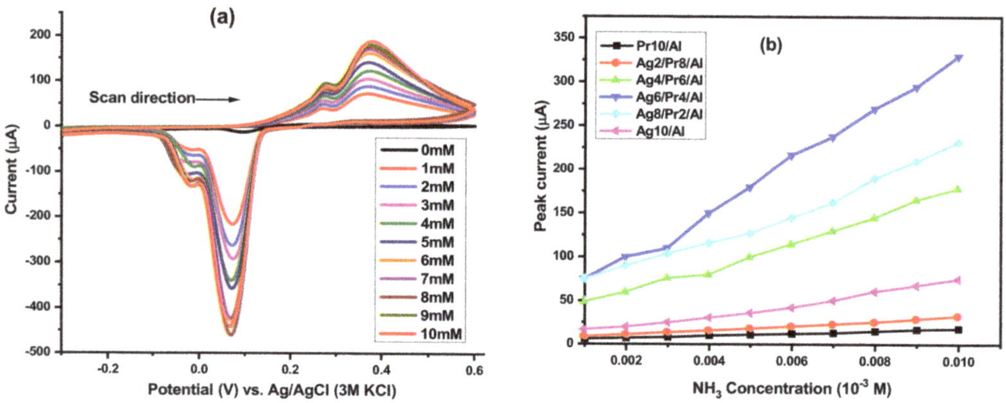

Figure 11. (**a**) Cyclic voltammograms with $Ag_2O_{(8)}$–$PrO_{2(2)}/\gamma$-Al_2O_3 electrode at different concentrations of NH_3 and (**b**) Reinmuth plots between Ip_a and C_{NH3}, showing variation of peak current with C_{NH3} for and for all modified electrodes.

Table 5. Kinetic parameters for NH_3 electro-oxidation at the electrode surfaces.

Electrocatalysts	α	$(D^{\circ})/10^{-9}$ $cm^2 \cdot s^{-1}$	$(m_T)/cm \cdot s^{-1}$	$k^{\circ}/10^{-3}$ $cm \cdot s^{-1}$
$PrO_{2(10)}/\gamma$-Al_2O_3	0.5	0.063	0.0005	0.40
$Ag_2O_{(2)}$–$PrO_{2(8)}/\gamma$-Al_2O_3	0.5	0.140	0.0007	0.72
$Ag_2O_{(4)}$–$PrO_{2(6)}/\gamma$-Al_2O_3	0.7	11.00	0.0070	6.10
$Ag_2O_{(6)}$–$PrO_{2(4)}/\gamma$-Al_2O_3	0.6	36.50	0.0120	7.40
$Ag_2O_{(8)}$–$PrO_{2(2)}/\gamma$-Al_2O_3	0.4	4.040	0.0040	5.20
$Ag_2O_{(10)}/\gamma$-Al_2O_3	0.4	0.300	0.0011	1.60

5.3. Thermodynamic Studies for Ammonia Electro-Oxidation

Cyclic voltammograms were observed by varying the temperature of the system from 10 to 50 °C with the $Ag_2O_{(6)}$–$PrO_{2(4)}$/γ-Al_2O_3 catalyst, as shown in Figure 12a. The observed thermodynamic parameters tabulated below show that electro-oxidation of ammonia in alkaline solution is exothermic and non-spontaneous reaction. The entropy of ammonia electro-oxidation is negative because it is an adsorption-controlled process. Before adsorption, the molecules are free to move in three dimensions; when these get adsorbed on the surface of catalyst, their motion is restricted. As a result, disorderliness decreases, and entropy of the system decreases. The free energy of activation is less for $Ag_2O_{(6)}$–$PrO_{2(4)}$/γ-Al_2O_3 modified electrode, which reveals that this composition exhibits superior catalytic properties among this series because it lowers the activation energy of system to a greater extent.

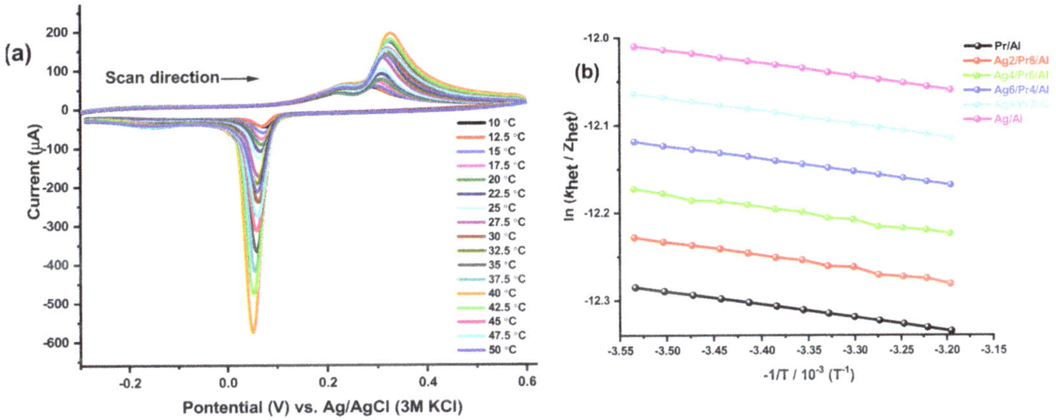

Figure 12. (**a**) Peak current output varied with increase in temperature and (**b**) Marcus plots between temperature and rate constants.

All the compositions in the series exhibit a similar response towards the increase in temperature of the system. Different thermodynamic parameters enlisted in Table 6 are deduced for the NH_3 electro-oxidation, with the help of the Marcus equation. It is a straight-line equation and gives the value of enthalpy of reaction (ΔH) from the slope of the Marcus plot, i.e., Figure 12b between ln $[k^\circ/Z_{het}]$ vs. $-1/T$, while change in entropy (ΔS) can be estimated from the intercept.

$$k^\circ = Z_{het} \cdot \exp\left[\frac{-\Delta G}{RT}\right] \quad (8)$$

Equations (5) and (6) may be expressed in a simpler form:

$$\ln \frac{k^\circ}{Z(het)} = \left[\frac{-\Delta G}{RT}\right] \quad (9)$$

Here, ΔG is free energy of activation, k° heterogeneous rate constant and Z_{het} is collision number whose values Z_{het} at different temperatures (T) can be calculated by Equation (9) [66].

$$Z_{het} = \sqrt{\frac{RT}{2\pi M}} \quad (10)$$

Table 6. Thermodynamic parameters calculated from the Marcus equation.

Electrocatalysts	$(\Delta H)/kJ\cdot mol^{-1}$	$(\Delta S)/kJ\cdot mol^{-1}K^{-1}$	$(\Delta G)/kJ\cdot mol^{-1}$
$PrO_{2(10)}/\gamma\text{-}Al_2O_3$	−1.24	−0.119	32.0
$Ag_2O_{(2)}\text{–}PrO_{2(8)}/\gamma\text{-}Al_2O_3$	−1.30	−0.100	29.0
$Ag_2O_{(4)}\text{–}PrO_{2(6)}/\gamma\text{-}Al_2O_3$	−1.23	−0.080	23.6
$Ag_2O_{(6)}\text{–}PrO_{2(4)}/\gamma\text{-}Al_2O_3$	−1.22	−0.082	23.0
$Ag_2O_{(8)}\text{–}PrO_{2(2)}/\gamma\text{-}Al_2O_3$	−1.23	−0.085	24.0
$Ag_2O_{(10)}/\gamma\text{-}Al_2O_3$	−1.24	−0.094	27.0

M is the molar concentration of NH_3, according to thermodynamic equation.

$$\Delta G = \Delta H - T\Delta S \quad (11)$$

$$\ln \frac{k^{\circ}}{Z(het)} = -\frac{\Delta H}{RT} + \frac{\Delta S}{R} \quad (12)$$

5.4. Mechanism of Ammonia Oxidation

The independent behavior of cathodic and anodic reactions was further confirmed by comparing the linear-sweep voltammograms (LSV) with cyclic voltammogram (CV) observed for 5 mM NH_3 in 0.1 M KOH at 100 $mV\cdot s^{-1}$, as represented in Figure 13. It reveals that the anodic peak (black curve) is purely for ammonia oxidation, and the cathodic peak (blue curve) is for the reduction of silver oxide into metal silver. The comparison of LSV with CV will help to propose a mechanism for ammonia oxidation on surface of electrocatalysts.

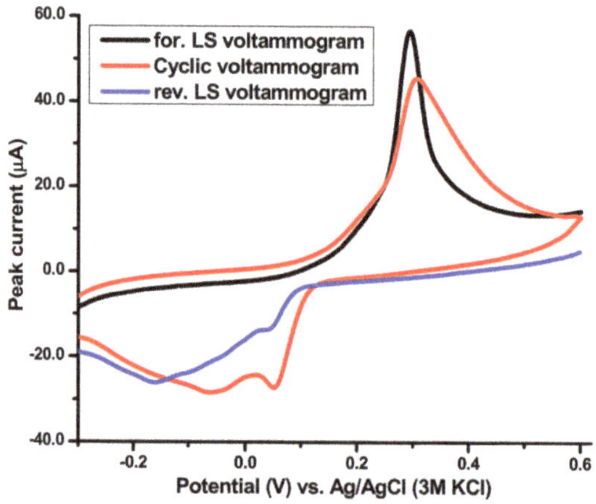

Figure 13. Linear-sweep voltammograms (LSV) vs. cyclic voltammogram (CV) for oxidation of ammonia in 0.1 M KOH at 100 $mV\cdot s^{-1}$ on the surface of optimum composition/GC.

These responses correspond to the mechanism of ammonia oxidation. NH_3 is firstly adsorbed on electrode surface from bulk solution. After adsorption, it undergoes the $3e^-$ oxidation process, forming water and N_2 as by-products [67].

Overall reaction

$$NH_3 + 3OH^- \rightarrow \frac{1}{2} N_2 + 3H_2O + 3e^- \quad (13)$$

Mechanism

$$NH_3(aq) \rightarrow NH_3(ad) \quad (14)$$

$$NH_3(ad) + OH^- \rightarrow NH_2^-(ad) + H_2O + e^- \quad (15)$$

$$NH_2^-(ad) + OH^- \rightarrow NH^-(ad) + H_2O + e^- \quad (16)$$

$$NH^-(ad) + OH^- \rightarrow N(ad) + H_2O + e^- \quad (17)$$

$$2N(ad) \rightarrow N_2 \quad (18)$$

6. Conclusions

The γ-Al_2O_3 supported $Ag_2O_{(x)}$-$PrO_{2(y)}$ nanocatalysts were synthesized via simple synthetic routes. Moreover, γ-Al_2O_3 was synthesized by precipitation method, and metal oxide promotors were loaded on γ-Al_2O_3 by co-impregnation method. All the as-synthesized materials were subjected to characterization techniques, including FTIR, XRD, EDX, SEM, CV and EIS. The sizes of the proposed nanocatalysts were calculated by XRD and SEM, which confirms that synthesized materials are in nano-range and are well dispersed on the support. The successful loading of xAg_2O-$yPrO_2$ nanocomposites on support was confirmed by XRD, FTIR and EDX. The surface morphology of samples was examined by SEM, which shows that metal oxide precursors are homogeneously dispersed on the surface of the support. The active surface area and electron transfer constant were calculated by CV and EIS. All the compositions gave appreciable current output in ammonia electro-oxidation reaction, owing to their estimated ECSA values and electron transfer properties. Among all the catalysts in the series, $Ag_2O_{(6)}$–$PrO_{2(4)}/\gamma$-Al_2O_3 gave the higher values for both, making it an optimum composition for electrochemical applications.

The cyclic voltammetric investigations were employed for analyzing the electrochemical properties of materials. The catalytic activity of all synthesized samples was investigated towards oxidation of ammonia, considering scan rate, concentration and temperature effects in focus. All modified electrodes showed adequate catalytic behavior towards ammonia oxidation, whereby modified electrodes behaved as anodic material. The calculated kinetic parameters revealed that ammonia oxidation on these catalysts was of diffusion control nature and irreversible. The thermodynamic parameters declared that the overall process of ammonia on catalysts surface was exothermic (negative enthalpy) and non-spontaneous (positive Gibbs free energy). The entropy of the system was negative, as disorderliness decreases in the system due to adsorption of intermediate species on the catalyst surface before oxidation, which reveals that ammonia electro-oxidation is an adsorption-controlled process. Among all modified electrodes, the $Ag_2O_{(6)}-/PrO_{2(4)}/\gamma$-$Al_2O_3$ modified electrode exhibited better catalytic response, owing to its higher diffusion coefficient, mass transport coefficient, heterogeneous rate constant and lowest free energy of activation.

Author Contributions: Conceptualization, M.K. and N.K.J.; Methodology, M.K. and N.K.J.; Validation, S.K.; Formal Analysis, T.S.A.; Investigation, N.K.J.; Resources, S.A.; Data Curation, N.K.J.; Writing—Original Draft Preparation, M.K.; Writing—Review and Editing, N.K.J.; Visualization, S.K.; Supervision, N.K.J.; Project Administration, N.K.J.; SEM characterization, I.Q.; TEM characterization, S.A.; Funding Acquisition, T.S.A. All authors have read and agreed to the published version of the manuscript.

Funding: This research received no external funding.

Data Availability Statement: Data is contained within article.

Acknowledgments: The authors acknowledge the Fuel Cell Lab, Department of Chemistry, Quaid-I-Azam University, Islamabad, which furnished this novel research, and the researchers collaborative Project Number (RSP-2020/254), King Saud University, Riyadh, Saudi Arabia.

Conflicts of Interest: The authors declare no conflict of interest.

References

1. Nowotny, J.; Hoshino, T.; Dodson, J.; Atanacio, A.J.; Ionescu, M.; Peterson, V.; Prince, K.E.; Yamawaki, M.; Bak, T.; Sigmund, W.; et al. Towards sustainable energy. Generation of hydrogen fuel using nuclear energy. *Int. J. Hydrogen Energy* **2016**, *41*, 12812–12825. [CrossRef]
2. Pervaiz, M.; Ahmad, I.; Yousaf, M.; Kirn, S.; Munawar, A.; Saeed, Z.; Adnan, A.; Gulzar, T.; Kamal, T.; Ahmad, A.; et al. Synthesis, spectral and antimicrobial studies of amino acid derivative Schiff base metal (Co, Mn, Cu, and Cd) complexes. *Spectrochim. Acta Part A Mol. Biomol. Spectrosc.* **2019**, *206*, 642–649. [CrossRef] [PubMed]
3. Kashif, M.; Ngaini, Z.; Harry, A.V.; Vekariya, R.L.; Ahmad, A.; Zuo, Z.; Sahari, S.K.; Hussain, S.; Khan, Z.A.; Alarifi, A. An experimental and DFT study on novel dyes incorporated with natural dyes on titanium dioxide (TiO_2) towards solar cell application. *Appl. Phys. A* **2020**, *126*, 1–13. [CrossRef]
4. Ahmad, A.; Jini, D.; Aravind, M.; Parvathiraja, C.; Ali, R.; Kiyani, M.Z.; Alothman, A. A novel study on synthesis of egg shell based activated carbon for degradation of methylene blue via photocatalysis. *Arab. J. Chem.* **2020**, *13*, 8717–8722. [CrossRef]
5. Hussain, S.; Khan, A.J.; Arshad, M.; Javed, M.S.; Ahmad, A.; Shah, S.S.A.; Khan, M.R.; Akram, S.; Zulfiqar; Ali, S.; et al. Charge storage in binder-free 2D-hexagonal CoMoO4 nanosheets as a redox active material for pseudocapacitors. *Ceram. Int.* **2020**. [CrossRef]
6. Afif, A.; Radenahmad, N.; Cheok, Q.; Shams, S.; Kim, J.H.; Azad, A.K. Ammonia-fed fuel cells: A comprehensive review. *Renew. Sustain. Energy Rev.* **2016**, *60*, 822–835. [CrossRef]
7. Zhang, X.Z.; Xu, P.H.; Liu, G.W.; Ahmad, A.; Chen, X.H.; Zhu, Y.L.; Alothman, A.; Hussain, S.; Qiao, G.J. Synthesis, Characterization and Wettability of Cu-Sn Alloy on the Si-Implanted 6H-SiC. *Coatings* **2020**, *10*, 906. [CrossRef]
8. Aravind, M.; Ahmad, A.; Ahmad, I.; Amalanathan, M.; Naseem, K.; Mary, S.M.M.; Parvathiraja, C.; Hussain, S.; Algarni, T.S.; Pervaiz, M.; et al. Critical green routing synthesis of silver NPs using jasmine flower extract for biological activities and photocatalytical degradation of methylene blue. *J. Environ. Chem. Eng.* **2021**, *9*, 104877. [CrossRef]
9. Naseem, K.; Rehman, M.Z.U.; Ahmad, A.; Dubal, D.; Algarni, T.S. Plant Extract Induced Biogenic Preparation of Silver Nanoparticles and Their Potential as Catalyst for Degradation of Toxic Dyes. *Coatings* **2020**, *10*, 1235. [CrossRef]
10. Zhan, M.; Hussain, S.; Algarni, T.S.; Shah, S.; Liu, J.; Zhang, X.; Ahmad, A.; Javed, M.S.; Qiao, G.; Liu, G. Facet controlled polyhedral ZIF-8 MOF nanostructures for excellent NO2 gas-sensing applications. *Mater. Res. Bull.* **2021**, *136*, 111133. [CrossRef]
11. Khan, F.S.A.; Mubarak, N.M.; Khalid, M.; Walvekar, R.; Abdullah, E.C.; Ahmad, A.; Karri, R.R.; Pakalapati, H. Functionalized multi-walled carbon nanotubes and hydroxyapatite nanorods reinforced with polypropylene for biomedical application. *Sci. Rep.* **2021**, *11*, 1–10. [CrossRef] [PubMed]
12. Kashif, M.; Jafaar, E.; Sahari, S.K.; Low, F.W.; Hoa, N.D.; Ahmad, A.; Abbas, A.; Ngaini, Z.; Shafa, M.; Qurashi, A. Organic sensitization of graphene oxide and reduced graphene oxide thin films for photovoltaic applications. *Int. J. Energy Res.* **2021**. [CrossRef]
13. Vidal-Iglesias, F.J.; Solla-Gullón, J.; Montiel, V.; Feliu, J.; Aldaz, A. Screening of electrocatalysts for direct ammonia fuel cell: Ammonia oxidation on PtMe (Me: Ir, Rh, Pd, Ru) and preferentially oriented Pt(100) nanoparticles. *J. Power Sources* **2007**, *171*, 448–456. [CrossRef]
14. Thou, C.Z.; Khan, F.S.A.; Mubarak, N.; Ahmad, A.; Khalid, M.; Jagadish, P.; Walvekar, R.; Abdullah, E.; Khan, S.; Khan, M.; et al. Surface charge on chitosan/cellulose nanowhiskers composite via functionalized and untreated carbon nanotube. *Arab. J. Chem.* **2021**, *14*, 103022. [CrossRef]
15. Saleem, M.; Irfan, M.; Tabassum, S.; Albaqami, M.D.; Javed, M.S.; Hussain, S.; Pervaiz, M.; Ahmad, I.; Ahmad, A.; Zuber, M. Experimental and theoretical study of highly porous lignocellulose assisted metal oxide photoelectrodes for dye-sensitized solar cells. *Arab. J. Chem.* **2021**, *14*, 102937. [CrossRef]
16. Zou, P.; Chen, S.; Lan, R.; Tao, S. Investigation of Perovskite Oxide $SrCo_{0.8}Cu_{0.1}Nb_{0.1}O_{3-\delta}$ as a Cathode Material for Room Temperature Direct Ammonia Fuel Cells. *ChemSusChem* **2019**, *12*, 2788–2794. [CrossRef]
17. Adli, N.M.; Zhang, H.; Mukherjee, S.; Wu, G. Review—Ammonia Oxidation Electrocatalysis for Hydrogen Generation and Fuel Cells. *J. Electrochem. Soc.* **2018**, *165*, J3130–J3147. [CrossRef]
18. Suzuki, S.; Muroyama, H.; Matsui, T.; Eguchi, K. Fundamental studies on direct ammonia fuel cell employing anion exchange membrane. *J. Power Sources* **2012**, *208*, 257–262. [CrossRef]
19. Li, L.; Zhu, Z.; Yan, Z.; Lu, G.; Rintoul, L. Catalytic ammonia decomposition over Ru/carbon catalysts: The importance of the structure of carbon support. *Appl. Catal. A: Gen.* **2007**, *320*, 166–172. [CrossRef]
20. Schaper, H.; Doesburg, E.; Van Reijen, L. The influence of lanthanum oxide on the thermal stability of gamma alumina catalyst supports. *Appl. Catal.* **1983**, *7*, 211–220. [CrossRef]
21. Mardkhe, M.K.; Huang, B.; Bartholomew, C.H.; Alam, T.M.; Woodfield, B.F. Synthesis and characterization of silica doped alumina catalyst support with superior thermal stability and unique pore properties. *J. Porous Mater.* **2015**, *23*, 475–487. [CrossRef]
22. Jo, J.-O.; Trinh, Q.H.; Kim, S.H.; Mok, Y.S. Plasma-catalytic decomposition of nitrous oxide over γ-alumina-supported metal oxides. *Catal. Today* **2018**, *310*, 42–48. [CrossRef]
23. Yang, X.; Sun, M.; Liu, B.; Lin, W. NO_x removal by oxidation on alumina supported metal oxide catalysts followed by alkaline absorption. *China Pet. Process. Petrochem. Technol.* **2019**, *21*, 1–9.

24. Sandupatla, A.S.; Ray, K.; Thaosen, P.; Sivananda, C.; Deo, G. Oxidative dehydrogenation of propane over alumina supported vanadia catalyst–Effect of carbon dioxide and secondary surface metal oxide additive. *Catal. Today* **2020**, *354*, 176–182. [CrossRef]
25. Adil, S.F.; Assal, M.E.; Khan, M.; Shaik, M.R.; Kuniyil, M.; Sekou, D.; Dewidar, A.Z.; Al-Warthan, A.; Siddiqui, M.R.H. Eco-Friendly Mechanochemical Preparation of Ag_2O–MnO_2/Graphene Oxide Nanocomposite: An Efficient and Reusable Catalyst for the Base-Free, Aerial Oxidation of Alcohols. *Catalysts* **2020**, *10*, 281. [CrossRef]
26. Yin, Z.; Xie, L.; Cao, S.; Xiao, Y.; Chen, G.; Jiang, Y.; Wei, W.; Wu, L. Ag/Ag_2O confined visible-light driven catalyst for highly efficient selective hydrogenation of nitroarenes in pure water medium at room temperature. *Chem. Eng. J.* **2020**, *394*, 125036. [CrossRef]
27. Tankov, I.; Arishtirova, K.; Bueno, J.M.; Damyanova, S. Surface and structural features of Pt/PrO_2–Al_2O_3 catalysts for dry methane reforming. *Appl. Catal. A: Gen.* **2014**, *474*, 135–148. [CrossRef]
28. Wachs, I.E. Recent conceptual advances in the catalysis science of mixed metal oxide catalytic materials. *Catal. Today* **2005**, *100*, 79–94. [CrossRef]
29. Sharma, R.V.; Kumar, P.; Dalai, A.K. Selective hydrogenolysis of glycerol to propylene glycol by using Cu:Zn:Cr:Zr mixed metal oxides catalyst. *Appl. Catal. A Gen.* **2014**, *477*, 147–156. [CrossRef]
30. Berenguer Betrián, R.; Quijada, C.; Morallón, E. The nature of the electro-catalytic response of mixed metal oxides: Pt-and Ru-doped SnO_2 anodes. *ChemElectroChem* **2019**, *6*, 1057–1068. [CrossRef]
31. Gangwar, J.; Gupta, B.K.; Kumar, P.; Tripathi, S.K.; Srivastava, A.K. Time-resolved and photoluminescence spectroscopy of θ-Al_2O_3nanowires for promising fast optical sensor applications. *Dalton Trans.* **2014**, *43*, 17034–17043. [CrossRef] [PubMed]
32. Domínguez-Crespo, M.; Díaz-García, L.; De La Paz, M.C.; Zárate-Ramos, R.; Santes, V.; Pérez, G.L.; Arce-Estrada, E.; Torres-Huerta, A. Influence of Alumina Crystal Size on the Hydrotreating Activity of Supported NiMo Catalysts Using Real Feedstock. *Pet. Sci. Technol.* **2006**, *24*, 485–506. [CrossRef]
33. Xu, L.; Wei, B.; Liu, W.; Zhang, H.; Su, C.; Che, J. Flower-like ZnO-Ag_2O composites: Precipitation synthesis and photocatalytic activity. *Nanoscale Res. Lett.* **2013**, *8*, 536. [CrossRef]
34. Zhu, L.; Wei, B.; Xu, L.; Lu, Z.; Zhang, H.; Gao, H.; Che, J. Ag_2O–Bi_2O_3 composites: Synthesis, characterization and high efficient photocatalytic activities. *CrystEngComm* **2012**, *14*, 5705–5709. [CrossRef]
35. Dhoondia, Z.H.; Chakraborty, H. Lactobacillus mediated synthesis of silver oxide nanoparticles. *Nanomater. Nanotechnol.* **2012**, *2*, 15. [CrossRef]
36. Vorokh, A. Scherrer formula: Estimation of error in determining small nanoparticle size. *Nanosyst. Physics, Chem. Math.* **2018**, *9*. [CrossRef]
37. Ghosh, M.; Dilawar, N.; Bandyopadhyay, A.K.; Raychaudhuri, A.K. Phonon dynamics of Zn(Mg,Cd)O alloy nanostructures and their phase segregation. *J. Appl. Phys.* **2009**, *106*, 084306. [CrossRef]
38. Mahan, G.D.; Lucas, A.A. Collective vibrational modes of adsorbed CO. *J. Chem. Phys.* **1978**, *68*, 1344–1348. [CrossRef]
39. Cheng, B.; Xiao, Y.; Wu, G.; Zhang, L. The vibrational properties of one-dimensional ZnO: Ce nanostructures. *Appl. Phys. Lett.* **2004**, *84*, 416–418. [CrossRef]
40. Tripathi, S.; Mehrotra, G.K.; Dutta, P.K. Chitosan–silver oxide nanocomposite film: Preparation and antimicrobial activity. *Bull. Mater. Sci.* **2011**, *34*, 29–35. [CrossRef]
41. Waterhouse, G.I.N.; Bowmaker, G.A.; Metson, J.B. The thermal decomposition of silver (I, III) oxide: A combined XRD, FT-IR and Raman spectroscopic study. *Phys. Chem. Chem. Phys.* **2001**, *3*, 3838–3845. [CrossRef]
42. González Hernández, N.N.; Contreras, J.L.; Pinto, M.; Zeifert, B.; Flores Moreno, J.L.; Fuentes, G.A.; Hernández-Terán, M.E.; Vázquez, T.; Salmones, J.; Jurado, J.M. Improved NO_x reduction using C_3H_8 and H_2 with Ag/Al_2O_3 catalysts promoted with Pt and WO_x. *Catalysts* **2020**, *10*, 1212. [CrossRef]
43. Yiğitalp, A. Nafion Modified NCA Cathode Synthesis for Superior Cycling Performance of Lithium-Ion Batteries. Ph.D. Thesis, Sabancı University, Istanbul, Turkey, 2019.
44. Kashif, M.; Jaafar, E.; Bhadja, P.; Low, F.W.; Sahari, S.K.; Hussain, S.; Loong, F.K.; Ahmad, A.; Algarni, T.S.; Shafa, M.; et al. Effect of potassium permanganate on morphological, structural and electro-optical properties of graphene oxide thin films. *Arab. J. Chem.* **2021**, *14*, 102953. [CrossRef]
45. Sharifi, E.; Salimi, A.; Shams, E. Electrocatalytic activity of nickel oxide nanoparticles as mediatorless system for NADH and ethanol sensing at physiological pH solution. *Biosens. Bioelectron.* **2013**, *45*, 260–266. [CrossRef]
46. Sharifi, E.; Salimi, A.; Shams, E.; Noorbakhsh, A.; Amini, M.K. Shape-dependent electron transfer kinetics and catalytic activity of NiO nanoparticles immobilized onto DNA modified electrode: Fabrication of highly sensitive enzymeless glucose sensor. *Biosens. Bioelectron.* **2014**, *56*, 313–319. [CrossRef]
47. Guo, D.J.; Li, H.L. Highly dispersed Ag nanoparticles on functional MWNT surfaces for methanol oxidation in alkaline solution. *Carbon* **2005**, *43*, 1259–1264. [CrossRef]
48. Droog, J.M.; Huisman, F. Electrochemical formation and reduction of silver oxides in alkaline media. *J. Electroanal. Chem.* **1980**, *115*, 211–224. [CrossRef]
49. Sun, Z.; Han, Y.; Gao, M.; Wei, X.; Hu, X. Cetyl trimethyl ammonium bromide-assisted electrochemical preparation of palladium-nickel bimetallic electrode. *Int. J. Electrochem. Sci.* **2011**, *6*, 5626–5638.

50. Matinise, N.; Mayedwa, N.; Ikpo, C.O.; Hlongwa, N.W.; Ndipingwi, M.M.; Molefe, L.; Dywili, N.; Yonkeu, A.L.D.; Waryo, T.; Baker, P.G.L.; et al. Bimetallic Nanocomposites of Palladium (100) and Ruthenium for Electrooxidation of Ammonia. *J. Nano Res.* **2016**, *44*, 100–113. [CrossRef]
51. Lomocso, T.L.; Baranova, E.A. Electrochemical oxidation of ammonia on carbon-supported bi-metallic PtM (M=Ir, Pd, SnOx) nanoparticles. *Electrochim. Acta* **2011**, *56*, 8551–8558. [CrossRef]
52. Matsui, T.; Suzuki, S.; Katayama, Y.; Yamauchi, K.; Okanishi, T.; Muroyama, H.; Eguchi, K. In Situ Attenuated Total Reflection Infrared Spectroscopy on Electrochemical Ammonia Oxidation over Pt Electrode in Alkaline Aqueous Solutions. *Langmuir* **2015**, *31*, 11717–11723. [CrossRef]
53. Katayama, Y.; Okanishi, T.; Muroyama, H.; Matsui, T.; Eguchi, K. Electrochemical oxidation of ammonia over rare earth oxide modified platinum catalysts. *J. Phys. Chem. C* **2015**, *119*, 9134–9141. [CrossRef]
54. Zhang, H.; Wang, Y.; Wu, Z.; Leung, D.Y. An ammonia electrolytic cell with NiCu/C as anode catalyst for hydrogen production. *Energy Procedia* **2017**, *142*, 1539–1544. [CrossRef]
55. Chrzescijanska, E.; Wudarska, E.; Kusmierek, E.; Rynkowski, J. Study of acetylsalicylic acid electroreduction behavior at platinum electrode. *J. Electroanal. Chem.* **2014**, *713*, 17–21. [CrossRef]
56. Mujtaba, A.; Janjua, N.K. Fabrication and electrocatalytic application of CuO@Al_2O_3 hybrids. *J. Electrochem. Soc.* **2015**, *162*, H328. [CrossRef]
57. Castilho, M.; Almeida, L.E.; Tabak, M.; Mazo, L.H. Voltammetric oxidation of dipyridamole in aqueous acid solutions. *J. Braz. Chem. Soc.* **2000**, *11*, 148–153. [CrossRef]
58. Holze, R.; Brett, C.M.A.; Brett, A.M.O. *Electrochemistry—Principles, Methods and Applications*; Oxford University Press Inc.: New York, NY, USA, 1993.
59. Soleymani, J.; Hasanzadeh, M.; Shadjou, N.; Jafari, M.K.; Gharamaleki, J.V.; Yadollahi, M.; Jouyban, A. A new kinetic–mechanistic approach to elucidate electrooxidation of doxorubicin hydrochloride in unprocessed human fluids using magnetic graphene based nanocomposite modified glassy carbon electrode. *Mater. Sci. Eng. C* **2016**, *61*, 638–650. [CrossRef]
60. Masek, A.; Chrzescijanska, E.; Zaborski, M. Electrooxidation of morin hydrate at a Pt electrode studied by cyclic voltammetry. *Food Chem.* **2014**, *148*, 18–23. [CrossRef] [PubMed]
61. Mazloum-Ardakani, M.; Taleat, Z. Investigation of electrochemistry behavior of hydroxylamine at glassy carbon electrode by indigocarmine. *Int. J. Electrochem. Sci.* **2009**, *4*, 694–706.
62. Janjua, N.K.; Jabeen, M.; Islam, M.; Yaqub, A.; Sabahat, S.; Mehmood, S.; Abbas, G. Electrochemical properties of barium cerate doped with zinc for methanol oxidation. *J. Chem. Soc. Pak.* **2015**, *37*.
63. Mu, Y.; Jia, D.; He, Y.; Miao, Y.; Wu, H.-L. Nano nickel oxide modified non-enzymatic glucose sensors with enhanced sensitivity through an electrochemical process strategy at high potential. *Biosens. Bioelectron.* **2011**, *26*, 2948–2952. [CrossRef] [PubMed]
64. Kapałka, A.; Cally, A.; Neodo, S.; Comninellis, C.; Wächter, M.; Udert, K.M. Electrochemical behavior of ammonia at Ni/Ni(OH)2 electrode. *Electrochem. Commun.* **2010**, *12*, 18–21. [CrossRef]
65. Diaz, L.A.; Valenzuela-Muñiz, A.; Muthuvel, M.; Botte, G.G. Analysis of ammonia electro-oxidation kinetics using a rotating disk electrode. *Electrochim. Acta* **2013**, *89*, 413–421. [CrossRef]
66. Marcus, R.A. On the Theory of Electron-Transfer Reactions. VI. Unified Treatment for Homogeneous and Electrode Reactions. *J. Chem. Phys.* **1965**, *43*, 679–701. [CrossRef]
67. Oswin, H.G.; Salomon, M. The anodic oxidation of ammonia at platinum black electrodes in aqueous koh electrolyte. *Can. J. Chem.* **1963**, *41*, 1686–1694. [CrossRef]

coatings

MDPI

Article

Effects of Cr_2O_3 Content on Microstructure and Mechanical Properties of Al_2O_3 Matrix Composites

Kunkun Cui [1], Yingyi Zhang [1,2,*], Tao Fu [1], Shahid Hussain [3,*], Tahani Saad Algarni [4], Jie Wang [1], Xu Zhang [1] and Shafaqat Ali [5,6,*]

1 School of Metallurgical Engineering, Anhui University of Technology, Maanshan 243002, China; 15613581810@136.com (K.C.); 13716853174@163.com (T.F.); mrwang0101@ahut.edu.cn (J.W.); zx13013111171@163.com (X.Z.)
2 College of Material Science and Engineering, Chongqing University, Chongqing 400030, China
3 School of Materials Science and Engineering, Jiangsu University, Zhenjiang 212013, China
4 Chemistry Department, College of Science, King Saud University, Riyadh 11451, Saudi Arabia; tahanis@ksu.edu.sa
5 Department of Environmental Science and Engineering, Government College University, Faisalabad 38000, Pakistan
6 Department of Biological Sciences and Technology, China Medical University, Taichung 40402, Taiwan
* Correspondence: zhangyingyi@cqu.edu.cn (Y.Z.); shahid@ujs.edu.cn (S.H.); shafaqataligill@yahoo.com (S.A.)

Abstract: Al_2O_3-Cr_2O_3 refractories are completely substitution solid solutions and can effectively resist slag erosion when used as an industrial furnace lining. In order to provide suitable chromium corundum refractory with excellent slag resistance and mechanical properties for smelting reduction ironmaking, Al_2O_3-Cr_2O_3 samples with different mass percentages (0, 10, 20, 30, 40 wt.%) of Cr_2O_3 were prepared by a normal pressure sintering process to study its sintering properties, mechanical properties, thermal shock resistance, and microstructure. The results of densification behavior showed that the introduction of Cr_2O_3 deteriorates the compactness, the relative density and volume shrinkage rate of the composite material decrease with the increase of the Cr_2O_3 content, and the apparent porosity increases accordingly. In terms of mechanical properties, the hardness, compressive strength, and flexural strength of Al_2O_3-Cr_2O_3 material decrease gradually with the increase of Cr_2O_3. After 10 and 20 thermal shock cycles, the flexural strengths of the samples all decreased. With the increase of Cr_2O_3 in these samples, the loss rate of flexural strength gradually increased. Considering the slag resistance and mechanical properties of the composite material, the Al_2O_3-Cr_2O_3 composite refractory with Cr_2O_3 content of 20–30% can meet the requirements of smelting reduction iron making kiln lining.

Keywords: Al_2O_3-Cr_2O_3 composite; consolidation behavior; microstructure; mechanical properties; thermal shock resistance

Citation: Cui, K.; Zhang, Y.; Fu, T.; Hussain, S.; Saad Algarni, T.; Wang, J.; Zhang, X.; Ali, S. Effects of Cr_2O_3 Content on Microstructure and Mechanical Properties of Al_2O_3 Matrix Composites. *Coatings* **2021**, *11*, 234. https://doi.org/10.3390/coatings11020234

Academic Editors: Deepak Dubal and Awais Ahmad

Received: 13 January 2021
Accepted: 10 February 2021
Published: 16 February 2021

1. Introduction

The smelting reduction ironmaking process has a wide adaptability for raw material sources, which can avoid the use of coke. This process has little environmental pollution, so it is easy to build green factories. In recent years, the smelting reduction ironmaking process is regarded as the development direction of ironmaking in the future [1–3]. Nowadays, the COREX (smelting reduction) and HIsmelt processes, (direct smelting process) as new smelting reduction technologies, have been industrialized [4]. At present, the lining materials used in smelting reduction ironmaking reaction furnace mainly include Al_2O_3-C, MgO-C, MgO-Cr_2O_3, and Al_2O_3-Cr_2O_3 refractories [5–8]. Carbon-containing refractories have excellent slag resistance and thermal shock resistance, but they are easy to be oxidized. MgO-Cr_2O_3 materials have good slag resistance and erosion resistance, but the performance of MgO-Cr_2O_3 on thermal shock stability is poor, so they are easy to peel off [9,10]. Al_2O_3-Cr_2O_3 bricks with excellent slag resistance and suitable for use in oxidation/reduction atmosphere are selected for the HIsmelt equipment in Kwinana Plant, Australia. However, the lining needs to be replaced as a whole every two years for its short service life. Meanwhile, the furnace lining was found with serious damage on its sidewall bricks. Therefore, the short service life of the refractory material at the side wall slag line becomes a bottleneck to improve the service life of the HIsmelt furnace body [9,11].

Al_2O_3 and Cr_2O_3 have the same hexagonal crystal structure; they can form a complete substitutional solid solution $(Al_{1-x}Cr_x)_2O_3$ ($0 \leq x \leq 1$) at high temperature in the whole composition range without any sign of eutectic [12,13]. The solubility of Cr_2O_3 in coal gasification slag (SiO_2-CaO system) and various glass melts is much smaller than that of other oxide materials, so Cr_2O_3 or refractories containing Cr_2O_3 have good corrosion resistance to steel slag, non-ferrous smelting slag, coal gasification slag, oil–gas furnace slag, solid incinerator, and various glass melts [14–20]. However, the environment in the furnace of the smelting reduction ironmaking is very harsh with the high temperature, high oxidation degree, high FeO content, large fluctuation, etc. The service status of chromium corundum brick cannot reach the expectation, and the service life is still far from that of blast furnace lining [21]. At present, some people suggest that the slag resistance of Al_2O_3-Cr_2O_3 bricks in the smelting reduction ironmaking furnace can be improved when the content of Cr_2O_3 is increased on the basis of the original Al_2O_3-Cr_2O_3 bricks. Li et al. [22] reported the slag resistance to smelting reduction of Al_2O_3-Cr_2O_3 refractories with different Cr_2O_3 contents. They found that the slag permeability resistance of the material with low Cr_2O_3 content is poor; the slag resistance characteristic of the Al_2O_3-Cr_2O_3 refractory material will become better with the increase of Cr_2O_3 content. The slag resistance of the composites has been effectively improved due to the addition of Cr_2O_3. However, so far, there are few studies on the mechanical properties, thermal shock resistance, and microstructure of the composites due to the addition of Cr_2O_3. Kafkaslıoğlu et al. [23] reported the mechanical properties of an Al_2O_3-Cr_2O_3 ceramic system as armor material, but they only reported the effect of low Cr_2O_3 (0.5, 1, 5 vol.%) content on the microstructure and mechanical properties of Al_2O_3-Cr_2O_3 ceramic. Therefore, it is particularly important to study the effect of high percentage of Cr_2O_3 on the mechanical properties and microstructure of Al_2O_3-Cr_2O_3 composite refractory.

In this work, the Al_2O_3-Cr_2O_3 composite refractories with various Cr_2O_3 content were prepared by a normal pressure sintering process at 1600 °C for 4 h. The effect of Cr_2O_3 content on the microstructure and phase composition of Al_2O_3-Cr_2O_3 refractory was investigated. In addition, the effects of Cr_2O_3 content on the porosity, hardness, linear shrinkage, density, flexural strength, compressive strength, and thermal shock resistance of composite refractories have been widely discussed. Finally, some reasonable suggestions on the selection of lining refractories in smelting reduction ironmaking are presented.

2. Materials and Methods

2.1. Materials Preparation

Al_2O_3 and Cr_2O_3 powders were used as raw materials. These powders were selected from highly-purity materials in the market. The physical properties and chemical compositions of these raw materials are shown in Table 1. Five kinds of cuboid composites containing different content of Cr_2O_3 and Al_2O_3 particles were sintered, namely, monolithic Al_2O_3 (A), Al_2O_3-10 wt.% Cr_2O_3 (A-10C), Al_2O_3-20 wt.% Cr_2O_3 (A-20C), Al_2O_3-30 wt.% Cr_2O_3 (A-30C), and Al_2O_3-40 wt.% Cr_2O_3 (A-40C). Firstly, the mixture of Cr_2O_3 and Al_2O_3 was carefully weighed, mixed according to the ratio of Cr_2O_3 and Al_2O_3 above, fully mixed and crushed by a high-energy ball milling analyzer (F-P4000) with a grinding time of 30 min, and put it in an oven (DHG-9030A) to dry for 24 h. Secondly, the dried materials were forced to pass through a 200 mesh sieve to break the agglomerated materials; then, they were mixed by deionized water with 5% methylcellulose binder for the preparation of compacts. We weighed 20 g for each sample and poured them respectively into a 60×15 mm^2 mold for pressing, obtaining the rectangular slab with an approximate size of $60 \times 15 \times 7$ mm^3 by compression through hydraulic equipment under 20 MPa. Finally, the green blank was dried at 160 °C for 24 h in an oven, heated to 1600 °C at a heating rate of 20 °C min^{-1} in a programmable resistance furnace, kept at 1600 °C for 4 h, and then naturally cooled to room temperature.

Table 1. Powder composition, particle size, purity, and production place of the materials used in this study.

Material	Particle Diameter (μm)	Purity (wt.%)	Resource
Al_2O_3	1–8	>99.5%	Henan, China/Ou shang
Cr_2O_3	0.1–1	>99.5%	Hebei, China/Qing guang

2.2. Characterization

The phase evolution of the samples was investigated by X-ray diffractometer (D8ADV ANCE, Bruker, Billerica, MA, USA) with Cu Kα (λ = 1.5406 Å) and a maximum current and voltage of 45 kV and 40 mA, respectively, and the diffraction angle was from 10° to 90° with a scanning speed of 2°·min^{-1}. The surface morphology, structure, and grain size of the samples were observed by scanning electron microscopy (SEM, JSM 6510LV, JEOL Ltd., Akishima, Tokyo, Japan), and the surface composition of the samples was analyzed by X-ray energy spectroscopy (EDS). The experimental density (ED) and apparent porosity (AP) of sintered samples were measured by a conventional liquid displacement method based on the Archimedes drainage principle, and the densification behavior of the sintered samples was evaluated by measuring the experimental density, apparent porosity, and firing shrinkage of sintered samples. In the mechanical test, Vickers microhardness (HV1.0) was measured with a load of 9.81 N, and five indentations were pressed on the sample by a Vickers hardness tester (HVS-1000Z, Zhenjiang, China) at room temperature to test the hardness of the sample; the experimental facility of flexural strength is shown in Figure 1a, and the schematic diagram of the span of the flexural strength test and the size of the sample is presented in Figure 1c. The size of the sample is about 60 × 15 × 7 mm^3, and the span of the three-point bending test is 30 mm. The indenter at the upper end of the sample applies stress to the sample at a loading speed of 50 N/s ± 10 N/s until the sample breaks; then, it calculates the flexural strength of the sample. The implementation standard of the test operation is the flexural strength test method for ceramic materials (GB/T 4741-1999). The equipment for testing the compressive strength of the sample (STS520K, Zhenjiang China) is shown in Figure 1b, and the schematic diagram of the compressive strength test process along with the size of the sample is shown in Figure 1d. A cylindrical sample with a size of Φ 20 mm × 20 mm is placed on a lower plate of the testing machine. The upper pressure plate on the testing machine applies stress to the sample at a rate of 1 ± 0.1 MPa·s^{-1} until the sample is crushed, and the compressive strength at normal temperature is calculated according to the maximum load and the average compressive cross-sectional area when the sample is crushed. The implementation standard of the test operation is the compressive strength test method for ceramic materials (GB/T 4740-1999). In the thermal shock experiment, the samples were heated from room temperature to 1100 °C at a rate of 20 °C/min in a heating furnace; then, they were taken out after 15 min and cooled from 1100 °C to room temperature in air. We repeated the process for two groups 10 and 20 times, respectively. The flexural strength loss rate of the sample was calculated by testing the residual strength of the sample to characterize its thermal shock resistance. The calculation method of the flexural strength loss rate of the sample is as follows [24]:

$$S_\sigma = \frac{\sigma_1 - \sigma_2}{\sigma_1} \times 100\% \quad (1)$$

where S_σ is the loss rate of flexural strength, σ_1 (MPa) is the flexural strength before thermal shock, and σ_2 (MPa) is the flexural strength after thermal shock. The morphology, structure, and grain size of the surface and cross-section of the samples after thermal shock were observed by field emission scanning electron microscope.

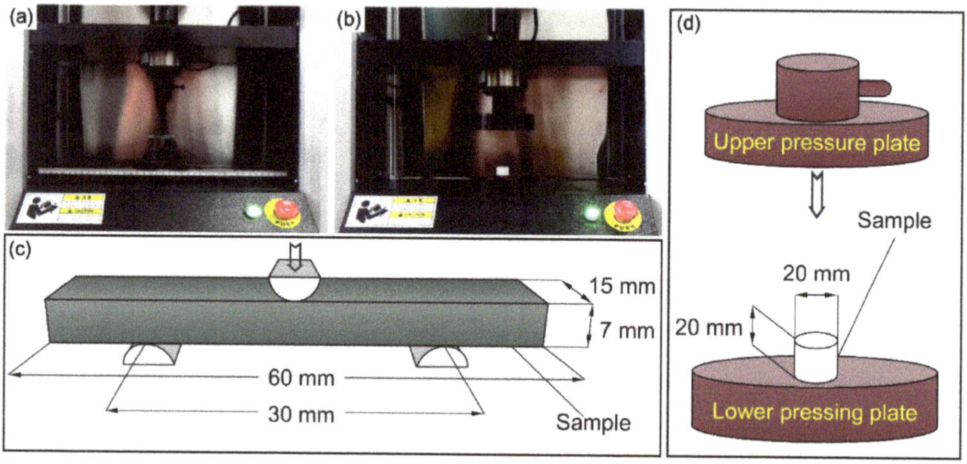

Figure 1. Real diagram (**a**,**b**) and schematic diagram (**c**,**d**) of the flexural strength and compressive strength experiment.

3. Results and Discussion

3.1. Phase Identification

Figure 2a shows XRD patterns of sintered Al_2O_3-Cr_2O_3 ceramic materials with different Cr_2O_3 contents. Al_2O_3 and Cr_2O_3 have the same crystal structure, and all components will transfer completely to substitutional solid solutions at 1600 °C without other compounds forming. It is pointed out in reference [25] that the lattice parameters of corundum structures "a" and "c" increase linearly with the increment of Cr_2O_3 content, which conforms to Vegard's law. When the radius of Cr^{3+} ions (0.076 nm) is larger than that of Al^{3+} ions (0.068 nm), the dissolution of Cr_2O_3 increases the lattice size of Al_2O_3 and decreases the diffraction peak angle, which conforms to Bragg's law [26]. As shown in Figure 2a, the diffraction peak intensity of Al_2O_3-Cr_2O_3 solid solution phase decreased significantly with the increase of Cr_2O_3 content. The diffraction peak of the (104) crystal plane is the sharpest with the highest in intensity, and the corresponding diffraction peaks are shown in Figure 2b. It can be seen that the (104) peak moves to a lower angle from 36° to 34° with the increase of Cr_2O_3 content [23].

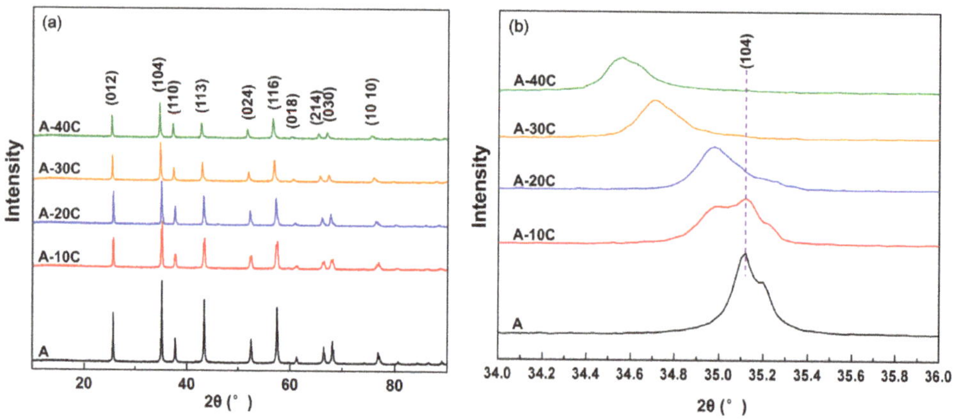

Figure 2. XRD patterns of Al_2O_3-Cr_2O_3 composites with different Cr_2O_3 content.

Macroscopic photos and SEM microstructure images of Al_2O_3-Cr_2O_3 composites with different Cr_2O_3 ratios are shown in Figure 3. The macroscopic photos of the Al_2O_3-Cr_2O_3

composite are shown in Figure 3a. With the increase of Cr_2O_3 content, the appearance of the composite changes from white (sample A) to pink (sample A-20C). Finally, the sample color changes from purplish gray (sample A-30C) to grayish green (sample A-40C). Under the condition of low chromium content (10–20 wt.%), the sample shows a pink color caused by Cr^{3+} ions separated or interacted in pairs. On the other side, when the chromium oxide content grows higher, the color changes gray because the spin–spin interaction between Cr^{3+} ions along the crystal axis increases. Finally, in the condition of high chromium oxide content (40%), the color turns green due to the interaction of Cr^{3+}–O^{2-}–Cr^{3+} [27].

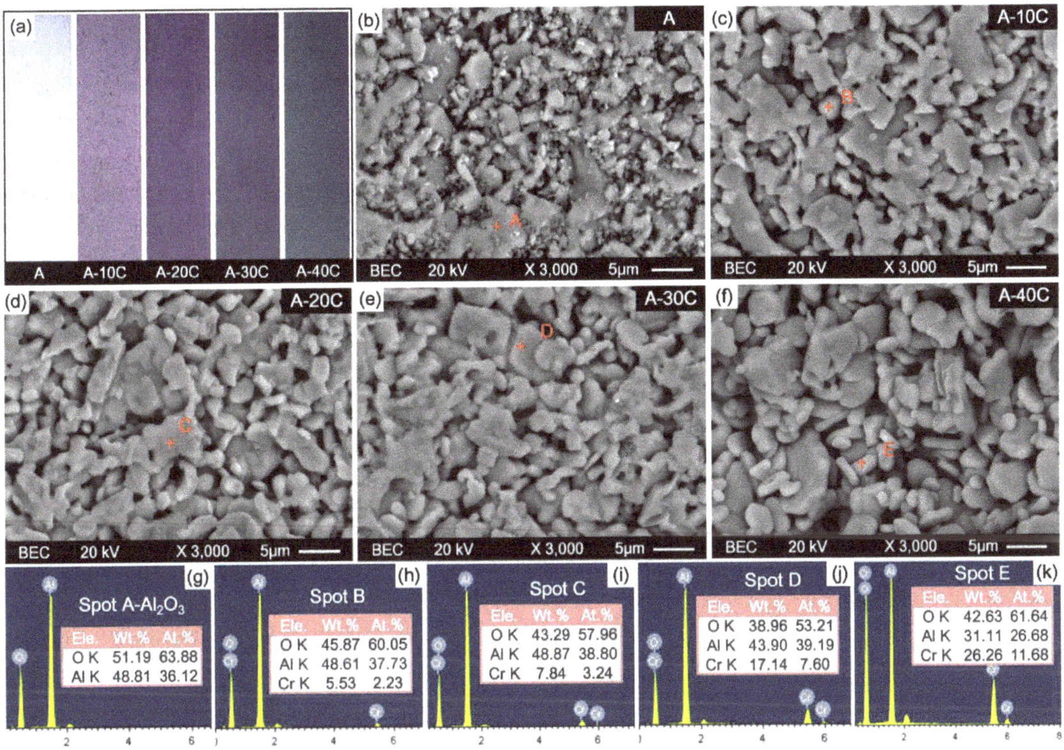

Figure 3. Macroscopic photo, backscatter (BSE) images (**a–f**) and X-ray energy spectrum (EDS) spectrums of Al_2O_3-Cr_2O_3 composites (**g–k**).

The backscatter images (BSE) of the samples with different Cr_2O_3 content are shown in Figure 3 b–f. It can be seen that the Al_2O_3 ceramic is very compact with a fine grain size, as presented in Figure 3b. With the increase of Cr_2O_3 content, the grain size and porosity of the Al_2O_3-Cr_2O_3 composites increased slightly. The composites are almost entirely composed of Al_2O_3-Cr_2O_3 solid solution, and the Cr_2O_3 grains cannot be distinguished from Al_2O_3 grains due to the formation of solid solution. Figure 3g–k shows the X-ray energy spectrum (EDS) images of the samples with different Cr_2O_3 content. The results show that sample A is only composed of the Al_2O_3 phase. A substitutional solid solution of Al_2O_3 and Cr_2O_3 with granular is observed, and the particle size of the Al_2O_3-Cr_2O_3 solid solution phase is about 2 to 10 µm, as shown in Figure 3h–k. The atomic percentage and weight percentage of Cr gradually increase with the increasing of Cr_2O_3 content, which is consistent with the XRD analysis results.

3.2. Consolidation Behaviour

The sintering properties of Al_2O_3-Cr_2O_3 composites are shown in Figure 4. The relative density (RD) of sample A without adding Cr_2O_3 powder is 95.71%. As shown in Figure 4a, the RD of the Al_2O_3-Cr_2O_3 composites decreases gradually with the increase of Cr_2O_3 content. When the content of Cr_2O_3 is 10 wt.%, the relative density of the Al_2O_3-Cr_2O_3 composite decreased sharply, and the relative density of sample A-10C is 90.43%. When the addition amount of Cr_2O_3 increases to 30 wt.%, the relative density of the Al_2O_3-Cr_2O_3 composite decreases slightly, and the relative density of sample A-30C is 88.18%. The lowest relative density is observed when the Cr_2O_3 content is 40 wt.%, and the relative density of the sample A-40C is 87.23%. Firing shrinkage is the combined effect of sintering and thermal transformation/reaction [13]. A certain amount of volume shrinkage occurs after sintering during the preparation of the composite material. The change law of the length shrinkage ratio and volume shrinkage ratio is similar to that of relative density, as shown in Figure 4b. The length shrinkage ratio and volume shrinkage ratio of the Al_2O_3-Cr_2O_3 composites gradually decreases with the increase of Cr_2O_3 content. Sample A has the highest length shrinkage ratio and volume shrinkage ratio, and the length shrinkage ratio and volume shrinkage ratio are 8.1% and 22.2%, respectively. The lowest length shrinkage ratio and volume shrinkage ratio are observed when the Cr_2O_3 content is 40 wt.%, and the length shrinkage ratio and volume shrinkage ratio of the sample A-40C are 1.4% and 3.4%, respectively.

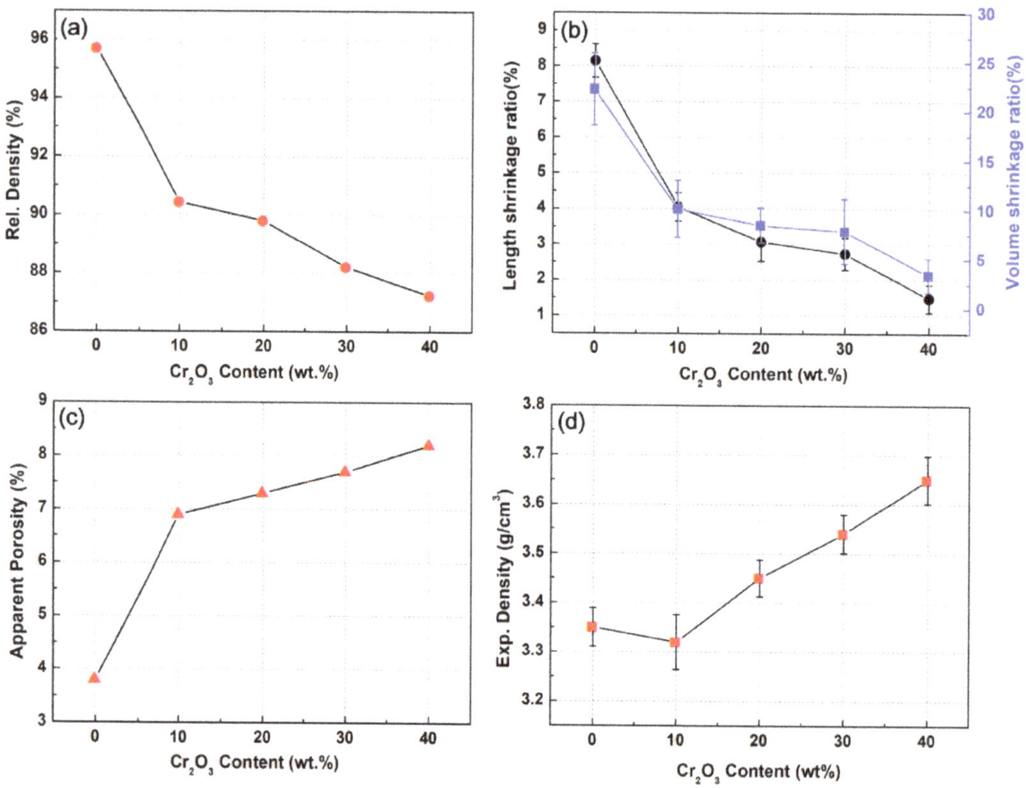

Figure 4. Parameters of prepared Al_2O_3-Cr_2O_3 composites: (**a**) relative density; (**b**) firing shrinkage; (**c**) apparent porosity; (**d**) experimental density.

The change of bulk density of the ceramic system is related to the formation of a new phase, interactions between phases, or particle migration by thermal forces [28]. The apparent porosity (AP) ratio of Al_2O_3-Cr_2O_3 composites increases gradually with the increase of Cr_2O_3 content, as shown in Figure 4c. When the addition of Cr_2O_3 is 10 wt.%, the AP ratio of Al_2O_3-Cr_2O_3 composite increased sharply, and the AP ratio of sample A-10C is 6.90%. When the Cr_2O_3 content ranging from 20 to 40 wt.%, the AP ratio of the Al_2O_3-Cr_2O_3 composite increases slightly, and the AP ratios of samples A-20C, A-30C, and A-40C are 7.25%, 7.69%, and 8.22%, respectively. As shown in Figure 3, the Al_2O_3-Cr_2O_3 solid solution phase presents an obvious granular structure with the increase of Cr_2O_3 content, and the grain size and porosity also gradually increase, which is the main reason for the decrease of the relative density and firing shrinkage of Al_2O_3-Cr_2O_3 composites. The experimental density also presented a gradual increase law with the increase of Cr_2O_3 content. Theoretically, since the specific gravity of Cr_2O_3 (5.15) is higher than that of Al_2O_3 (3.95), the density of the sample should increase when the Cr_2O_3 content goes up. However, it can be seen from the figure that the ED of sample A-10 decreases obviously with the increase of Cr_2O_3 content and then increases steadily with the further increase of Cr_2O_3 content. The trend of AP can be observed from the figure that increased from 3.8% of the original sample A to 8.2% of the sample A-40C, which indicates that the addition of Cr_2O_3 will gradually increase AP. This is because Cr_2O_3 is easy to vaporize even at a very low partial pressure of oxygen. In addition, the formation of the solid solution absorbs heat energy additionally, so that the ceramic density is reduced [23].

3.3. Effect of Cr_2O_3 on Mechanical Properties and Thermal Shock Properties of Al_2O_3-Cr_2O_3 Composites

3.3.1. Hardness

Figure 5 shows the hardness curve of Al_2O_3-Cr_2O_3 composites with different Cr_2O_3 content measured by a Vickers hardness tester. It can be seen that the hardness of Al_2O_3-Cr_2O_3 composites is lower than that of monolithic alumina (18.31 $\pm$ 0.5 GPa). With the increase of Cr_2O_3 content, the hardness of Al_2O_3-Cr_2O_3 composites gradually decreases. When the proportion of Cr_2O_3 is between 0 and 30 wt.%, the hardness decreases slowly. When the content of Cr_2O_3 reaches 40 wt.%, the hardness value decreases sharply, and the minimum hardness value of the sample A-40C composite is only 8.2 GPa. Kafkaslıoğlu et al. [23] reported the mechanical properties of Al_2O_3-Cr_2O_3 ceramics as armor materials. However, they only studied the effect of low addition of Cr_2O_3 (0.5, 1 and 5 vol.%) on the microstructure and mechanical properties of Al_2O_3-Cr_2O_3 ceramics. In their experiment, the hardness of Al_2O_3-Cr_2O_3 ceramic is 20.2 $\pm$ 1.1 GPa when the content of Cr_2O_3 is 0 vol.%. When the content of Cr_2O_3 is 5 vol.%, the hardness of the Al_2O_3-Cr_2O_3 ceramic is 19.0 $\pm$ 2.0 GPa. The hardness value presented a gradually decreasing trend with the increase of Cr_2O_3 content, which is similar to the hardness change rule in this study. The typical diamond indentation images of samples A-10C, A-30C, and A-40C are shown in Figure 6a–c, respectively. It can be seen that the indentation area gradually increases with the increase of Cr_2O_3 addition, which is consistent with the calculation results of the hardness tester. Since the hardness of Cr_2O_3 itself is lower than that of Al_2O_3, adding excessive Cr_2O_3 will reduce the hardness of the material. In addition, the decrease of hardness is closely related to the consolidation behavior of the Al_2O_3-Cr_2O_3 composites. With the increase of the solid solution content of Cr_2O_3, the grain size gradually increases, the material becomes loose and porous, and the hardness gradually decreases.

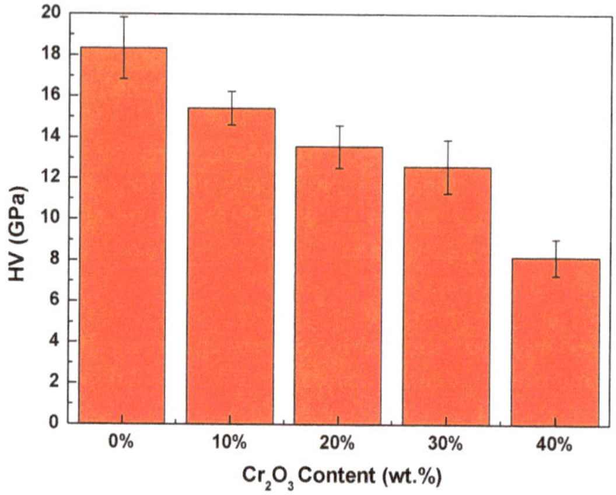

Figure 5. Effect of Cr_2O_3 content on the hardness of composites.

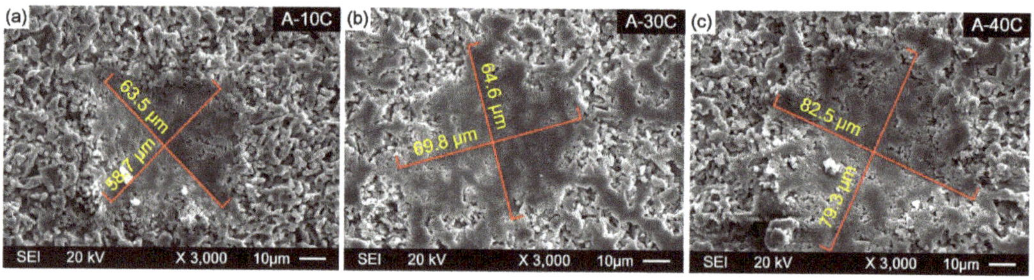

Figure 6. Typical diamond indentation images: (**a**) A-10C; (**b**) A-30C; (**c**) A-40C.

3.3.2. Compressive Strength

Figure 7 shows the compressive strength of Al_2O_3-Cr_2O_3 composites with different Cr_2O_3 content measured by the universal mechanical testing machine. The result shows that the compressive strength of Al_2O_3-Cr_2O_3 composites is lower than that of monolithic alumina (406 MPa). With the increase of Cr_2O_3 content, the hardness of Al_2O_3-Cr_2O_3 composites decreases gradually. The lowest compressive strength is observed when the Cr_2O_3 content is 40 wt.%, and the compressive strength of the sample A-40C is 181 MPa. Although the increase of Cr_2O_3 content has a negative impact on the compressive strength of Al_2O_3-Cr_2O_3 composites, the material prepared in this experiment still has great advantages over the refractories used in general metallurgical kilns. Shu et al. [29] reported that Al_2O_3-SiC composites were prepared by sintering kaolin as the main raw material, and its compressive strength is only 54.3 MPa. Xiao et al. [30] reported that the normal temperature compressive strength of the Al_2O_3-C refractory used in the blast furnace they prepared was about 40 MPa. Yi et al. [31] reported that the compressive strength of the MgO-C refractories prepared by them at room temperature is about 70 MPa, which is also much lower than the compressive strength of the refractory material prepared in this experiment.

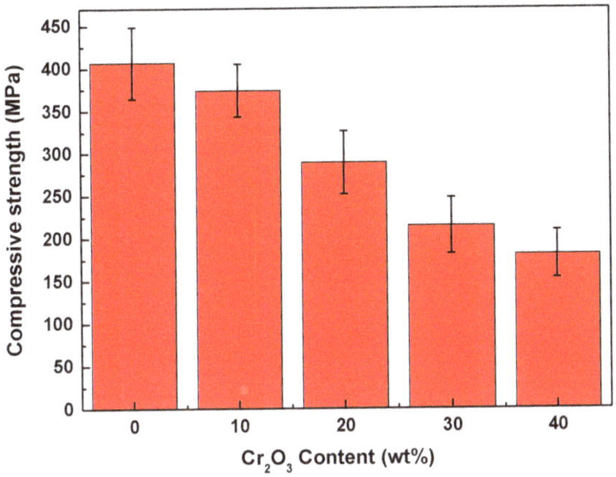

Figure 7. Effect of Cr_2O_3 content on compressive strength of composites.

3.3.3. Thermal Shock Resistance and Flexural Strength

Chromium corundum multiphase ceramics are mainly used at high-temperature environments, and the ceramics may be subjected to thermal shock while working. Thermal shock will generate thermal stress, and ceramics will be destroyed when the thermal stress exceeds the fracture energy [32]. Therefore, thermal shock resistance is one of the most important properties of high-temperature structural materials. The thermal shock resistance of ceramic materials is a combination of mechanical properties and thermal properties, which can be expressed by the loss rate of bending strength. On the other hand, the thermal shock resistance of materials is determined by their thermal expansion coefficient, fracture energy, elastic modulus, and thermal conductivity coefficient. The main factors affecting physical quantities are the phase composition and microstructure of materials. The comprehensive properties of crystal phase determine the thermal shock resistance of the composite materials. Based on the thermal shock theory:

$$\Delta T_C \propto \left(\lambda^2 G / \alpha E_0\right)^{1/2} \tag{2}$$

where λ, α, G, and E_0 are the thermal conductivity coefficient, thermal expansion coefficient, fracture energy, and elastic modulus of ceramic materials, respectively. G is determined by the flexural strength of the material, and a higher flexural strength is required in order to achieve a higher ΔT_C.

As shown in Figure 8a, with the increase of Cr_2O_3 content, the flexural strength of the composite decreases gradually. The flexural strength of the monolithic alumina material is 108 MPa, and the flexural strength of the sample A-40C decreases to 78 MPa. With the increase of Cr_2O_3 addition amount and thermal shock times, the flexural strength of the samples decreased to varying degrees, being the flexural strength of sample A-40 is the lowest (50 MPa) after 20 thermal shocks. As shown in Figure 8b, after 10 and 20 thermal shock cycles, the flexural strength loss rates of sample A are 11.11% and 26.85%, respectively. The flexural strength loss rate is the smallest, which indicates that sample A obtains the best thermal shock resistance. With the increase of Cr_2O_3 addition in the sample, the bending strength loss rate of the sample gradually increases, and the flexural strength loss rate of the sample also gradually increases with the increase of thermal shock times. This shows that the addition of Cr_2O_3 has a negative effect on the thermal shock resistance of the sample. Kafkaslolu et al. [23] showed that adding 0.5 vol.% Cr_2O_3 to Al_2O_3 significantly increased the flexural strength of the composites from 199 to 286 MPa. They believed that the grain boundary modification caused by the larger size of the Cr^{3+} ions replacing Al^{3+} ions results in localized compressive stresses and hinders the propagation of cracks through grain boundaries, thus improving the fracture strength. However, with the increase of Cr_2O_3 content, localized compressive stresses caused by ion size misfit begin to overlap with increasing Cr_2O_3 content and its effect decreases, and the fracture strength of Al_2O_3-Cr_2O_3 composites with 1 vol.% Cr_2O_3 and 5 vol.% Cr_2O_3 content begins to decrease. In this experiment, the effect of the increase of more Cr_2O_3 content (10–40 wt.%) on the composites was investigated. With the increase of Cr_2O_3, the existence of Cr^{3+} ions causes the growth rate of Al_2O_3 to increase and promotes the grain size growth of Al_2O_3. Moreover, the ion radius of Cr^{3+} is larger than that of Al^{3+}, and the grain size gradually increases, so the grain boundary gradually decreases, and the smaller the area, the smaller the grain boundary contact area, the more favorable the crack propagation, so the fracture strength of the composite material gradually decreases.

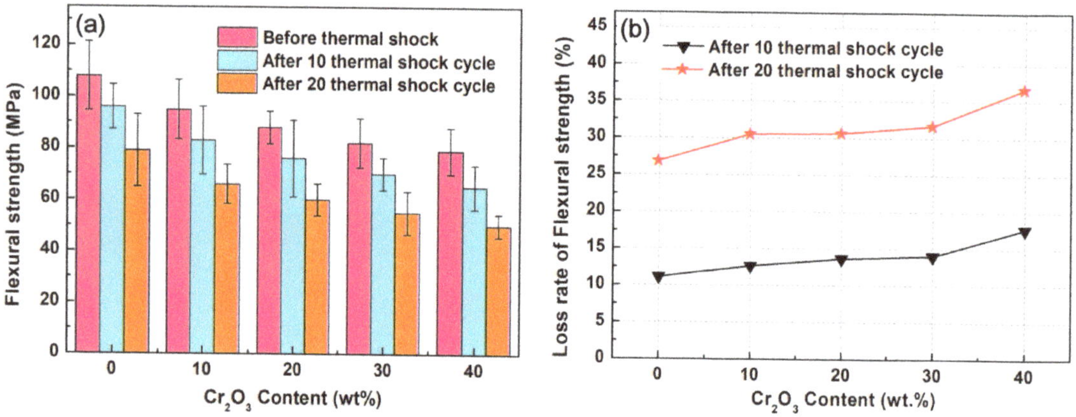

Figure 8. (**a**) Flexural strength before and after thermal shock; (**b**) the flexural strength loss rate of the sample after thermal shock.

In order to study the influence of phase composition, microstructure, and physical properties on thermal shock resistance, SEM analysis was carried out to study the microstructure evolution of the samples before and after thermal shock. Figure 9 is a schematic diagram of a broken sample after three-point bending, and the area marked in the figure is the position of the SEM tests.

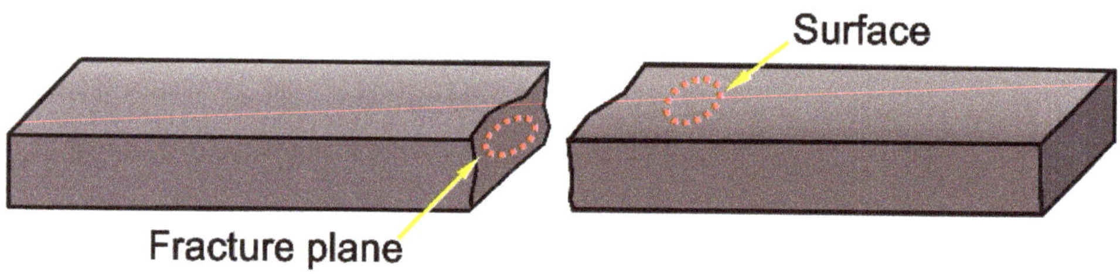

Figure 9. Schematic diagram of fractured sample.

The surface SEM images of the sintered samples before and after thermal shock are shown in Figure 10. The surface secondary-electron (SE) images of the sintered sample are shown in Figure 10a–e. It can be seen that the proportion of solid solution increases gradually with the increase of Cr_2O_3 addition. Meanwhile, the grain size and the porosity of the material also increase gradually, which causes the composites to become more and more loose and porous. It perfectly verified the results of the sintering behavior mentioned above. The surface SEI images of the sintered sample after 20 thermal shocks are shown in Figure 10f–j. It can be seen that the surfaces of the samples after thermal shock have different degrees of pulverization, which is due to the growth of crystal particles in the material under high-temperature atmosphere. The stress concentration is formed locally and eventually leads to the surface pulverization of the material. As shown in Figure 10f–j, the proportion of the surface pulverization area of the composite material gradually increases with the increase of Cr_2O_3 addition.

The fracture SEM images of samples with different Cr_2O_3 content after a three-point bending strength test are shown in Figure 11. The fracture morphologies at low magnification are presented in Figure 11a–e. It can be seen that the fracture surface of the samples presents a state of discontinuous lamellar, and the material has different degrees of cleavage fracture. This fracture morphology is formed by the propagation of cracks along a family of dissociation planes parallel to each other but with different heights. The liquid phase bonding between Al_2O_3 particles has a certain intensity, and the fracture morphology becomes flat with the gradual increase of Cr_2O_3 addition. The SE images of the fracture under high magnification are shown in Figure 11f–j. The "cleavage step" in the sample diagram can be clearly observed; intergranular and transgranular mixed fracture is the fracture mode of the sample, and transgranular fracture dominates. With the increase of Cr_2O_3 addition, the fracture mode changed from intergranular and transgranular mixed fracture mode to intergranular fracture mode. The energy required for crack propagation through grains is higher than that required for crack propagation through grain boundaries. Therefore, with the gradual increase of Cr_2O_3 addition, the fracture mode changes and the material structure gradually becomes loose, resulting in the gradual deterioration of the bending strength of the samples.

Figure 12 shows the fracture morphologies of samples with different Cr_2O_3 contents after 20 thermal shocks. Figure 12a–e shows the SE images of the fracture at low magnification. With the gradual increase of Cr_2O_3 addition, the morphology of the fracture becomes flat. The fracture morphologies under high magnification are presented in Figure 12f–j. It can be seen that the Al_2O_3-Cr_2O_3 composites present a crystalline fracture section, and the intergranular fracture is a dominant mode of failure in Al_2O_3-Cr_2O_3 composites. After 20 thermal shocks, the samples have different degrees of pulverization, and the grain size is smaller than that before thermal shock, but the fracture mode of the material is mainly intergranular fracture. With the increase of Cr_2O_3 addition, the pulverization of the sample becomes more and more serious, and the transgranular fracture is hardly found, so the flexural strength of the Al_2O_3-Cr_2O_3 composites gradually decreases.

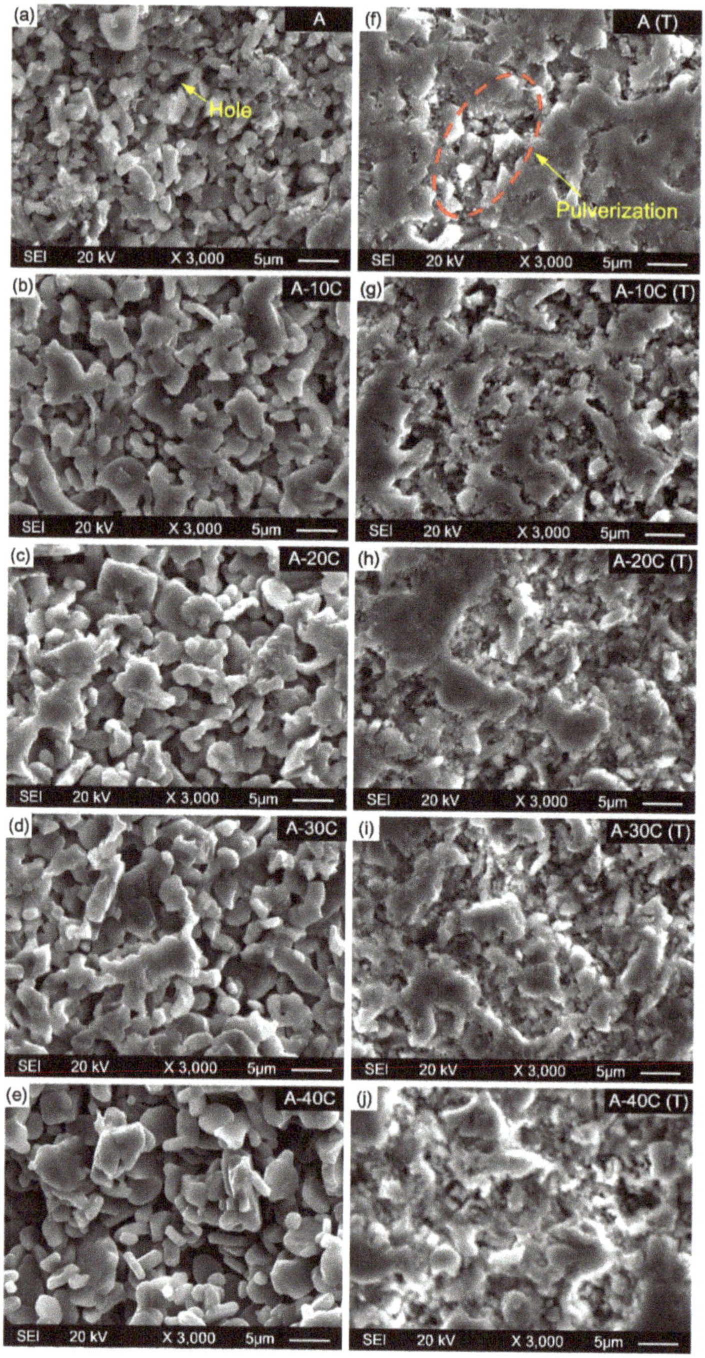

Figure 10. Secondary-electron (SE) images of the samples surface before and after thermal shock (**a**–**j**).

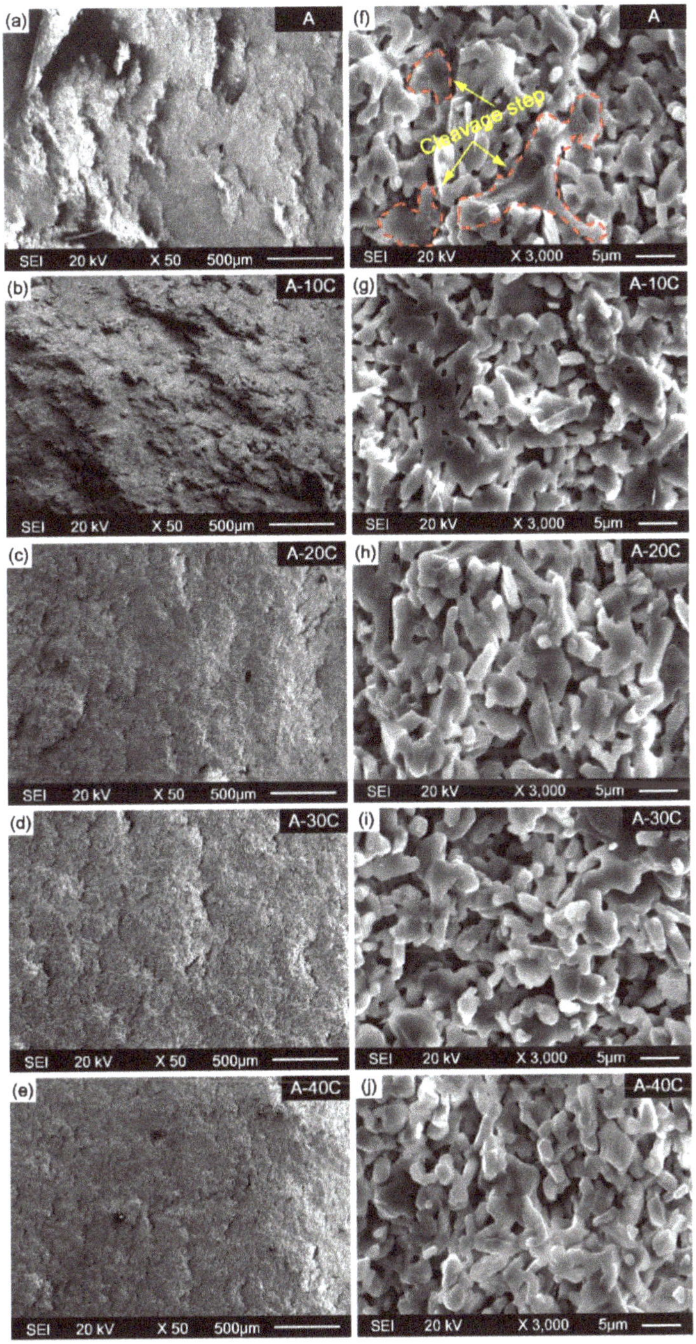

Figure 11. Fracture morphology of the samples after three-point bending (**a**–**j**).

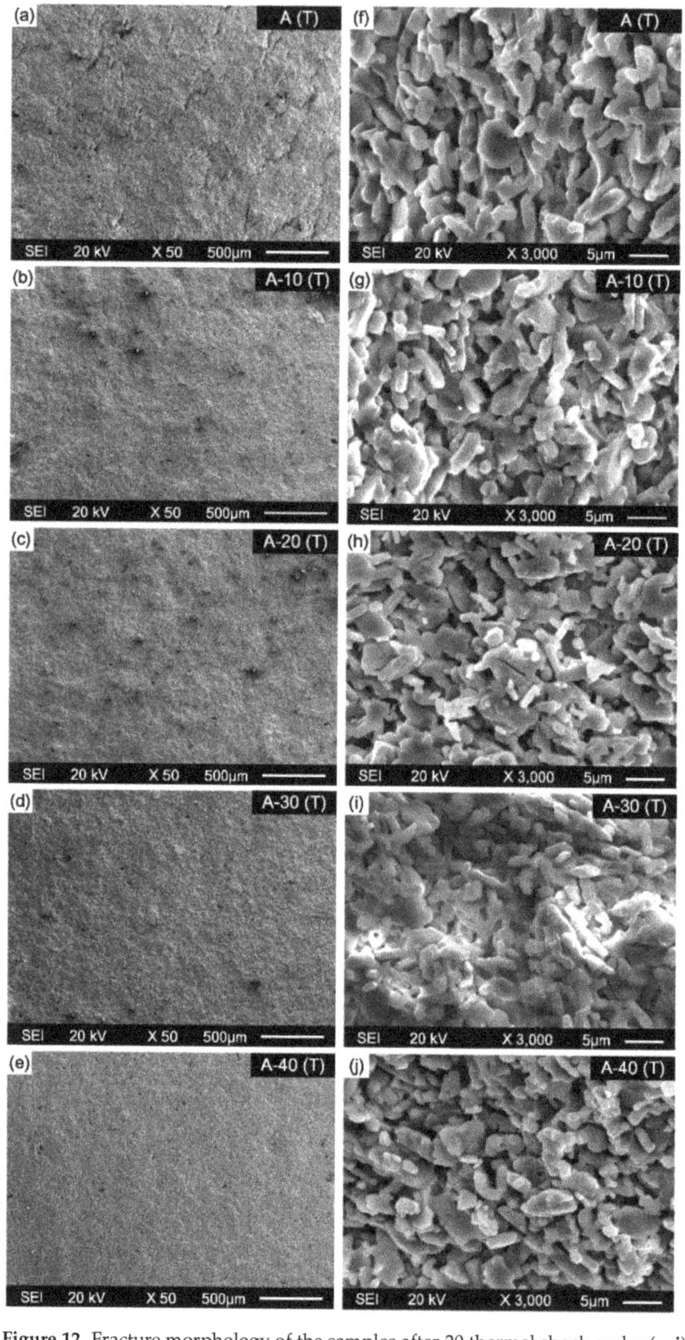

Figure 12. Fracture morphology of the samples after 20 thermal shock cycles (**a**–**j**).

4. Conclusions

Al_2O_3-Cr_2O_3 composites were prepared by reaction sintering with Al_2O_3 and Cr_2O_3 powder at 1600 °C. The effects of Cr_2O_3 on microstructure, consolidation behavior, hardness, bending strength, and thermal shock resistance of Al_2O_3-Cr_2O_3 composites were investigated, and the conclusions are as follows:

1. The grain size and porosity increase gradually due to the formation of Al_2O_3-Cr_2O_3 solid solution phases; thus, the densification behavior of materials Al_2O_3-Cr_2O_3 gradually get worse along with the increase of content of Cr_2O_3. When the Cr_2O_3 content is 40 wt.%, the relative density and volume shrinkage rate of the Al_2O_3-Cr_2O_3 system achieve the minimum combined with the maximum porosity.
2. Due to the reduction of densification degree, the material becomes progressively porous, and the fracture mode of the material changes from transgranular and intergranular mixed fracture to intergranular fracture mode. Meanwhile, the hardness, compressive strength, and flexural strength of Al_2O_3-Cr_2O_3 composites all decreased. When the content of Cr_2O_3 in the system exceeds 30 wt.%, the mechanical properties of the Al_2O_3-Cr_2O_3 material decrease significantly.
3. After 10 and 20 cyclic thermal shocks, the flexural strength of the sample is reduced to varying degrees, and the fracture mode of composites is dominated by intercrystalline fracture. The flexural strength loss rate of samples gradually increases with the increase of Cr_2O_3 content. The maximum bending strength loss rate was observed when the Cr_2O_3 content is 40 wt.% after 10 and 20 thermal shock cycles; the flexural strength loss rates of sample A-40C were 17.72% and 36.71%, respectively. In addition, with the increase of Cr_2O_3 addition, the surface pulverization after thermal shock gradually becomes serious.
4. Although the increase of Cr_2O_3 content deteriorates the mechanical properties of Al_2O_3-Cr_2O_3 composites, the composites still have better mechanical properties when the Cr_2O_3 content is 20–30% and can meet the service requirements of molten reduction ironmaking.

Author Contributions: Conceptualization, Y.Z.; Methodology, Y.Z. and K.C.; Validation, Y.Z. and K.C.; Formal Analysis, Y.Z. and K.C.; Investigation, Y.Z., K.C., X.Z. and T.F.; Resources, Y.Z.; Data Curation, Y.Z., K.C., and S.H.; Writing—Original Draft Preparation, Y.Z. and K.C.; Writing—Review and Editing, Y.Z. and K.C.; Visualization, Y.Z., S.H. and J.W.; Supervision, Y.Z.; Project Administration, Y.Z., S.A.; Funding Acquisition, Y.Z., T.S.A. All authors have read and agreed to the published version of the manuscript.

Funding: This work was supported by the National Natural Science Foundation Grant Nos. 51604049, 51950410596, the National Key R&D Program of China (2017YFB0603800 & 2017YFB0603802) and Researchers Supporting Project Number (RSP-2020/254) King Saud University, Riyadh, Saudi Arabia.

Data Availability Statement: No new data were created or analyzed in this study. Data sharing is not applicable to this article.

Conflicts of Interest: The authors declare no conflict of interest.

References

1. Meijer, K.; Zeilstra, C.; Teerhuis, C.; Ouwehand, M.; Van Der Stel, J. Developments in Alternative Ironmaking. *Trans. Indian Inst. Met.* **2013**, *66*, 475–481. [CrossRef]
2. Kurunov, I.F. The direct production of iron and alternatives to the blast furnace in iron metallurgy for the 21st century. *Metallurgist* **2010**, *54*, 335–342. [CrossRef]
3. Pervaiz, M.; Ahmad, I.; Yousaf, M.; Kirn, S.; Munawar, A.; Saeed, Z.; Rashid, A. Synthesis, spectral and antimicrobial studies of amino acid derivative Schiff base metal (Co, Mn, Cu, and Cd) complexes. *Spectrochim. Acta Part A Mol. Biomol. Spectrosc.* **2019**, *206*, 642–649. [CrossRef] [PubMed]
4. Hasanbeigi, A.; Arens, M.; Price, L. Alternative emerging ironmaking technologies for energy-efficiency and carbon dioxide emissions reduction: A technical review. *Renew. Sustain. Energy Rev.* **2014**, *33*, 645–658. [CrossRef]
5. Kashif, M.; Ngaini, Z.; Harry, A.V.; Vekariya, R.L.; Ahmad, A.; Zuo, Z.; Alarifi, A. An experimental and DFT study on novel dyes incorporated with natural dyes on titanium dioxide (TiO_2) towards solar cell application. *Appl. Phys. A* **2020**, *126*, 1–13. [CrossRef]
6. Zhang, Q.; Guo, L.; Chen, X.D. Analysis of Improving COREX-3000 Competence. *J. Iron Steel Res. Int.* **2009**, *16*, 1225–1227.
7. Hussain, S.; Khan, A.J.; Arshad, M.; Javed, M.S.; Ahmad, A.; Shah, S.S.A.; Khan, M.R.; Akram, S.; Zulfiqar, S.; Ali, S.; et al. Charge storage in binder-free 2D-hexagonal $CoMoO_4$ nanosheets as a redox active material for pseudocapacitors. *Ceram. Int.* **2020**. [CrossRef]

8. Li, J.Q.; Wang, W.W.; Gan, F.F.; Wu, J.G. Study on Corrosion of magnesium chromium materials by iron bath smelting reduction slag. *Refractory* **2011**, *45*, 1–5. (In Chinese)
9. Zhang, X.Z.; Xu, P.H.; Liu, G.W.; Ahmad, A.; Chen, X.H.; Zhu, Y.L.; Qiao, G.J. Synthesis, characterization and wettability of Cu-Sn alloy on the Si-implanted 6H-SiC. *Coatings* **2020**, *10*, 906. [CrossRef]
10. Ahmad, A.; Jini, D.; Aravind, M.; Parvathiraja, C.; Ali, R.; Kiyani, M.Z.; Alothman, A. A novel study on synthesis of egg shell based activated carbon for degradation of methylene blue via photocatalysis. *Arab. J. Chem.* **2020**, *13*, 8717–8722. [CrossRef]
11. Aravind, M.; Ahmad, A.; Ahmad, I.; Amalanathan, M.; Naseem, K.; Mary, S.M.M.; Parvathiraja, C.; Hussain, S.; Algarni, T.S.; Pervaiz, M.; et al. Critical green routing synthesis of silver NPs using jasmine flower extract for biological activities and photocatalytical degradation of methylene blue. *J. Environ. Chem. Eng.* **2021**, *9*, 104877. [CrossRef]
12. Naseem, K.; Rehman, M.Z.U.; Ahmad, A.; Dubal, D.; Algarni, T.S. Plant Extract Induced Biogenic Preparation of Silver Nanoparticles and Their Potential as Catalyst for Degradation of Toxic Dyes. *Coatings* **2020**, *10*, 1235. [CrossRef]
13. Ahmad, A.; Mubharak, N.; Naseem, K.; Tabassum, H.; Rizwan, M.; Najda, A.; Kashif, M.; Bin-Jumah, M.; Hussain, A.; Shaheen, A.; et al. Recent advancement and development of chitin and chitosan-based nanocomposite for drug delivery: Critical approach to clinical research. *Arab. J. Chem.* **2020**, *13*, 8935–8964. [CrossRef]
14. Zhan, M.; Hussain, S.; Algarni, T.S.; Shah, S.; Liu, J.; Zhang, X.; Ahmad, A.; Javed, M.S.; Qiao, G.; Liu, G. Facet controlled polyhedral ZIF-8 MOF nanostructures for excellent NO_2 gas-sensing applications. *Mater. Res. Bull.* **2021**, *136*, 111133. [CrossRef]
15. Chen, J.; Xiao, J.; Zhang, Y.; Wei, Y.; Han, B.; Li, Y.; Zhang, S.; Li, N. Corrosion mechanism of Cr_2O_3-Al_2O_3-ZrO2 refractories in a coal-water slurry gasifier: A post-mortem analysis. *Corros. Sci.* **2020**, *163*, 108250. [CrossRef]
16. Ahmad, A.; Mubarak, N.; Jannat, F.T.; Ashfaq, T.; Santulli, C.; Rizwan, M.; Najda, A.; Bin-Jumah, M.; Abdel-Daim, M.M.; Hussain, S.; et al. A Critical Review on the Synthesis of Natural Sodium Alginate Based Composite Materials: An Innovative Biological Polymer for Biomedical Delivery Applications. *Processes* **2021**, *9*, 137. [CrossRef]
17. Khan, F.S.A.; Mubarak, N.M.; Khalid, M.; Walvekar, R.; Abdullah, E.C.; Ahmad, A.; Karri, R.R.; Pakalapati, H. Functionalized multi-walled carbon nanotubes and hydroxyapatite nanorods reinforced with polypropylene for biomedical application. *Sci. Rep.* **2021**, *11*, 1–10. [CrossRef]
18. KKashif, M.; Jafaar, E.; Sahari, S.K.; Low, F.W.; Hoa, N.D.; Ahmad, A.; Abbas, A.; Ngaini, Z.; Shafa, M.; Qurashi, A. Organic sensitization of graphene oxide and reduced graphene oxide thin films for photovoltaic applications. *Int. J. Energy Res.* **2021**. [CrossRef]
19. Wang, X.; Zhao, P.; Chen, J.; Zhao, H.; He, K. Corrosion resistance of Al–Cr-slag containing chromium–corundum refractories to slags with different basicity. *Ceram. Int.* **2018**, *44*, 12162–12168. [CrossRef]
20. Zhang, Y.; Li, Y.; Bai, C. Microstructure and oxidation behaviour of Si–$MoSi_2$ functionally graded coating on Mo substrate. *Ceram. Int.* **2017**, *43*, 6250–6256. [CrossRef]
21. Pan, D.; Zhao, H.; Zhang, H.; Zhao, P.; Li, Y.; Zou, Q. Corrosion mechanism of spray refractory in COREX slag with varying basicity. *Ceram. Int.* **2019**, *45*, 24398–24404. [CrossRef]
22. Li, P.T.; Sun, H.G.; Li, J.Q.; Yan, S.Z.; Du, Y.H. Different compositions of Al_2O_3-Cr_2O_3 Study on the resistance to smelting reduction slag of No.3 brick. *Refractory* **2016**, *50*, 352–358. (In Chinese)
23. Yıldız, B.K.; Yılmaz, H.; Tür, Y.K. Evaluation of mechanical properties of Al_2O_3–Cr_2O_3 ceramic system prepared in different Cr_2O_3 ratios for ceramic armour components. *Ceram. Int.* **2019**, *45*, 20575–20582. [CrossRef]
24. Xu, X.; Li, J.; Wu, J.; Tang, Z.; Chen, L.; Li, Y.; Lu, C. Preparation and thermal shock resistance of corundum-mullite composite ceramics from andalusite. *Ceram. Int.* **2017**, *43*, 1762–1767. [CrossRef]
25. Kirkaldy, J.S. Flux-independent theory of nonlinear diffusion for Vegard's law solutions. *Mater. Sci. Eng. A* **2007**, *444*, 104–111. [CrossRef]
26. Zhao, P.; Zhao, H.; Yu, J.; Zhang, H.; Gao, H.; Chen, Q. Crystal structure and properties of Al_2O_3-Cr_2O_3 solid solutions with different Cr_2O_3 contents. *Ceram. Int.* **2018**, *44*, 1356–1361. [CrossRef]
27. Bondioli, F.; Ferrari, A.M.; Leonelli, C.; Manfredini, T.; Linati, L.; Mustarelli, P. Reaction Mechanism in Alumina/Chromia (Al_2O_3-Cr_2O_3) Solid Solutions Obtained by Coprecipitation. *J. Am. Ceram. Soc.* **2000**, *83*, 2036–2040. [CrossRef]
28. Nath, M.; Kumar, P.; Maldhure, A.V.; Sinhamahapatra, S.; Dana, K.; Ghosh, A.; Tripathi, H.S. Anomalous densification behaviour of Al_2O_3–Cr_2O_3 system. *Mater. Charact.* **2016**, *111*, 8–13. [CrossRef]
29. Shu, X.M.; Sang, S.B.; Wu, S.J.; Xia, C.Y.; Wu, L.S. Preparation of lightweight mullite aggregate from kaolin and its effect on properties of mullite SiC refractories. *Refractory* **2020**, *54*, 19–23. (In Chinese)
30. Xiao, C.J.; Li, J.; Li, Z.X. Effect of carbon content on Al_2O_3 Effect of MgO-C refractories on properties. *Foshan Ceram.* **2015**, *25*, 9–11. (In Chinese)
31. Yi, J.G.; Zhu, B.Q.; Li, X.C. Effect of carbon embedded heat treatment temperature on phase composition, microstructure and mechanical properties of low carbon MgO-C materials. *Refractory* **2014**, *48*, 170–173. (In Chinese)
32. Tong, Y.; Zhu, W.; Bai, S.; Hu, Y.; Xie, X.; Li, Y. Thermal shock resistance of continuous carbon fiber reinforced ZrC based ultra-high temperature ceramic composites prepared via Zr-Si alloyed melt infiltration. *Mater. Sci. Eng. A* **2018**, *735*, 166–172. [CrossRef]

coatings

Article

Mechanical Characteristics and Adhesion of Glass-Kevlar Hybrid Composites by Applying Different Ratios of Epoxy in Lamination

Sajid Hussian Siyal [1,*], Subhan Ali Jogi [1], Salman Muhammadi [1], Zubair Ahmed Laghari [1], Sadam Ali Khichi [1], Khalida Naseem [2], Tahani Saad Algarni [3], Asma Alothman [3], Shahid Hussain [4,*] and Muhammad Sufyan Javed [5,*]

1 Department of Metallurgy and Materials Engineering, Dawood University of Engineering and Technology-Karachi, Sindh 74800, Pakistan; subhan.duet@gmail.com (S.A.J.); salmanmuhammadi1997@gmail.com (S.M.); zubair.laghari@duet.edu.pk (Z.A.L.); sadam.khichi@hotmail.com (S.A.K.)
2 Department of Chemistry, University of Central Punjab Lahore, Lahore 54000, Pakistan; omeboy007@gmail.com
3 Chemistry Department, College of Science, King Saud University, Riyadh 11451, Saudi Arabia; tahanis@ksu.edu.sa (T.S.A.); zaothman@ksu.edu.sa (A.A.)
4 School of Materials Science and Engineering, Jiangsu University, Zhenjiang 212013, China
5 School of Physical Science and Technology, Lanzhou University, Lanzhou 730000, China
* Correspondence: shahid@ujs.edu.cn (S.H.S.); sajid.hussain@duet.edu.pk (S.H.); safisabri@gmail.com (M.S.J.)

Citation: Hussian Siyal, S.; Ali Jogi, S.; Muhammadi, S.; Ahmed Laghari, Z.; Ali Khichi, S.; Naseem, K.; Saad Algarni, T.; Alothman, A.; Hussain, S.; Javed, M.S. Mechanical Characteristics and Adhesion of Glass-Kevlar Hybrid Composites by Applying Different Ratios of Epoxy in Lamination. *Coatings* **2021**, *11*, 94. https://doi.org/10.3390/coatings11010094

Received: 27 October 2020
Accepted: 9 January 2021
Published: 15 January 2021

Abstract: Hybrid composites have great potential for specific strength and specific stiffness, effective in aerospace industries, submarines, and light-weight automotives. The mechanical strength and adhesiveness of hybrid laminates can be enhanced by effective use of matrix materials in different ratios of epoxy resin and epoxy hardener. Gentle use of resin and hardener in the fabrication of hybrid composites can alter tensile modulus, the bonding strength between matrix and fabric. Spectacular progress has been achieved by the selection of appropriate amounts of resin and hardener in the hybridization of composite laminate. Hybridization was made by Kevlar inorganic/organic fabrics and glass fabrics stacked with epoxy matrix material. To achieve the combination of mechanical properties and bonding strength, transparent epoxy resin and hardener of commercial grades mixed in various ratios are incorporated as matrix material to fabricate laminate. Three different sheets, named A (3:2), B (4:1), and C (2:3), were embedded by the hand layup method to prepare a hybrid composite. Experimental tests, according to ASTM 3039, were performed to determine the tensile mechanical properties. Peel tests, according to ASTM 6862-11, were performed to investigate the interlaminar strength between Kevlar and glass layers. Shore A and Shore C hardness durometers were used to find out the hardness of the specimens at different spots using the ASTM D-2240 standard. Finally, physical testing, such as density and then water absorption, was carried out using the ASTM D-570 standard to check the swelling ratio of the different specimens. The results obtained highlight that the specimen of the glass/Kevlar hybrid embedded in the ratio 3:2 in lamination has the best mechanical properties (tensile strength and hardness) and the lowest swelling ratio, while the material system in the ratio 4:1 shows the best interlaminar properties and adhesion capabilities.

Keywords: glass/Kevlar; hybrid composites; hand layup; epoxy; hardener; tensile; hardness shore D; water absorption; density; peel; ratio

1. Introduction

Composites have distinct phases, made by dissimilar fibers and matrix materials. Reinforcement fibers or stacking sheets are employed in matrix materials for making composite structures at the macroscopic level. Matrix materials have a continuous phase and are used as adhesives, and fibers have a discontinuous phase in metal matrix composites (MMCs), ceramic matrix composites (CMCs), and polymer matrix composites (PMCs) [1]. Reinforcing synthetic fibers, such as Kevlar fibers, carbon fibers, and glass fibers, and natural fibers, such as jute, sisal, hemp, etc., serve as a dispersed phase in composite materials [2]. In aeronautical applications, light-weight, high-strength organic/inorganic stacked laminated hybrid composites are ideal candidates [3]. The organic fiber Kevlar is five times better than steel due to its specific weight ratio. Kevlar has improved interfacial fracture toughness and degree of symmetry in its internal structure [4]. High-strength aramid fiber is employed as core material in hybrid laminate to prepare light-weight, high-strength bulletproof vests and helmets [5]. In the last few decades, researchers have focused on enhancing the performance of composite materials by mixing synthetic and natural fibers to prepare hybridized structures [6]. E-glass fibers have better ultimate tensile strength and are placed in the outside layer in lamination to produce a flaw-free structure [7]. There is no significant difference in the density if layup placement changes in hybridization [8]. Aramid (Kevlar), a synthetic fiber, was developed in 1965 by two research scientists, Stephanie Kwolek and Herbert Blades. Kevlar belongs to the aromatic polyamide family and consists of extended chains of synthetic polyamide [9]. Aramid fibers have incredible ultimate tensile strength, achieved by intermolecular hydrogen bonds and the aromatic stacking interface of aromatic groups in the neighboring strands [10]. The interatomic bonding is considerably stronger than the van der Waals intermolecular connection in other manmade polymers [11]. Kevlar (aramid) is based on comparatively stiff molecules, which formulate the planar sheet-like arrangement, comparable to silk protein. The intermolecular relation among the molecular strands of Kevlar significantly enhances mechanical strength characteristics and provides superior heat and flame resistivity [12]. Kevlar, which is by now renowned as an elevated-performance fiber, has a remarkable strength-to-weight ratio and improved toughness [13]. As far as cost competitiveness is concerned, glass fiber is cheap and easily available in the market, whereas Kevlar fiber is considerably expensive [14]. To achieve a balance between the cost competitiveness and the strength-to-weight ratio of laminate, hybrid composites are fabricated by stacking E-glass/Kevlar fibers [15]. The amount of water absorption in laminate reduces the mechanical strength of the hybrid composite by weakening the fibers' strength [16,17]. Glass fibers are placed at the interface and Kevlar fibers are used in the core to make a sandwich structure [18]. Glass-fiber-reinforced polymers are used to make printed wire boards for electronic appliances [19]. The mechanical behavior of fiber-reinforced polymers was investigated by understanding ultimate tensile strength, stiffness, and fiber-matrix bond strength [20]. Kevlar fibers are used as core in the lamination sequence to suppress the fracture toughness of the hybrid composite [21]. The ratios of epoxy resin to hardener and the volume fraction are effective tools in the fabrication of composite materials [22]. Kevlar fibers reinforced with glass fibers have incredible capacity to produce less dense materials with superior strength, which attain high break and bear resistance characteristics [23]. Epoxy resin and epoxy hardener have good bonding strength and are positioned within layers of fibers to produce rigid and environment-resistant materials [24]. However, the mechanical properties of composite materials strongly depend on the adhesive material as well as the different ratios in which resin and hardener are mixed together to form a matrix phase [25]. The proper use of adhesive materials can supplement the mechanical properties, reduce component weight, increase durability, provide better design latitude, handle high levels of stress, and increase the strength of composites [26]. The performance of the continuous phase is related with the ratio of resin to hardener in the discontinuous phase in hybridization [27]. The interfacial bond strength of the glass/Kevlar hybrid composite is investigated by conducting a peeling test of the hybrid composite laminate [28]. In advanced technologies, materials are required with a combination of properties, such as strength, stiffness, impact, water absorption, and strength-to-weight ratio. Glass/Kevlar hybrid composites have an unusual combination of properties, which cannot be met by other conventional materials [29]. The strength of a composite depends not only on the assets of the matrix but also on how well it sticks to the particles and fibers of the dispersed phase [30]. It is vital that adhesive bonding forces between fiber and matrix be high to minimize fiber pull-out [31]. Sufficient bonding is essential to take advantage of the stress transfer from the weak matrix to the durable fibers.

2. Experimental Work

2.1. Preparation of Specimens

In E-glass fabrics with Kevlar 49, each layer has a thickness of 0.5 mm stacked with epoxy resin and a hardener of commercial grade as adhesive medium for the fabrication of hybrid composites. E-glass fabrics were placed at the top and the bottom, whereas the central layer was of Kevlar 49. In this study, laminates of hybrid composites (A, B, and C) were fabricated (Figure 1) with dissimilar amounts of epoxy resin and hardener using the hand layup method given in Table 1. Materials properties of E-glass, Kevlar, and epoxy given in Table 2.

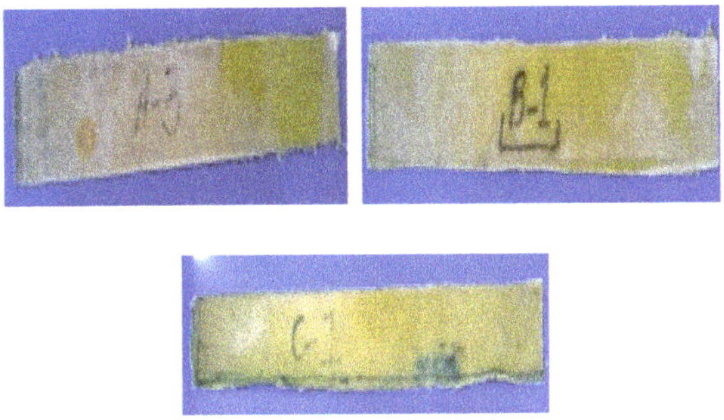

Figure 1. Specimens fabricated using different resin ratios (A = 3:2, B = 4:1, C = 2:3) respectively.

Table 1. Layup placement of glass fibers and Kevlar 49 by applying different ratios of resin and hardener.

Layer Formation	Layers	Glass Fibers	Kevlar 49	Ratio of Epoxy Resin	Ratio of Hardener
A	03	02	01	3	2
B	03	02	01	4	1
C	03	02	01	2	3

Table 2. Materials properties of E-glass, Kevlar, and epoxy.

Material	Fiber Strength	Laminate Strength	Density of Laminate (g/cc)	Strength-to-Weight Ratio
E-glass	3450	1500	2.66	564
Kevlar	2757	1430	1.44	993
Epoxy	N/A	12–40	1–1.15	28

2.2. Tensile Test

During a tension test, standard specimens are placed on the grip of a tensile testing machine (Cometech, Taichung, Taiwan) (Figure 2) so that the specimens undergo tension before fracture. In usual practice, the gauge length of the standard sample increases and the cross-section area reduces during operation. The stress-strain diagram results and the data are analyzed. The standard specimen of ASTM D-3039 was used to conduct the test for a hybrid laminate [32]. Three specimens (24 mm wide and 150 mm long) were subjected to the grip of a 5 kN universal tensile testing machine. The strain rate (s^{-1}) was set to conduct the tensile test for each laminate. The laminate with the same dimensions as those of the glass/Kevlar hybrid with different ratios of epoxy resin and hardener—A (3:2), B (4:1), and C (2:3)—was examined at room temperature. The results and statistical analysis of specimens A, B, and C were tabulated, and the peak stress and peak load of each specimen were investigated considering the stress (N/mm^2) and the strain (%).

Figure 2. A 5 kN universal tensile testing machine.

2.3. Peel Test

The peel adhesion test is used to determine the force required to debond two components joined by a strong adhesive. The test result, also known as bond strength, is generally represented as N (force to debond)/24 mm depending on specimen width. There are various types of peel tests. Examples include the 90° peel test, 180° peel test, single lap joint shear test, and T peel test. In this work, 90° peel tests were performed with two different test setups named type I and type II.

2.3.1. Sample Preparation

Three specimens, each being 24 mm wide and 150 mm long, was used to perform 90° peel tests, according to the ASTM D-6862-11 standard [33]. The samples were directly cut from the prepared sheets (27 cm × 9 cm), in which one of the two glued components was flexible enough to be bent 90° without breaking.

2.3.2. Type I: When Glass Fiber Is Clamped on Both Upper and Lower Jaws of the UTM for Debonding

One layer of glass fiber is clamped on the upper, movable jaw and another layer of glass fiber is fixed on the lower, fixed jaw of the universal tensile testing machine. The force is set at zero and then, on pushing the start button, the specimen from one end starts debonding, either mechanically or chemically, until the movable layer of glass fiber is completely peeled apart or destroyed.

2.3.3. Type II: When Glass Fiber Is Clamped on the Upper Jaw and Kevlar Is Clamped on the Lower Jaw of the UTM for Debonding

An attempt is made to prevent the glass fiber from breaking during peeling. Kevlar fiber is clamped on the lower, fixed jaw and glass fiber is clamped on the upper, movable jaw to peel apart to check the adhesion between glass fiber and Kevlar by peeling them apart completely.

3. Hardness Test

Hardness is resistance to indentation or penetration. Shore A and Shore C durometers are used as portable hardness testers to find the hardness of three specimens. The shore durometer is a device for measuring the hardness of polymers, polymer matrix composites, and elastomers. Higher numbers on the scale indicate a greater resistance to indentation and, thus, harder materials.

The hardness test is carried out as per the ASTM D2240 standard [34] using Shore A and Shore C durometers shown in Figure 3. An indentation is formed by a given force on a standardized indentation head. This test method is widely used for composite and polymer testing. As per ASTM standards, three samples are prepared with different resin and hardener ratios with the dimensions 24 mm × 24 mm.

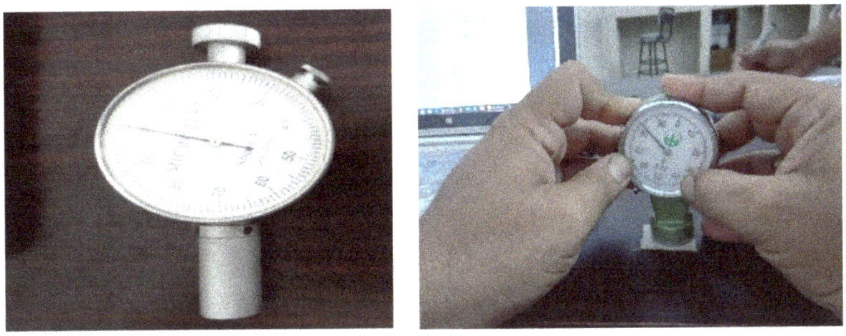

(a) (b)

Figure 3. Shore A (**a**) and Shore C hardness (**b**) tester, respectively.

The hardness test was carried out using two types of durometers, Shore A and Shore C. A load was applied, and the readings on the dial were noted down. The indentation was measured from different spots, and mean average values were calculated.

4. Density Test

A physical balance was used to evaluate the density of the hybrid composites. As per ASTM standards, three specimens were made. The specimens were placed onto the physical balance and the readings were obtained from the screen. The density was measured by putting the values in the formula Density $= \frac{\text{Mass}}{\text{Volume}}$. The dimensions of the specimen used for the density test were 24 mm × 24 mm × 0.2 mm.

5. Water Absorption Test

Water absorption is the amount of water uptake by a material under specific conditions. A physical balance was used to measure the amount of water absorbed by the specimens. As per the ASTM D-570 standard [35], three specimens were fabricated for water absorption, as shown in Figure 4. The samples were first weighed in dry condition and then put into distilled water at 31 °C for 24 h. The samples were then weighed again.

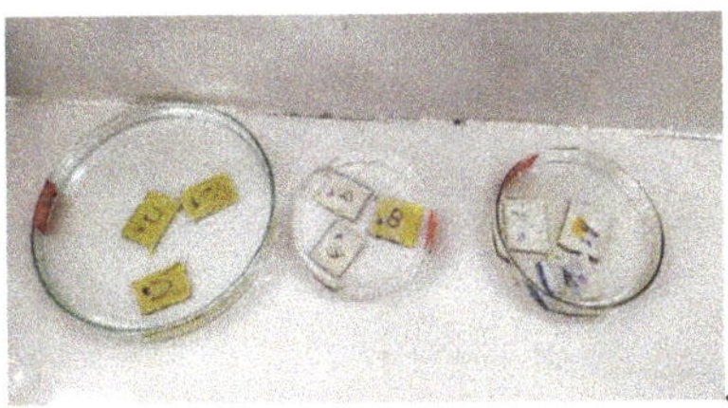

Figure 4. Specimens for the water absorption test using the ASTM D-570 standard.

Measurements

Water absorption is calculated as an increase in the weight percentage of specimens. Swelling ratio % $= \frac{\text{Wet laminate area} - \text{Dry laminate area}}{\text{Dry laminate area}} \times 100$ [36] Standard specimens are the same as in the density test (24 mm × 24 mm × 0.2 mm).

6. Results and Discussion

6.1. Tensile Test

A tensile test was carried out using the ASTM standard D-3039 [37] to analyze the mechanical behavior and strength of specimens under tension. The results are shown in the graphs below (Figure 5).

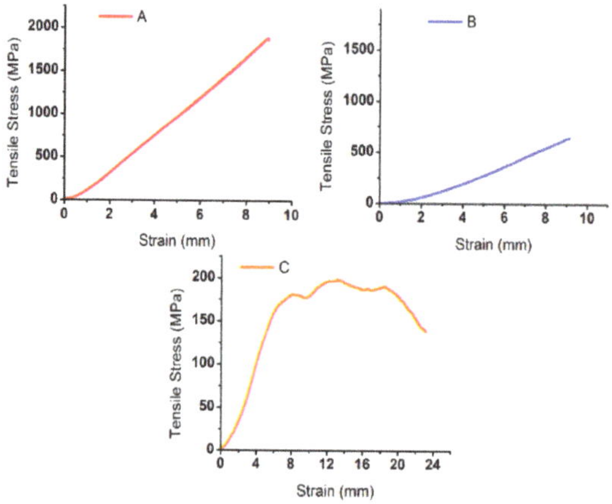

Figure 5. Tensile test results of each configuration: (**A–C**).

According to the graphs of the tensile test, specimen A, with 60% resin and 40% hardener, has the maximum value of stress, 1880 MPa, as shown in the graphs in Figure 5. There is a little fluctuation at stress 635 MPa in the graph of specimen B due to the prior breaking of glass fiber than Kevlar. As far as specimen C is concerned, it has the lowest value of tensile stress due to very poor adhesion between glass fiber and Kevlar with 40% epoxy resin and 60% hardener.

6.2. Peel Test

As the peel test is performed in two ways, in both conditions, the laminate of same dimensions as those of glass/Kevlar with different ratios of epoxy resin and epoxy hardener (A (3:2), B (4:1), and C (2:3)) was examined at room temperature.

Type I: When Glass Fiber is Clamped on Both Upper and Lower Jaws for Debonding.

Both the upper and lower layers of glass fabric are peeled off from Kevlar fabric, and a graph is plotted against the force required to debond/24 mm and the strain to examine the adhesiveness between glassy fiber (GF) and Kevlar.

The results of the peel test are shown in Figure 6 in the form of the force required to debond different specimens.

According to this figure, it is clear that specimen B (with 80% resin and 20% hardener) uses maximum force in debonding the laminate layers compared to other specimens with codes A and C. This force is expressed as 16.5 N/24 mm. Comparison values of the peel test shown in Figure 7.

Type II: When Glass Fabric is Clamped on the Upper Jaw and Kevlar is Clamped on the Lower Jaw for Debonding.

Kevlar fabric is clamped on the lower, fixed jaw and glass fabric is clamped on the upper, movable jaw to check the adhesion between the glass fabric and Kevlar by peeling them apart completely. Peel test examination of type II shown in Figure 8.

This is a special segment of the peel test, when glass fiber is clamped on the upper jaw and Kevlar is clamped on the lower jaw for debonding, in which an attempt is made to enhance the debonding force. The specimen with code B shows the maximum debonding force (38 N/24 mm) in this case also, as shown in the comparison graph in Figure 9.

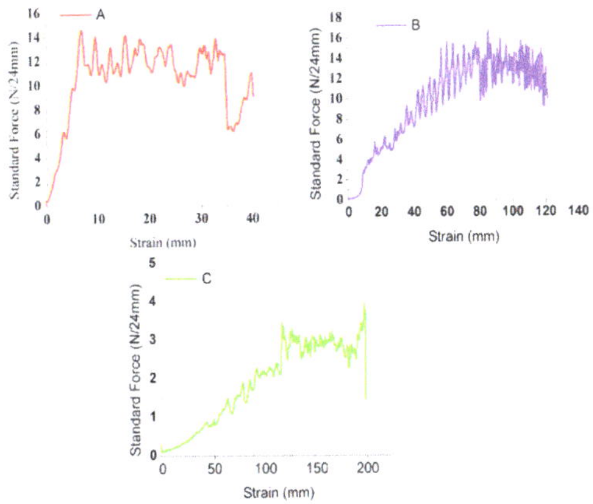

Figure 6. Peel test analysis of type I specimens (**A**–**C**).

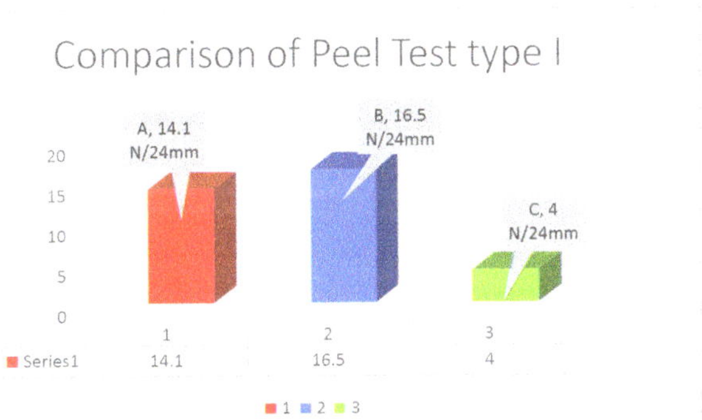

Figure 7. Comparison values of the peel test of type I specimens (**A**–**C**).

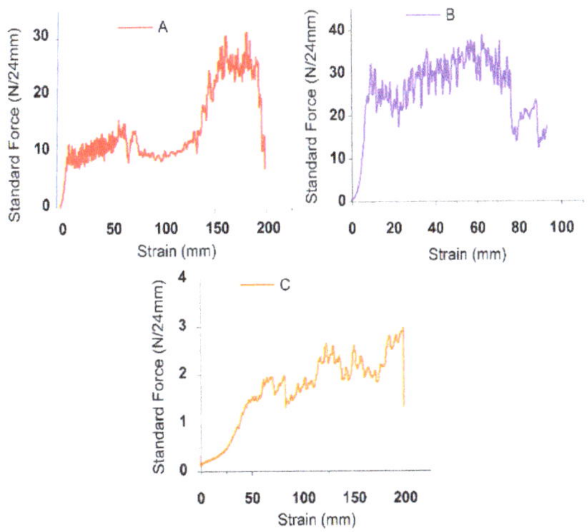

Figure 8. Peel test examination of type II specimens (**A**–**C**).

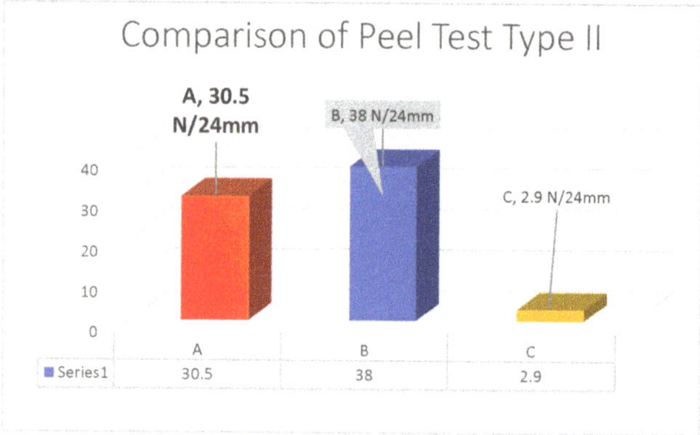

Figure 9. Difference between the peel test results of type II specimens (**A–C**).

6.3. Hardness Test

A portable hardness tester was used to perform the hardness test according to the ASTM A 1038-17 standard [38]. Three samples with size 24 mm × 24 mm were prepared with different epoxy resin and epoxy hardener ratios (A = 3:2, B = 4:1, and C = 2:3). A load was applied, and the readings on the dial were noted down.

Shore A Hardness Test (Table 3):

Table 3. Shore A hardness of three different specimens.

Readings	**A**			**B**			**C**		
	A1	**A2**	**A3**	**B1**	**B2**	**B3**	**C1**	**C2**	**C3**
1.	95	93	96	91	94	94	80	85	80
2.	96	95	97	93	92	85	85	86	84
3.	95	91	96	91	95	88	86	84	81
Mean of indentation	95.33	93	96.33	91.66	93.67	89	83.67	85	81.67
Mean hardness		94.887			91.443			83.44	

Shore C Hardness Test (Table 4):

Table 4. Data analysis of Shore C hardness of three distinct configurations.

Readings	**A**			**B**			**C**		
	A1	**A2**	**A3**	**B1**	**B2**	**B3**	**C1**	**C2**	**C3**
1.	82	78	85	72	76	75	68	74	65
2.	83	81	86	75	70	78	78	75	68
3.	78	79	86	70	84	77	75	78	66
Mean of indentation	81	79.33	85.67	72.33	76.67	76.67	73.67	75.67	66.33
Mean hardness		82			75.23			71.89	

Tables 3 and 4 report that there is a small variation in harness by applying dissimilar ratios in lamination. It is evident that hardness does not rely on the ratio of matrix material in hybridization [1].

6.4. Density Test

According to the results of the density test, specimen A has the best value of density compared to specimens B and C, as described in Table 5.

Table 5. Measured density of specimens A, B, and C at difference resin ratios.

Specimen	Weight (g)	Density (g/cm³)	Mean (g/cm³)
A	0.98	1.0	0.96
B	0.91	0.88	0.86
C	0.85	0.72	0.82

6.5. *Water Absorption Test*

Results of water absorption for specimens A, B, and C by applying different ratios of matrix in lamination given in Table 6.

Table 6. Results of water absorption for specimens A, B, and C by applying different ratios of matrix in lamination.

Specimen	W1 (g)	W2 (g)	Mean W1	Mean W2	Swelling Ratio %
A	0.98	1.0	0.96	1.18	1.23
B	0.91	0.95	0.86	1.20	1.08
C	0.85	0.78	0.82	1.32	1.46

The water absorption ratio is calculated as an increase in the weight percentage of specimens A, B, and C by using the formula Swelling ratio % $= \frac{\text{Wet laminate} - \text{Dry laminate}}{\text{Dry laminate}} \times 100$. By analyzing Table 5, it is observed that specimen A absorbs the minimal amount of water and, thus, has the minimum swelling ratio compared to specimens B and C. From the research findings, it is evident that specimen A has superior strength performance than samples B and C. The minimal swelling ratio is best suited for mechanical properties of hybrid composite materials.

7. Conclusions

Efforts were made to enhance the interlaminar adhesion of glass/Kevlar hybrid composites by using dissimilar proportions of epoxy resin and hardener. In this regard, mechanical and physical testing of hybrid composites was performed. Experimental verification shows that the resin and hardener ratio has a profound influence on both the mechanical and physical properties of hybrid composites. The laminate with 60% resin and 40% hardener has superior mechanical properties, such as tensile strength hardness and minimum swelling ratio, compared with 40% resin and 60% hardener in a matrix, whereas specimens made of 80% epoxy resin and 20% hardener have better bonding strength among the layers compared to other specimens. The specimens in which the ratio of hardener is suppressed rather than that of resin have mechanical properties such as adhesion and deduce.

Author Contributions: Conceptualization, S.H.S. and S.A.J.; methodology, S.M. and Z.A.L.; software, S.A.K.; validation, K.N., S.H., and M.S.J.; formal analysis, S.H.; investigation, T.S.A.; resources, T.S.A.; data curation, K.N.; writing—original draft preparation, S.H.S.; writing—review and editing, S.H.; visualization, M.S.J.; supervision, S.H.; project administration, A.A.; funding acquisition, T.S.A. All authors have read and agreed to the published version of the manuscript.

Funding: This work was funded by the Researchers Supporting Project No. (RSP-2020/254) King Saud University, Riyadh, Saudi Arabia.

Data Availability Statement: Data are available in the paper.

Conflicts of Interest: The authors declare no conflict of interest.

References

1. Chavali, M.; Palanisamy, P.; Nikolova, M.P.; Wu, R.-J.; Tadiboyina, R.; Rao, P.P. Inorganic composites in biomedical engineering. In *Materials for Biomedical Engineering*; Elsevier: Amsterdam, The Netherlands, 2019; pp. 47–80.
2. Atiqah, A.; Chandrasekar, M.; Kumar, T.S.M.; Senthilkumar, K.; Ansari, M.N. Characterization and interface of natural and synthetic hybrid composites. *Encycl. Renew. Sustain. Mater.* **2020**, 389–400. [CrossRef]
3. Kashif, M.; Ngaini, Z.; Harry, A.V.; Vekariya, R.L.; Ahmad, A.; Zuo, Z.; Alarifi, A. An experimental and DFT study on novel dyes incorporated with natural dyes on titanium dioxide (TiO_2) towards solar cell application. *Appl. Phys. A* **2020**, *126*, 1–13. [CrossRef]
4. Ahmad, A.; Jini, D.; Aravind, M.; Parvathiraja, C.; Ali, R.; Kiyani, M.Z.; Alothman, A. A novel study on synthesis of egg shell based activated carbon for degradation of methylene blue via photocatalysis. *Arab. J. Chem.* **2020**, *13*, 8717–8722. [CrossRef]
5. Zhang, X.Z.; Xu, P.H.; Liu, G.W.; Ahmad, A.; Chen, X.H.; Zhu, Y.L.; Qiao, G.J. Synthesis, characterization and wettability of Cu-Sn alloy on the Si-implanted 6H-SiC. *Coatings* **2020**, *10*, 906. [CrossRef]
6. Aravind, M.; Ahmad, A.; Ahmad, I.; Amalanathan, M.; Naseem, K.; Mary, S.M.M.; Zubair, M. Critical green routing synthesis of silver NPs using jasmine flower extract for biological activities and photocatalytical degradation of methylene blue. *J. Environ. Chem. Eng.* **2020**, *9*, 104877. [CrossRef]
7. Hussain, S.; Khan, A.J.; Arshad, M.; Javed, M.S.; Ahmad, A.; Shah, S.S.A.; Khan, M.R.; Akram, S.; Zulfiqar; Ali, S.; et al. Charge storage in binder-free 2D-hexagonal $CoMoO_4$ nanosheets as a redox active material for pseudocapacitors. *Ceram. Int.* **2020**. [CrossRef]
8. Saleem, M.; Irfan, M.; Tabassum, S.; Alothman, Z.; Javed, M.S.; Hussain, S.; Zubair, M. Experimental and theoretical study of highly porous lignocellulose assisted metal oxide photoelectrodes for dye-sensitized solar cells. *Arab. J. Chem.* **2020**, *14*, 102937. [CrossRef]
9. Naseem, K.; Zia Ur Rehman, M.; Ahmad, A.; Dubal, D.; AlGarni, T.S. Plant extract induced biogenic preparation of silver nanoparticles and their potential as catalyst for degradation of toxic dyes. *Coatings* **2020**, *10*, 1235. [CrossRef]
10. Madkour, L.H. *Nanoelectronic Materials: Fundamentals and Applications*; Springer: Berlin/Heidelberg, Germany, 2019; Volume 116.
11. Zhan, M.; Hussain, S.; AlGarni, T.S.; Shah, S.; Liu, J.; Zhang, X.; Liu, G. Facet controlled polyhedral ZIF-8 MOF nanostructures for excellent NO_2 gas-sensing applications. *Mater. Res. Bull.* **2021**, *136*, 111133. [CrossRef]
12. Kashif, M.; Jaafar, E.; Bhadja, P.; Low, F.W.; Sahari, S.K.; Hussain, S.; Al-Tamrah, S.A. Effect of potassium permanganate on morphological, structural and electro-optical properties of graphene oxide thin films. *Arab. J. Chem.* **2020**, *14*, 102953. [CrossRef]
13. Tang, L.; Dang, J.; He, M.; Li, J.; Kong, J.; Tang, Y.; Gu, J. Preparation and properties of cyanate-based wave-transparent laminated composites reinforced by dopamine/POSS functionalized Kevlar cloth. *Compos. Sci. Technol.* **2019**, *169*, 120–126. [CrossRef]
14. Pandey, J.; Nagarajan, V.; Mohanty, A.K.; Misra, M. Commercial potential and competitiveness of natural fiber composites. In *Biocomposites*; Elsevier: Amsterdam, The Netherlands, 2015; pp. 1–15.
15. Ahmad, A.; Mubharak, N.M.; Naseem, K.; Tabassum, H.; Rizwan, M.; Najda, A.; Hussain, S. Recent advancement and development of chitin and chitosan-based nanocomposite for drug delivery: Critical approach to clinical research. *Arab. J. Chem.* **2020**, *13*, 8935–8964. [CrossRef]
16. Reis, P.N.; Neto, M.A.; Amaro, A.M. Effect of the extreme conditions on the tensile impact strength of GFRP composites. *Compos. Struct.* **2018**, *188*, 48–54. [CrossRef]
17. Mustafa, E.H.B.; Dyadyura, K.; Jan, V.; Harničárová, M.; Zajac, J.; Modrák, V.; Pandová, I.; Vrábel, P.; Nováková-Marcinčínová, E.; Pavelek, Z. *Manufacturing Technology of Composite Material Structure*; Sudan University of Science and Technology: Khartoum, Sudan, 2017.
18. Xiong, J.; Du, Y.; Mousanezhad, D.; Asl, M.E.; Norato, J.; Vaziri, A. Sandwich structures with prismatic and foam cores: A review. *Adv. Eng. Mater.* **2019**, *21*, 1800036. [CrossRef]
19. Kwonpongsagoon, S.; Jareemit, S.; Kanchanapiya, P. Environmental impacts of recycled nonmetallic fraction from waste printed circuit board. *Int. J. Geomate* **2017**, *12*, 8–14. [CrossRef]
20. Qi, L.; Ju, L.; Zhou, J.; Li, S.; Zhang, T.; Tian, W. Tensile and fatigue behavior of carbon fiber reinforced magnesium composite fabricated by liquid-solid extrusion following vacuum pressure infiltration. *J. Alloys Compd.* **2017**, *721*, 55–63. [CrossRef]
21. Pervaiz, M.; Ahmad, I.; Yousaf, M.; Kirn, S.; Munawar, A.; Saeed, Z.; Rashid, A. Synthesis, spectral and antimicrobial studies of amino acid derivative Schiff base metal (Co, Mn, Cu, and Cd) complexes. *Spectrochim. Acta Part A Mol. Biomol. Spectrosc.* **2019**, *206*, 642–649. [CrossRef]
22. Rajesh Kumar, G.; Hariharan, V.; Saravanakumar, S.S. Enhancing the free vibration characteristics of epoxy polymers using sustainable phoenix Sp. fibers and nano-clay for machine tool applications. *J. Nat. Fibers* **2019**, 1–8. [CrossRef]
23. Kabir, S.M.F.; Mathur, K.; Seyam, A.M. A critical review on 3D printed continuous fiber-reinforced composites: History, mechanism, materials and properties. *Compos. Struct.* **2020**, *232*, 111476. [CrossRef]
24. Anjum, N.; Suresha, B.; Prasad, S.L.A. Influence of Water ageing on mechanical properties of $CaCO_3$ filler filled epoxy resin and sansevieria/carbon fiber reinforced composites. *Open J. Compos. Mater.* **2019**, *9*, 1–20. [CrossRef]
25. Liu, M.; Rohde, B.J.; Krishnamoorti, R.; Robertson, M.L.; Dawood, M. Bond behavior of epoxy resin–polydicyclopentadiene phase separated interpenetrating networks for adhering carbon fiber reinforced polymer to steel. *Polym. Eng. Sci.* **2019**, *60*, 104–112. [CrossRef]

26. de Souza, L.C.; Rodrigues, N.S.; Cunha, D.A.; Feitosa, V.P.; Santiago, S.L.; Reis, A.; Loguercio, A.D.; Paris, T.; Saboia, V.d.A.; Perdigao, J. Two-year clinical evaluation of proanthocyanidins added to a two-step etch-and-rinse adhesive. *J. Dent.* **2019**, *81*, 7–16. [CrossRef] [PubMed]
27. Braga, R.; Magalhaes, P., Jr. Analysis of the mechanical and thermal properties of jute and glass fiber as reinforcement epoxy hybrid composites. *Mater. Sci. Eng. C* **2015**, *56*, 269–273. [CrossRef] [PubMed]
28. Jusoh, M.S.M.; Yahya, M.Y.; Hussein, N.I.S. The effect of fibre layering pattern in resisting bending loads of natural fibre-based hybrid composite materials. In *MATEC Web of Conferences*; EDP Sciences: Ulis, France, 2016.
29. Xavier, J.; Rodney, K.D.; Prakash, S.J. Mechanical characterisation of epoxy polymer composite reinforced with ramie and synthetic fiber. *SSRN Electron. J.* **2019**. [CrossRef]
30. Kader, W.B. Physico-mechanical properties of typha angustata (elephant grass) fiber reinforced thermoplastic composites. *IJAR J.* **2019**. [CrossRef]
31. Dolan, G.K.; Cartwright, B.; Bonilla, M.R.; Gidley, M.J.; Stokes, J.R.; Yakubov, G. Probing adhesion between nanoscale cellulose fibres using AFM lateral force spectroscopy: The effect of hemicelluloses on hydrogen bonding. *Carbohydr. Polym.* **2019**, *208*, 97–107. [CrossRef]
32. Dia, A.; Dieng, L.; Gaillet, L.; Gning, P.B. Damage detection of a hybrid composite laminate aluminum/glass under quasi-static and fatigue loadings by acoustic emission technique. *Heliyon* **2019**, *5*, e01414. [CrossRef]
33. Phuong, P.T.M.; Won, H.J.; Oh, Y.J.; Lee, H.S.; Lee, K.D.; Park, S.Y. The chemistry and engineering of mussel-inspired glue matrix for tissue adhesive and hemostatic. *J. Ind. Eng. Chem.* **2019**, *80*, 749–756. [CrossRef]
34. Polyzois, G.L.; Tarantili, P.A.; Frangou, M.J.; Andreopoulos, A.G. Physical properties of a silicone prosthetic elastomer stored in simulated skin secretions. *J. Prosthet. Dent.* **2000**, *83*, 572–577. [CrossRef]
35. Ab Ghani, M.H.; Ahmad, S. The comparison of water absorption analysis between counterrotating and corotating twin-screw extruders with different antioxidants content in wood plastic composites. *Adv. Mater. Sci. Eng.* **2011**. [CrossRef]
36. Eisenreich, T.J.; Cox, D.S. Modification of the ASTM D 3039 tensile specimen for cryogenic applications. In *Materials*; Springer: Boston, MA, USA, 1992; pp. 437–444.
37. Jayaraman, T.; Murthy, A.P.; Elakkiya, V.; Chandrasekaran, S.; Nithyadharseni, P.; Khan, Z.; Ashokkumar, M. Recent development on carbon based heterostructures for their applications in energy and environment: A review. *J. Ind. Eng. Chem.* **2018**, *64*, 16–59. [CrossRef]
38. Mohsenizadeh, S.; Alipour, R.; Ahmad, Z.; Alias, A. Influence of auxetic foam in quasi-static axial crushing. *Int. J. Mater. Res.* **2016**, *107*, 916–924. [CrossRef]

coatings

MDPI

Article

Synthesis, Characterization and Wettability of Cu-Sn Alloy on the Si-Implanted 6H-SiC

Xiang Zhao Zhang [1], Pu Hao Xu [1], Gui Wu Liu [1,2,*], Awais Ahmad [3], Xiao Hui Chen [1,2], Ya Long Zhu [1], Asma Alothman [4,*], Shahid Hussain [1,*] and Guan Jun Qiao [1]

1 School of Materials Science and Engineering, Jiangsu University, Zhenjiang 212013, China; xzzhang2018@ujs.edu.cn (X.Z.Z.); 13667004282@163.com (P.H.X.); chenxiaohui058@163.com (X.H.C.); 18852897590@163.com (Y.L.Z.); gjqiao@ujs.edu.cn (G.J.Q.)
2 School of Mechanical and Electrical Engineering, Xinyu University, Xinyu 338000, China
3 Department of Chemistry, The University of Lahore, Lahore 54590, Pakistan; awaisahmed@gcuf.edu.pk
4 Department of Chemistry, College of Science, King Saud University, Riyadh 11451, Saudi Arabia
* Correspondence: gwliu76@ujs.edu.cn (G.W.L.); aaalothman@ksu.edu.sa (A.A.); shahid@ujs.edu.cn (S.H.)

Received: 22 August 2020; Accepted: 14 September 2020; Published: 21 September 2020

Abstract: The wettability of the metal/SiC system is not always excellent, resulting in the limitation of the widespread use of SiC ceramic. In this paper, three implantation doses of Si ions (5×10^{15}, 1×10^{16}, 5×10^{16} ions/cm^2) were implanted into the 6H-SiC substrate. The wetting of Cu-(2.5, 5, 7.5, 10) Sn alloys on the pristine and Si-SiC were studied by the sessile drop technique, and the interfacial chemical reaction of Cu-Sn/SiC wetting couples was investigated and discussed. The Si ion can markedly enhance the wetting of Cu-Sn on 6H-SiC substrate, and those of the corresponding contact angles (θ) are raised partly, with the Si ion dose increasing due to the weakening interfacial chemical reactions among four Cu-Sn alloys and 6H-SiC ceramics. Moreover, the θ of Cu-Sn on (Si-)SiC substrate is first decreased and then increased from ~62° to ~39°, and ~70° and ~140°, with the Sn concentration increasing from 2.5%, 5% and 7.5% to 10%, which is linked to the reactivity of Cu-Sn alloys and SiC ceramic and the variation of liquid-vapor surface energy. Particularly, only a continuous graphite layer is formed at the interface of the Cu-10Sn/Si-SiC system, resulting in a higher contact angle (>40°).

Keywords: 6H-SiC; Cu-Sn alloy; ion implantation; wettability; interface

1. Introduction

Silicon carbide (SiC) has been widely applied in the field of electronics industries and metal-ceramic composites, due to its sublime properties such as high strength, high modulus, high melting point and erosion resistance [1–3]. In fact, good wettability between liquid metals and SiC ceramic plays a vital role in these areas. Up to now, two critical challenges for the SiC applications are the low wettability and the undesired interfacial reaction of metals and SiC ceramic [4]. Commonly, the equilibrium value of θ, used to explore the wetting behavior of liquid on a flat and chemically homogeneous solid surface, observes the traditional Young's equation [5] $\cos\theta = \sigma_{SV} - \sigma_{SL}/\sigma_{LV}$ (where σ_{SV} and σ_{LV} describe the surface tension of the solid and liquid, respectively, and the σ_{SL} defines the solid/liquid interfacial energy). Based on this equation, two leading technologies can be performed to reduce the θ of metal on the SiC substrate.

One is increasing the σ_{SV} by changing the SiC surface (i.e., the ion implantation, Ref. [6] sintering metallization [7,8] and plasma pulses [9]. Compared with other surface modification technologies, ion implantation is a non-thermal and non-equilibrium process, where no new interface is introduced and the surface crystal structure can be altered in this process [10]. Recently, our group has proved that Mo, Ref. [11] Pd [12] and Si [13] ion implantation can increase the σ_{SV} of 6H-SiC monocrystal substrate by producing lattice imperfections and point defects, and thus the ion implantation is regarded as one of the key factors affecting the wettability of Al, Ref. [13] Al-Cu and [13] Al-Si-(Cu, Mg or Zn) [6,12] on SiC monocrystal substrate. For instance, Zhu et al. [13] studied the wetting of Al-Cu/Si-implanted SiC wetting couples, indicating that the contact angles of Al-Cu alloys on Si-SiC substrates were raised partly with an increase of the Si ion dose, and the Si ion can evidently improve the wetting of Cu-Al/SiC when the content of Al is no more than 42.9% in Cu.

Another is adding elements such as (Ti, Ref. [14,15] Cr, Ref. [16] Ni [17]) into the metal to reduce the σ_{SL} by enhancing the interfacial interactions of metal/SiC systems and/or to simultaneously reduce the σ_{LV}. For the pure Cu/SiC system, a high contact angle was detected due to the formation of the graphite layer at the wetting interface [18]. Various elements such as Al [13], Si [18], and Zr [19] were added into Cu to adjust the wettability of the Cu/SiC system. As reported by Zhou et al. [19], excellent wettability ($\theta \approx 5°$) was observed in the Cu-Zr/SiC system at 1200 °C, and the interfacial reaction layer was transformed from a graphite layer to metallic ZrC and Zr_2Si compounds after adding Zr into Cu. Furthermore, the other elements added into the pure metal can also bring about the changes of heat transfer performance compared with the pure metal [20–22], which can significantly affect the thermodynamic and kinetics of the chemical reaction between molten metal and ceramic during the wetting process. Therefore, the wettability of metal/ceramic system is further influenced accordingly.

As discussed above, both the surface modification of SiC and other metal element additions are the most promising methods to improve the wetting of metal/SiC systems. However, the influence of Sn addition on the wettability of the Cu/6H-SiC system was rarely reported, especially for the 6H-SiC ceramic after Si ion implantation. Herein, Si ion was implanted into the 6H-SiC monocrystal substrate, and the wettability of Cu-(2.5, 5, 7.5, 10) Sn (all in at.% in this text) alloys on 6H-SiC monocrystal substrate were investigated. We found that the Si ion can markedly enhance the wetting of Cu-Sn on 6H-SiC substrate, and the θ of Cu-Sn on (Si-)SiC substrate basically decreases first, and then increases alongside the Sn concentration.

2. Experimental Details

The double polished C-terminated 6H-SiC monocrystal (CAS: 409-21-2), with sizes of 10 mm × 10 mm × 0.33 mm, was employed as the wetting substrate. The Cu-(2.5, 5, 7.5, 10)Sn alloys used for the wetting experiments were fabricated by arc melting with non-consumable tungsten electrode under a purified Ar (99.99%, CAS: 7440-37-1) atmosphere, and being remelted five times to ensure a uniform composition, combining appropriate amounts of the Cu sheet (99.9 wt.% purity, CAS: 7440-50-8) and Sn granular (99.8 wt.% purity, CAS: 7440-31-5). In order to prevent the oxygen contamination, a Ti getter (CAS: 7440-32-6) was melted first, before the Cu-Sn alloy. The microstructure and phase composition of a typical Cu-7.5Sn alloy were analyzed and identified by scanning electron microscopy (SEM, FEI NovaNano450, back-scattered electron (BSE) mode) and X-ray diffraction (XRD). The scanning range of XRD examination was 20°–90°, with a speed of 5°/min. The Cu-Sn alloys featured a wire electrode cutting into the block, with sizes of 3 mm × 3 mm × 3 mm, which were carefully cleaned in acetone and ethanol successively before wetting experiments. The ion implantation was carried out in an ion implanter (MEVVA-36), and three doses of Si ions (5×10^{15}, 1×10^{16}, 5×10^{16} ions/cm^2) were implanted into the 6H-SiC substrate at 20 keV at an ambient temperature under the vacuum of ~5×10^{-3} Pa.

The sessile drop tests of Cu-Sn alloys on a Si-implanted SiC substrate were performed at a contact angle computing instrument (OCA15LHT-SV, Dataphysics, Filderstadt, Germany). The wetting samples were performed at 1373 K for 240 min in a vacuum of ~6×10^{-4} Pa, with a heating rate of 5 K/min. After wetting experiments, the cross-sectioned wetting systems were polished using diamond polishing fluid, and the interfacial microstructure was observed by SEM (BSE mode), coupled with energy dispersive spectroscopy (EDS). It was noted that the images obtained from BSE mode can be used to evidently present the composition contrast for the polished samples, while the images obtained from the secondary electron (SE) model mainly reflect the contrasting information of the surface geometry.

3. Results and Discussion

Figure 1 shows the typical microstructure of Cu-7.5Sn alloy combined with its XRD pattern. The Cu-7.5Sn alloy presents a typical dendritic microstructure, and is composed of two phases (dark and grey phase). Based on the EDS analysis, both phases are mainly Cu-rich solid solutions, where the Sn concentration in the grey phase is higher than that in the dark phase. According to the Cu-Sn binary phase diagram [23], the dark and grey phases can be confirmed as a Cu-rich solid solution (Cu) and (Cu) plus Cu-Sn compounds. However, only the (Cu) phase emerges in the XRD pattern due to the low content of the Cu-Sn compound (Figure 1b).

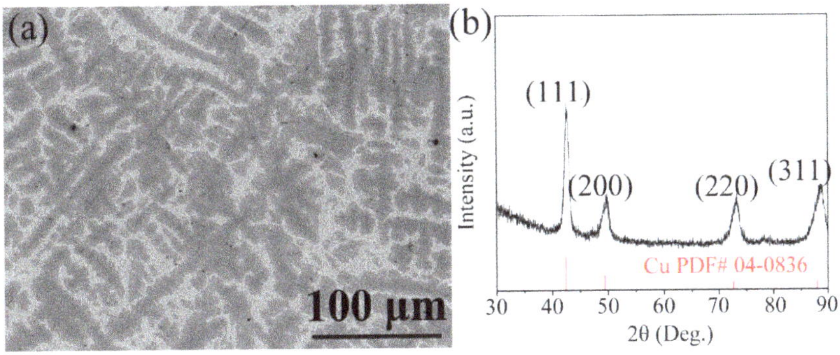

Figure 1. (**a**) Back-scattered electron (BSE) image; and (**b**) X-ray diffraction (XRD) pattern of Cu-7.5Sn alloy.

Figure 2 shows the wetting curves of molten Cu-(2.5, 5, 7.5, 10) Sn alloys on the (Si-)SiC substrates at 1373 K. In Figure 2, two Y-axes are drawn to present the information of the contact angle (the *Y*-axis on the left). The temperature (the *Y*-axis on the right) in Figure 2a–d and the line corresponding to the wetting temperature is 1373 K, showing a straight line in the figure. For the Cu-Sn/pristine SiC systems, the contact angle (θ) is first decreased, and then increased from ~62° to ~39° and ~70°, with the Sn concentration rising from 2.5% to 5% and 7.5%. In particular, a non-wetting phenomenon with a high contact angle of ~140° is observed in the Cu-10Sn/SiC system. On the other hand, the Si ion implantation has a pleasurable influence on the wetting of Cu-Sn/SiC couples. The θ of Cu-Sn on the Si-SiC substrate decreases markedly after Si ion implantation, especially for the Cu-10Sn/Si-SiC systems transforming from non-wetting to wetting. However, the θ of Cu-Sn/Si-SiC systems is increased partly with the Si ion dose increase, rising up to 5×10^{16} ions/cm^2. Those experimental results illustrate that the Si ion can blatantly improve the wettability of Cu-Sn/SiC systems. In contrast, a higher Si ion dose can weaken this phenomenon, showing an increased contact angle. In these cases, the σ_{SV} keeps invariant (before ion implantation), or is increased due to the presence of the lattice imperfection and point defects after ion implantation [13], while the σ_{LV} is decreased with the Sn concentration increasing, according to Amore's results [24]. Theoretically speaking, the increased σ_{SV} or/and decreased σ_{LV} can reduce the wettability of Cu-Sn/(Si-)SiC systems based on Young's equation. Thus, the abnormally increased contact angle with the rise of the Si ion dose can be ascribed to the increased σ_{LV} derived from the decreasing interfacial interactions between Cu-Sn alloys and Si-SiC ceramic.

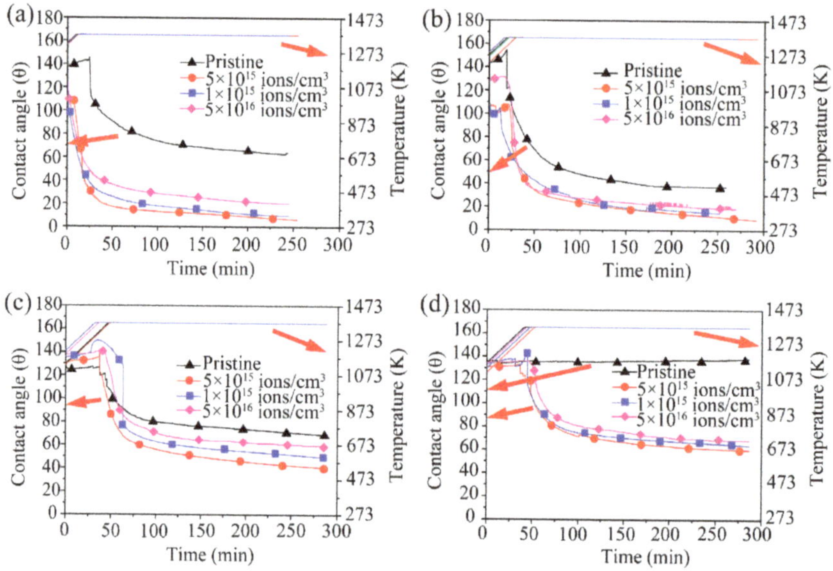

Figure 2. Curves of Cu-Sn/SiC systems before and after Si ion implantation (5 × 10^{15}, 1 ×10^{16} and 5 × 10^{16} ions/cm^2) at 1373 K: (**a**) Cu-2.5Sn; (**b**) Cu-5Sn; (**c**) Cu-7.5Sn; and (**d**) Cu-10Sn.

Figure 3 shows the interfacial BSE images of Cu-2.5Sn/SiC couples before and after Si ion implantation. From Figure 3a, a prominent graphite layer and numerous Cu_xSi_y compounds are formed at the surface of Cu-2.5Sn solidified drop due to the serious chemical reaction between Cu and SiC, which is also observed in the pure Cu and SiC systems [13]. After Si ion implantation, a visible graphite film can be observed at the triple line region, and its width decreases with the Si ion dose increasing (Figure 3b–d). Moreover, a laminated graphite layer is located on the upper part of the Cu-2.5Sn drop after Si ion implantation, as shown in Figure 3f–h. The laminated graphite layer gradually becomes thin, with the Si ion implantation dose rising, indicating that the interfacial interactions are weakened, resulting in an increase of σ_{SL} and a contact angle, either more or less (as shown in Figures 2 and 3i–l). Meanwhile, the 6H-SiC substrate is more or less consumed with the formation of uneven scallops (Figure 3e–h). Similarly, the solidified Cu-2.5Sn drop consists of a dark phase marked A and a small amount of grey phase marked B, as shown in the insert image in Figure 3f. Based on the EDS results, the chemical compositions of the two phases are 92.09Cu + 5.90Si + 2.01Sn and 66.48Cu + 33.52Sn, respectively. According to the Cu-Si and Cu-Sn binary phase diagrams [23,25], the dark and grey phases can be confirmed as Cu-rich solid solutions, containing few Cu-Si and Cu-Sn compounds, respectively.

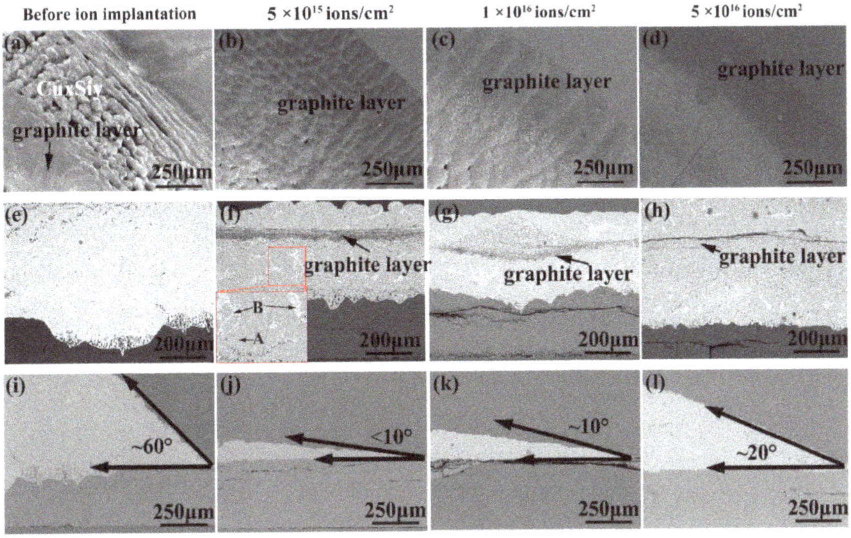

Figure 3. BSE images of Cu-2.5Sn/(Si-)SiC systems (**a**,**e**,**i**) before and (**b**–**d**,**f**–**h**,**j**–**l**) after Si ion implantation: (**a**–**d**) at the triple line region, cross-section (**e**–**h**) at the central interfaces and (**j**–**l**) at the triple line region.

Figures 4 and 5 exhibit the top-view and interfacial BSE images of Cu-(5, 7.5)Sn/Si-SiC systems with different Si ion implantation doses, respectively. From Figure 4a,c, a thin graphite film emerges at the triple line region of Cu-(5, 7.5)Sn/Si-SiC systems after the Si ion implantation dose of 5×10^{15} ions/cm^2. However, a higher Si ion implantation dose (5×10^{16} ions/cm^2) cannot induce the formation of the graphite layer in Cu-(5, 7.5)Sn/Si-SiC systems (Figure 4b,d).

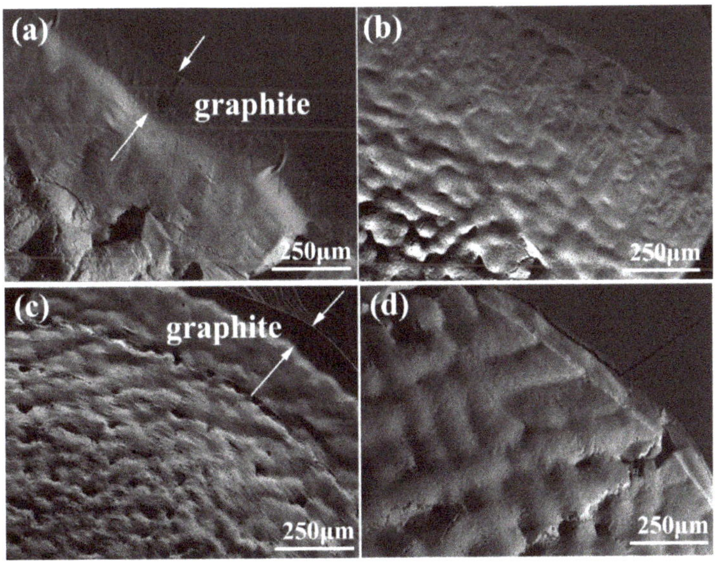

Figure 4. BSE images of (**a**,**b**) Cu-5Sn/Si-SiC and (**c**,**d**) Cu-7.5Sn/Si-SiC systems at the triple line region after Si ion implantation of (**a**,**c**) 5×10^{15} and (**b**,**d**) 5×10^{16} ions/cm^2.

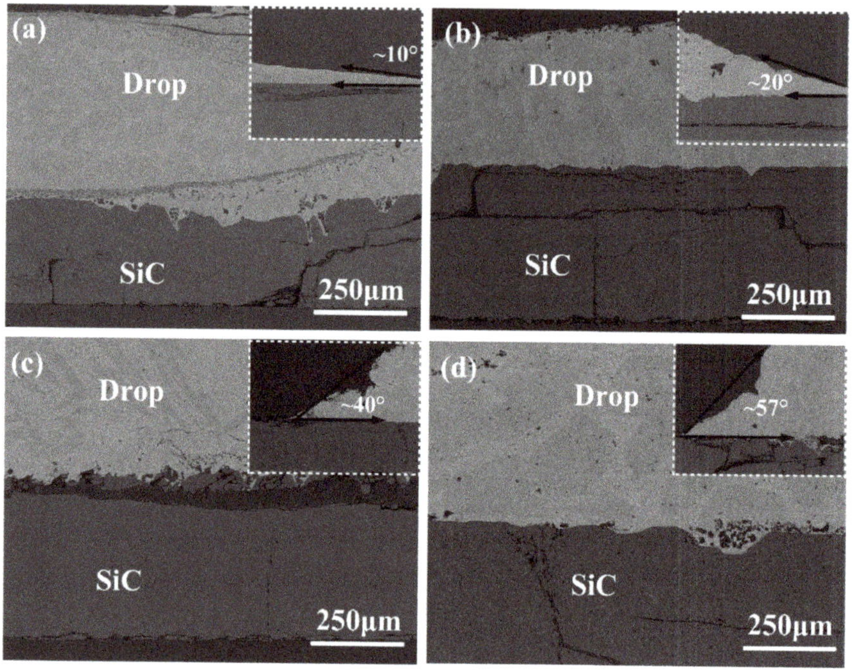

Figure 5. Interfacial BSE images of (**a**,**b**) Cu-5Sn/Si-SiC and (**c**,**d**) Cu-7.5Sn/Si-SiC systems at the central interfaces after Si ion implantation of (**a**,**c**) 5×10^{15} and (**b**,**d**) 5×10^{16} ions/cm^2.

Compared to the Cu-2.5Sn/Si-SiC system, the Cu-(5, 7.5) Sn/Si-SiC presents relatively weak interfacial reactions, and the formed graphite layer moves closer to the SiC substrate (Figure 5). This phenomenon is mainly attributed to the low activity of Cu, due to the relatively high concentration of the Sn element. As we know, the wettability of Cu-Sn/Si-SiC couple can be comprehensively determined by the σ_{SL} originating from interfacial interactions, and the σ_{LV} on the condition of the same substrate. In these cases, the increase in σ_{SL} and the decrease of σ_{LV} correspondingly dominates the Cu-5Sn/Si-SiC and Cu-7.5Sn/Si-SiC systems, so the θ of Cu-Sn/Si-SiC decreases firstly, and then increases, with the Sn concentration rising from 2.5% to 7.5%. Similarly, two kinds of phases (Figure 5) are also observed in the solidified Cu-(5, 7.5) Sn alloy drops, and the content of grey phase increases gradually (Figures 3–5). Accordingly, the Sn concentration in the grey phase is also raised.

Figure 6 exhibits the interfacial microstructure of Cu-10Sn/Si-SiC systems after Si ion implantation. Compared with Cu-(2.5, 5, 7.5) Sn/Si-SiC systems, the Cu-10Sn/Si-SiC presents relatively high contact angles (Figure 2). However, it is puzzling that the molten Cu-10Sn alloy has the lowest surface energy among the four Cu-Sn alloys, and thus theoretically the Cu-10Sn/Si-SiC systems should present excellent wettability, according to Young's equation. From Figure 6, only a continued graphite layer is closely attached to the SiC substrate, which is due to the fact that the high Sn concentration in Cu alloys can reduce the activity of Cu element, and thus the chemical reaction is limited at the interface. According to the reactive production control (RPC) model [26,27], wetting in the reactive system is predominated by the final interfacial production at the interface and triple region. Because the molten drop has a poor wettability on the graphite layer, there is relatively poor wettability of Cu-10Sn on Si-SiC substrates. Moreover, the interfacial chemical reaction between Cu-10Sn and Si-SiC becomes weaker and weaker as the Si ion dose rises, resulting in a thinner graphite layer and higher contact angle.

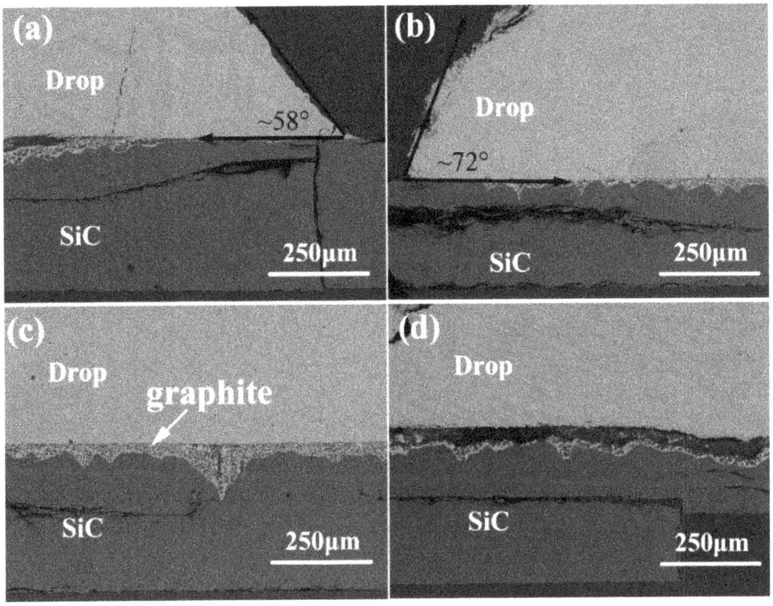

Figure 6. Interfacial BSE images of Cu-10Sn/SiC systems after the Si ion implantation of (**a**,**c**) 5×10^{15} and (**b**,**d**) 5×10^{16} ions/cm^2.

4. Conclusions

The wetting of molten Cu-(2.5, 5, 7.5, 10) Sn on Si-implanted SiC substrates was studied, and the effects of Si ion doses and Sn concentration in Cu alloys on the wettability were analyzed. The Si ion can markedly enhance the wetting of Cu-Sn/SiC systems; however, the contact angle of Cu-Sn on Si-SiC substrate partly increases, with the Si implantation dose increasing from 5×10^{15} to 5×10^{16} ions/cm^2. The wetting of Cu-Sn/(Si-)SiC systems is closely related to the increasing solid-liquid interfacial energy originated from the decreasing interfacial chemical reaction and the decreasing liquid-vapor surface energy, with the Sn concentration increasing from 2.5%, 5% and 7.5% to 10%. In particular, a higher Sn concentration of ≥7.5% can obviously reduce the activity of Cu, resulting in a relatively weak interfacial reaction and a higher contact angle (>40°). The above work and conclusions provide a novel way to change the surface properties of ceramic and improve the wettability of the metal/SiC ceramic system, which can further expand the prospective application area of SiC ceramic. However, changes of the surface state of SiC after ion implantation in an atomic scale, i.e., the ion site in SiC lattice and the lattice distortion of SiC, were less straightforward. In the future, the first-principle calculations based on density functional theory (DFT) and *Ab-initio* molecular dynamics (AIMD) simulation can be performed to study the changes in atomic scale after the ion implantation.

Author Contributions: Conceptualization, X.Z.Z. and A.A. (Awais Ahmad); methodology, A.A. (Asma Alothman), P.H.X., and X.Z.Z.; software, S.H.; validation, A.A. (Awais Ahmad) and S.H.; formal analysis, S.H.; investigation, G.J.Q.; resources, A.A. (Asma Alothman).; data curation, G.W.L.; writing—original draft preparation, A.A. (Awais Ahmad).; writing—review and editing, visualization, X.H.C.; supervision, S.H.; project administration, Y.L.Z.; funding acquisition, X.H.C. All authors have read and agreed to the published version of the manuscript.

Funding: This work was funded by the Researchers Supporting Project Number (RSP-2020/243) King Saud University, Riyadh, Saudi Arabia; National Key R&D Program of China (2017YFB0310400); the National Natural Science Foundation of China (51572112); the National Natural Science Foundation of Jiangxi (20192BAB206009, 20192BAB206030); the Key Research and Development Plan (BE2019094); the Six Talent Peaks Project (TD-XCL-004); 333 talents project (BRA2017387); and the Qing Lan Project ([2016]15) of Jiangsu Province providing financial support.

Conflicts of Interest: The authors declare no conflict of interest.

References

1. Fitriani, P.; Septiadi, A.; Hyuk, J.D.; Yoon, D.H. Joining of SiC monoliths using a thin MAX phase tape and the elimination of joining layer by solid-state diffusion. *J. Eur. Ceram. Soc.* **2018**, *38*, 3433–3440. [CrossRef]
2. Liu, J.W.; Zhou, X.B.; Tatarko, P.; Yuan, Q.; Huang, Q. Fabrication, microstructure, and properties of SiC/Al_4SiC_4 multiphase ceramics via an in-situ formed liquid phase sintering. *J. Adv. Ceram.* **2020**, *9*, 193–203. [CrossRef]
3. Zhang, W.; Yamashita, S.; Kita, H. Progress in tribological research of SiC ceramics in unlubricated sliding—A review. *Mater. Des.* **2020**, *190*, 108528. [CrossRef]
4. Pervaiz, M.; Ahmad, I.; Yousaf, M.; Kirn, S.; Munawar, A.; Saeed, Z.; Rashid, A. Synthesis, spectral and antimicrobial studies of amino acid derivative Schiff base metal (Co, Mn, Cu, and Cd) complexes. *Spectrochim. Acta Part A Mol. Biomol. Spectrosc.* **2019**, *206*, 642–649.
5. Eustathopoulos, N.; Nicholas, M.G.; Drevet, B. (Eds.) *Wettability at High Temperatures*; Elsevier Science Ltd.: Kidlington, UK, 1999; Volume 3.
6. Huang, Z.K.; Liu, H.; Liu, G.W.; Wang, T.T.; Zhang, X.Z.; Wu, J.; Wan, Y.G.; Qiao, G.J. Influences of surface polarity and Pd ion implantation on the wettability of Al-12Si(-2Mg)/SiC systems. *Mater. Chem. Phys.* **2018**, *211*, 329–334.
7. Song, Y.Y.; Liu, D.; Hu, S.P.; Song, X.G.; Lei, Y.Z.; Cao, J. Brazing of metallized SiC ceramic to GH99 superalloy using graphene nanoplatelets reinforced AgCuTi composite filler. *Ceram. Int.* **2019**, *45*, 8962–8970. [CrossRef]
8. Kashif, M.; Ngaini, Z.; Harry, A.V.; Vekariya, R.L.; Ahmad, A.; Zuo, Z.; Alarifi, A. An experimental and DFT study on novel dyes incorporated with natural dyes on titanium dioxide (TiO_2) towards solar cell application. *Appl. Phys.* **2020**, *126*, 1–13. [CrossRef]
9. Barlak, M.; Piekoszewski, J.; Sartowska, B.; Waliś, L.; Starosta, W.; Kierzek, J.; Pochrybniak, C.; Kowalska, E. Wettability of carbon and silicon carbide ceramics induced by their surface alloying with Zr and Cu elements using high intensity pulsed plasma beams. *Nukleonika* **2012**, *57*, 477–483.
10. McCafferty, E. Effect of Ion Implantation on the Corrosion Behavior of Iron, Stainless Steels, and Aluminum—A Review. *Corrosion* **2001**, *57*, 1011–1029. [CrossRef]
11. Zhao, S.T.; Valenza, F.; Liu, G.W.; Muolo, M.L.; Qiao, G.J.; Passerone, A. Surface characterization of Mo-implanted 6H-SiC by high temperature non-reactive wetting tests with the Ni-56Si alloy. *Ceram. Int.* **2014**, *40*, 7227–7234. [CrossRef]
12. Huang, Z.K.; Xu, W.L.; Liu, G.W.; Wang, T.T.; Zhang, X.Z.; Qiao, G.J. Wetting and interfacial behavior of molten Al-Si alloys on SiC monocrystal substrates: Effects of Cu or Zn addition and Pd ion implantation. *J. Mater. Sci. Mater. Electron.* **2018**, *29*, 17416–17424. [CrossRef]
13. Zhu, Y.L.; Zhang, M.F.; Zhang, X.Z.; Huang, Z.K.; Liu, G.W.; Qiao, G.J. Wetting and interfacial behavior of Cu–Al/SiC systems: Influences of Si ion implantation and Al concentration. *J. Alloys Compd.* **2020**, *824*, 153972. [CrossRef]
14. Fu, W.; Song, X.G.; Tian, R.C.; Lei, Y.Z.; Long, W.M.; Zhong, S.J.; Feng, J.C. Wettability and joining of SiC by Sn-Ti: Microstructure and mechanical properties. *J. Mater. Sci. Technol.* **2020**, *40*, 15–23. [CrossRef]
15. Yang, J.; Huang, J.H.; Ye, Z.; Chen, S.H.; Ji, R.; Zhao, Y. Influence of interfacial reaction on reactive wettability of molten Ag-Cu-X wt.%Ti filler metal on SiC ceramic substrate and mechanism analysis. *Appl. Surf. Sci.* **2018**, *436*, 768–778. [CrossRef]
16. Hussain, S.; Yang, X.; Aslam, M.K.; Shaheen, A.; Javed, M.S.; Aslam, N.; Qiao, G. Robust TiN nanoparticles polysulfide anchor for Li–S storage and diffusion pathways using first principle calculations. *Chem. Eng. J.* **2020**, *391*, 123595.
17. Wang, T.T.; Yang, Y.; Ren, Y.B.; Zhu, D.Y.; Zhang, T. Reactive wetting of Ni-Si alloys on graphite substrates: Effects of Si and Ni. *RSC Adv.* **2015**, *5*, 90866–90870. [CrossRef]
18. Rado, C.; Drevet, B.; Eustathopoulos, N. The role of compound formation in reactive wetting: The Cu/SiC system. *Acta Mater.* **2000**, *48*, 4483–4491. [CrossRef]
19. Zhou, B.F.; Wang, J.F.; Feng, K.Q. Study on the wetting interface of Zr-Cu alloys on the SiC ceramic surface. *RSC Adv.* **2020**, *10*, 3487–3492. [CrossRef]
20. Sarafraz, M.M.; Safaei, M.R.; Goodarzi, M.; Yang, B.; Arjomandi, M. Heat transfer analysis of Ga-In-Sn in a compact heat exchanger equipped with straight micro-passages. *Int. J. Heat Mass Transf.* **2019**, *139*, 675–684. [CrossRef]

21. Sarafraz, M.M.; Jafarian, M.; Arjomandi, M.; Nathan, G.J. Potential use of liquid metal oxides for chemical looping gasification: A thermodynamic assessment. *Appl. Energy* **2017**, *195*, 702–712. [CrossRef]
22. Sarafraz, M.M.; Tran, N.N.; Pourali, N.; Rebrov, E.V.; Hessel, V. Thermodynamic potential of a novel plasma-assisted sustainable process for co-production of ammonia and hydrogen with liquid metals. *Energy Convers. Manag.* **2020**, *210*, 112709. [CrossRef]
23. Okamoto, H. Supplemental Literature Review of Binary Phase Diagrams: Ag-Ho, Ag-Tb, Ag-Y, Cd-Na, Ce-Sn, Co-Dy, Cu-Dy, Cu-Sn, Ir-Pt, Mg-Pb, Mo-Ni, and Sc-Y. *J. Phase Equilibria Diffus.* **2014**, *35*, 208–219. [CrossRef]
24. Amore, S.; Ricci, E.; Lanata, T.; Novakovic, R. Surface tension and wetting behaviour of molten Cu–Sn alloys. *J. Alloys Compd.* **2008**, *452*, 161–166. [CrossRef]
25. Okamoto, H. Supplemental literature review of binary phase diagrams: Ag-Ni, Ag-Zr, Au-Bi, B-Ni, Co-Sb, Cu-Mn, Cu-Si, Cu-Zn, Fe-Zr, Li-Sb, Mg-Pu, and Si-Zr. *J. Phase Equilibria Diffus.* **2018**, *39*, 87–100. [CrossRef]
26. Eustathopoulos, N. Progress in understanding and modeling reactive wetting of metals on ceramics. *Curr. Opin. Solid State Mater. Sci.* **2005**, *9*, 152–160. [CrossRef]
27. Eustathopoulos, N. Wetting by liquid metals—Application in materials processing: The contribution of the grenoble group. *Metals* **2015**, *5*, 350–370. [CrossRef]

coatings

MDPI

Article

Radical Scavenging and Catalytic Activity of Fe-Cu Bimetallic Nanoparticles Synthesized from *Ixora finlaysoniana* Extract

Umer Younas [1,*], Syed Taimoor Hassan [1], Faisal Ali [1], Faiza Hassan [1], Zohaib Saeed [2], Muhammad Pervaiz [1], Safia Khan [3,*], Fakiha Tul Jannat [4], Sidra Bibi [3], Ayesha Sadiqa [1], Zahid Ali [1,5], Shahid Iqbal [6,*], Ayman A. Ghfar [7], Mohamed Ouladsmane [7], Murefah Mana AL-Anazy [8] and Shafaqat Ali [9,10,*]

1 Department of Chemistry, The University of Lahore, Lahore 54590, Pakistan; syedtaimoor91@gmail.com (S.T.H.); faisal.alimsc@gmail.com (F.A.); faiza.hassan@chem.uol.edu.pk (F.H.); mpbhatti786@gmail.com (M.P.); a.sadiqa046@gmail.com (A.S.); zahidchemist521@gmail.com (Z.A.)
2 Department of Chemistry, Government College University, Lahore 54000, Pakistan; zohaibsaeed46@gmail.com
3 Department of Chemistry, Quaid-I-Azam University Islamabad, Islamabad 45320, Pakistan; sidrabibi@chem.qau.edu.pk
4 Department of Chemistry, Government College University, Faisalabad 38000, Pakistan; jannat98@yahoo.com
5 Laboratory of Functional Material, College of Material Science and Engineering, Beijing University of Chemical and Technology, Beijing 100029, China
6 Department of Chemistry, University of Education Jauharabad Campus, Jauharabad 41200, Pakistan
7 Advanced Materials Research Chair, Chemistry Department, College of Science, King Saud University, Riyadh 11451, Saudi Arabia; aghafr@ksu.edu.sa (A.A.G.); mouladsmane@ksu.edu.sa (M.O.)
8 Department of Chemistry, College of Science, Princess Nourah Bint Abdulrahman University, Riyadh 11671, Saudi Arabia; mmalanazy@pnu.edu.sa
9 Department of Environmental Science and Engineering, Government College University, Faisalabad 38000, Pakistan
10 Department of Biological Sciences and Technology, China Medical University, Taichung 40402, Taiwan
* Correspondence: umer.younas@chem.uol.edu.pk (U.Y.); safiakhan@chem.qau.edu.pk (S.K.); ranashahid313@gmail.com (S.I.); shafaqataligill@yahoo.com (S.A.)

Citation: Younas, U.; Hassan, S.T.; Ali, F.; Hassan, F.; Saeed, Z.; Pervaiz, M.; Khan, S.; Jannat, F.T.; Bibi, S.; Sadiqa, A.; et al. Radical Scavenging and Catalytic Activity of Fe-Cu Bimetallic Nanoparticles Synthesized from *Ixora finlaysoniana* Extract. *Coatings* **2021**, *11*, 813. https://doi.org/10.3390/coatings11070813

Academic Editor: Natalia V. Kamanina

Received: 29 May 2021
Accepted: 15 June 2021
Published: 5 July 2021

Abstract: Iron–copper bimetallic nanoparticles (Fe-Cu BNPs) were prepared via a green synthesis route. *Ixora finlaysoniana* has been used in this study as a capping and stabilizing agent in the modification of Fe-Cu BNPs. As-synthesized BNPs were characterized using different techniques including UV/Vis spectrophotometry, FTIR, XRD and SEM. A particle size analyzer and SEM studies indicated the particle size to be in the range of 50–200 nm. In addition, degradation of MB dye in an aqueous system and radical-scavenging potential in a DPPH assay were also examined using BNPs. Methylene blue dye degradation in 17 min was monitored with UV/Vis spectrophotometry, which exhibited the efficiency of Fe-Cu BNPs. Bimetallic nanoparticles were also found to be efficient in neutralizing DPPH free radicals. Furthermore, kinetic studies of both dye degradation and radical scavenging potential are reported in this article. Subsequently, Fe-Cu BNPs synthesized via a green and sustainable method can be employed for dye degradation and free radical-scavenging activities.

Keywords: bimetallic nanoparticles; kinetics; antioxidant studies; catalytic activity

1. Introduction

Water pollution has always been a great concern over the years due to a decrease in water levels and an increase in pollution over time [1,2]. Among the many pollutants, dyes are one of the major contributors to water pollution [3,4]. Both cationic and anionic dyes as pollutants are a real threat to human life. Textile dyes can cause different diseases, such as cancer [5], cardiac problems [6] and lethal paraphernalia on cells [7]. These dyes stop sunlight reaching marine life, which results in a decrease in photosynthesis processes under water [8]. Therefore, the removal of such hazardous contaminants from water is of major interest [9]. For this purpose, a multitude of techniques have been employed for dye degradation, among which nanoparticles of various kinds have gained a lot of attention [10,11]. Many methods have been reported for the modification of nanoparticles, such as thermal and photochemical decomposition, chemical reduction, electrochemical reduction, the sol–gel method, sputtering, the micro-emulsion method, the hydrothermal method, the chemical precipitation method, the green method and the biological method, are frequently used routes for their synthesis [12,13]. Initially, the synthesis of monometallic nanoparticles was focused on by scientists and researchers. Recently, bimetallic nanocomposites have attracted attention due to their synergistic effects in many applications, especially dye degradation [14,15]. Osama Eljamal and coworkers reported the synthesis and characterization of iron–copper bimetallic nanoparticles (Fe-Cu BNPs) for an increase in the yield of methane and to enhance biogas production [16]. Recently, Jianlong Wang and Juntao Tang (2020) introduced novel Fe-Cu BNPs by the solvo-thermal method and used these BNPs in the reduction of sulfamethoxazole (a pharmaceutical pollutant) [17]. In addition, Fe-Cu BNPs have been used for the removal of chromium from waste water [18].

Moreover, copper and iron nanoparticles are both commercially important as their annual production amounts to millions of tons and, due to their uses in various industries, their production is likely to increase every year [19,20]. Unfortunately, other methods produce various pollutants as precursors that are highly damaging to the environment [21,22]. Therefore, it is necessary to find alternative ways under the umbrella of green chemistry to meet the demand for these nanoparticles in a better and safer way [18]. In green chemistry, there are various ways by which their production can be achieved, including biological waste [22] and plant extracts [23]. Looking at the vast and easy availability of plants, the plant-mediated method was considered for the production of these Fe-Cu BNPs [23].

Herein, *Ixora finlaysoniana*, also known as jungle flame, was used as a plant source in order to achieve our aim of synthesizing Fe-Cu BNPs in a natural and sustainable way. It is a shrub of the Rubiaceae family normally found in South East Asia, China, India and the Philippines [24].

This family is quite abundant, comprising more than 13,000 species found globally. Due to its medicinal impact, it has been used medically in various parts of the world. Variations in conditions and concentrations were made in order to optimize the most suitable method. This method is both economical and environmentally friendly with the most suitable condition [25]. 2,2-diphenyl-2-picrylhydrazyl hydrate (DPPH) is a nitrogen-centered radical with a maximum absorbance at 517 nm, which is converted to 1,1, diphenyl-2-picryl hydrazine when reacting with electron-accepting species. This hydrogen donation ability leads to the formation of a stable complex of free radicals, resulting in termination of free radical-based reactions [26]. The objectives of the present study were to synthesize Fe-Cu BNPs using a green method followed by characterization using UV/Vis spectrophotometry, XRD, FTIR and SEM. In addition, antioxidant and catalytic activities of the synthesized Fe-Cu BNPs were determined.

2. Materials and Methods

2.1. Chemicals and Reagents

Methylene blue (99.9%) was purchased from Fisher Scientific, Altrincham, UK and sodium borohydride (99%), DPPH (99.5%), $FeSO_4$ (99%) and $CuSO_4$ (99%) were acquired from Sigma-Aldrich, Taufkirchen, Germany and methanol (98%) was used as a solvent and was purchased from Unichem, Wuxi, China. All the chemicals and metal precursors were used as received without any further heating or purification treatment.

2.2. Preparation of Plant Extract

Ixora plants were collected from WAPDA Town, Model Town and Kalma Chowk nurseries of Lahore, Pakistan. After washing, plants were dried in the shade. In order to attain a constant weight, plants were placed in hot air oven for 3 h at 60 °C, followed by cutting and grinding. The powdered plant samples were then mixed with methanol and the extraction of bioactives was achieved using an orbital shaker for 3 h at 150 rpm. The filtrate obtained was dried using a rotary evaporator and stored at −4 °C prior to further use.

2.3. Nanoparticle Synthesis

Salt solution was prepared by mixing $FeSO_4$ and $CuSO_4$ in 25 mL of solvent. The solvent used was a mixture of methanol and water in a 1:3 molar ratio. Similarly, extract solution was prepared by mixing the extract in methanol. Both solutions were mixed at different concentrations to make various ppm solutions, ranging from 50 to 250 ppm. The purpose of making solutions at various concentrations was to find out the best concentration which gives satisfactory results. The best concentration was found to be 200 ppm after obtaining its UV/Vis spectra. The selected concentration was then centrifuged and filtered. The filtered sample containing Fe-Cu BNPs was dried at 65 °C in a vacuum oven for 2 h before further application. Figure 1 shows the Schematic diagram of Fe-Cu BNPs.

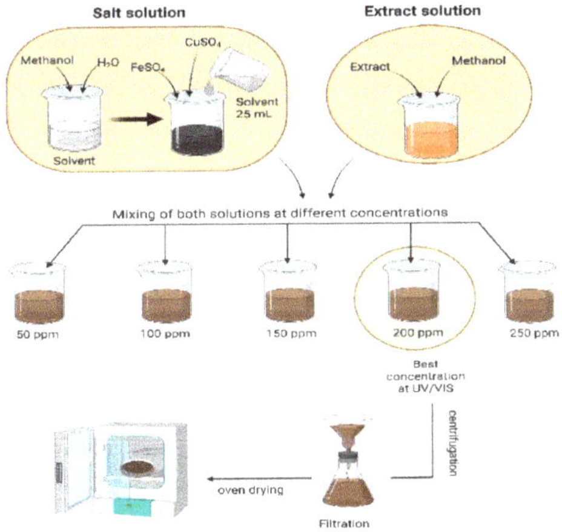

Figure 1. Schematic diagram of Fe-Cu BNPs.

2.4. Characterization of Nanoparticles

Synthesis of Fe-Cu BNPs was immediately confirmed using a UV/Vis spectrophotometer and lambda max was recorded on a CECIL-7400ce UV/Vis spectrophotometer (Cecil Instruments Ltd., Cambridge, UK). Particle size analyzer Lite-sizer 500 software version 1.8.1, Anton Paar, Graz, Austria) with a measurement time of 10 s at 30 °C with a maximum no of runs (60) was used. FTIR spectra of the Fe-Cu BNPs and aqueous plant extracts were recorded on an FTIR spectrophotometer (IR Prestige 21, Shimadzu, Kyoto, Japan). Additionally, the X-ray diffraction (XRD) studies were carried out at a scanning rate of 0.05 min^{-1} using a Bruker D8 Advanced (Bruker, Billerica, MA, USA), equipped with a scintillation counter using Cu Kα radiation (k = 1.5405 Å, nickel filter) at an acceleration voltage of 30 KV NOVA SEM 450 (FEI, Hillsboro, OR, USA) was utilized to obtain SEM images of synthesized BNPs and micrographs were obtained at 3 different magnifications.

2.5. DPPH Radical-Scavenging Potential

The antioxidant capacity of the BNPs was studied through the evaluation of the free radical-scavenging effect on the 1,1-diphenyl-2-picrylhydrazyl (DPPH) radical. The determination was carried out by following an already reported method [27]. Ten milliliters of BNPs (200, 400 and 600 ppm) was added to three separate flasks and 90 mL of distilled water was added followed by the addition of 3.9 mL, 25 mM DPPH methanol solution. The mixture was thoroughly vortexed and kept in the dark for 30 min. The absorbance was measured later, at 515 nm, against a blank of methanol without DPPH. Results were expressed as the percentage of inhibition of the DPPH radical. The percentage of inhibition of the DPPH radical was calculated according to Equation (1).

$$\text{DPPH scavenging effect } \% = [(\text{AD} - \text{AS})/\text{AD}] \times 100 \quad (1)$$

where AD control is the absorbance of DPPH solution without extracts and AS is the absorbance value for the sample.

2.6. Catalytic Activity

The catalytic activity for Fe-Cu BNPs was observed following an already published method [28]. Methylene blue 0.086 mM, $NaBH_4$ 26 mM and Fe-Cu BNPs with a 100 ppm concentration were prepared. In a cuvette of the UV/Vis spectrophotometer, 3 mL methylene blue, which acted as substrate, and 0.4 mL of 26 mM sodium borohydride acting as a reducing agent were mixed. In this solution containing the substrate and reducing agent, 0.5 mL Fe-Cu BNP solution was added, which behaved as a catalyst, and all the observations were recorded at 665 nm, the maximum absorbance (λ_{max}) for methylene blue.

3. Result and Discussion

3.1. UV/Visible Analysis

UV/visible spectrophotometric analysis can be an instant preliminary test for the confirmation of nanoparticle formation, as metallic nanoparticles show absorbance in the UV/Vis region. According to the literature, the absorbance range for iron nanoparticles is 280–350 nm [29], while copper nanoparticles show absorbance from 550–600 nm [30]. During the formation of Fe-Cu BNPs, the shift in absorbance values for iron is 292 nm and for copper it is 594 nm as compared to their monometallic counterparts that show absorbance at 325 and 589 nm, respectively, as shown in Figure 2a [31].

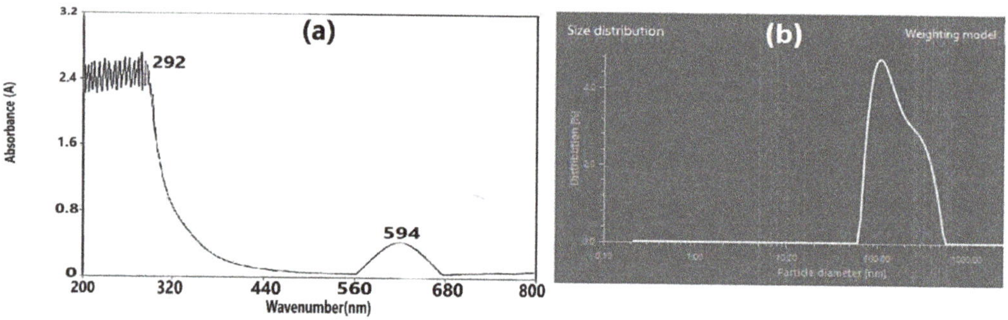

Figure 2. Absorbance peak of Fe-Cu BNPs (**a**) and particle size distribution of sample (**b**).

3.2. Particle Size Analysis

This technique is employed to find out about the size and distribution range of nanoparticles. It also confirms the presence of nanoparticles by describing the range of the particles. The sample used was in the form of solution and Figure 2b shows the particle size distribution that was obtained. It contains two parameters, particle diameter (horizontally) and distribution frequency (vertically). It clearly shows that nanoparticle ranges below and above 100 nm were present in the sample [32,33].

3.3. FTIR Analysis

The data obtained after FTIR analysis, as shown in Figure 3a, were then compared with an IR chart for the identification and confirmation of relevant components in the provided sample. The dip around 3169 cm^{-1} is due to C–H and hydroxyl and carboxylic group stretching in phenolic acid, gallic acid and protocatechuic acid present in the plant extract. The extract could possibly contain some other secondary metabolites or interfering compounds but successful extraction depends upon careful handling and preparation of plant samples. It would also minimize the incorporation of interfering components. FTIR did not indicate the presence of undesired species in the nanoparticles' final form. The dip around 1644 cm^{-1} is due to C=C stretching which is a basic component of most organic biomolecules [23]. All this evidence for the presence of phenolic compound peaks confirms the potential of the extract to reduce Fe/Cu, which is a strong indicator for the synthesis of target BNPs [26]. In addition, carboxylic acid present at the boundaries of Fe-NPs shows peaks at almost 800 to 850 cm^{-1}, whereas the peak at 1050 to 1111 cm^{-1} may represent C–O–C attached to Cu present in BNPs [34].

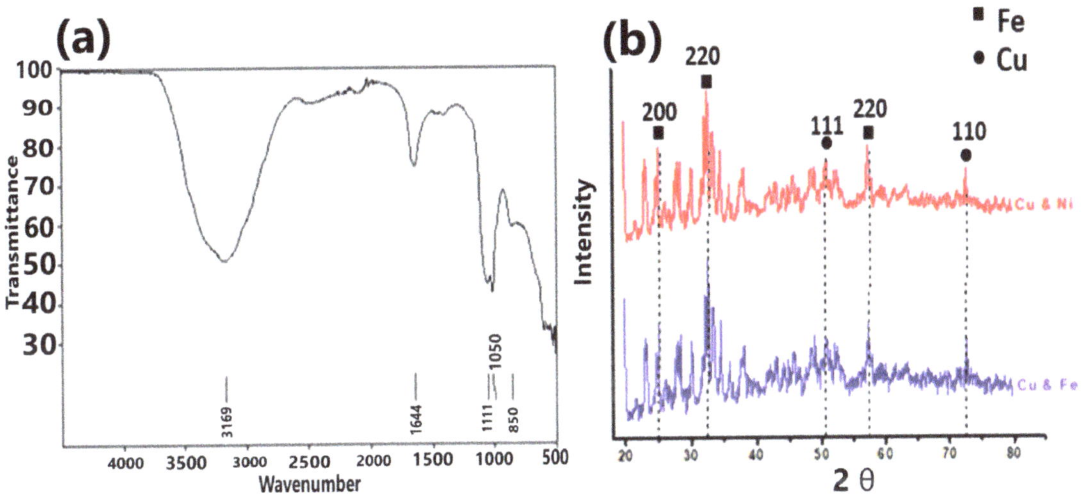

Figure 3. FTIR spectrum (**a**) and XRD patterns (**b**) of Fe-Cu BNPs.

3.4. XRD Analysis

X-ray diffraction is a very effective technique in determining the morphology and structure of nanoparticles. The peak in Figure 3b at 30.0° relates to γ-Fe_2O_3 and 57.39° relates to α-Fe_2O_3 (JCPDS 39-1346) while 50.24° and 74.21° correspond to Cu nanoparticles (JCPDS No: 04.0836) [35,36]. The presence of some minor peaks indicates impurity in the form of biomoieties. The sharp and intense peaks confirmed the presence of Fe/Cu nanoparticles and their crystalline nature [35]. Sharp peaks indicated the crystalline nature of nanoparticles. In conclusion, for as-synthesized Fe-Cu bimetallic nanoparticles, the existence of both the Fe and Cu peaks was recognized by a diffractogram.

3.5. Scanning Electron Microscopy (SEM)

SEM images of Fe-Cu BNPs were obtained through a NOVA Nano SEM 450. The sample in powder form was used for taking SEM images. The micrographs in Figure 4a–c were obtained at different magnifications (30, 50 and 100 nm, respectively). They provide the data regarding morphology of the nanoparticles and they appear to have a heterogeneous surface. The recorded micrographs display a mixed morphology of cubic and rectangular nanoparticles. The particle and sizes varied depending upon the clumping of nanoparticles. The surface of the particles did not remain smooth, which may be due to their interaction with extract biomolecules, however, these nanographs match those of synthesized Cu NPs [37–39]. However, keeping the SEM results in mind, the predicted size of the nanoparticles ranged from 50 to 100 nm.

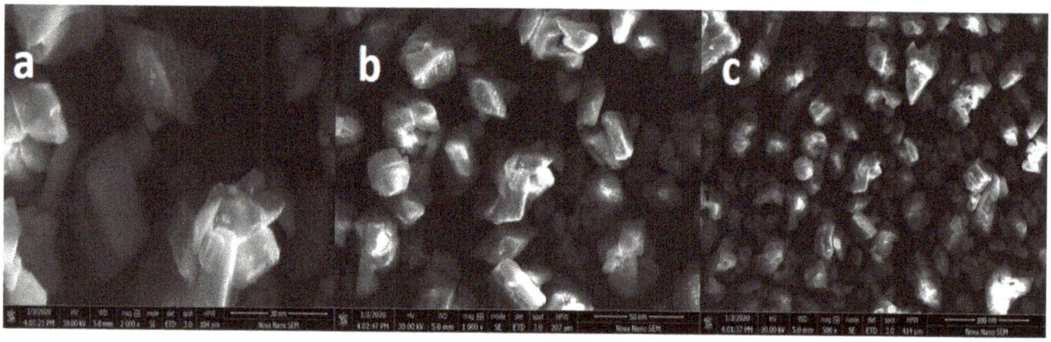

Figure 4. SEM micrographs of Fe-Cu BNPs at 30 nm (**a**), 50 nm (**b**) and 100 nm (**c**).

3.6. DPPH Radical-Scavenging Activity

UV–visible spectra for various plant extract concentrations with Fe-Cu BNPs taken after adding the BNP solution, at different time intervals, clearly show the potential of Fe-Cu BNPs, as presented in Figure 5. With the passage of time, absorbance of DPPH solution decreases and after 30 min, the DPPH solution becomes colorless [40,41]. By increasing the concentration of BNPs, a sharp decrease in the absorbance of DPPH solution was observed.

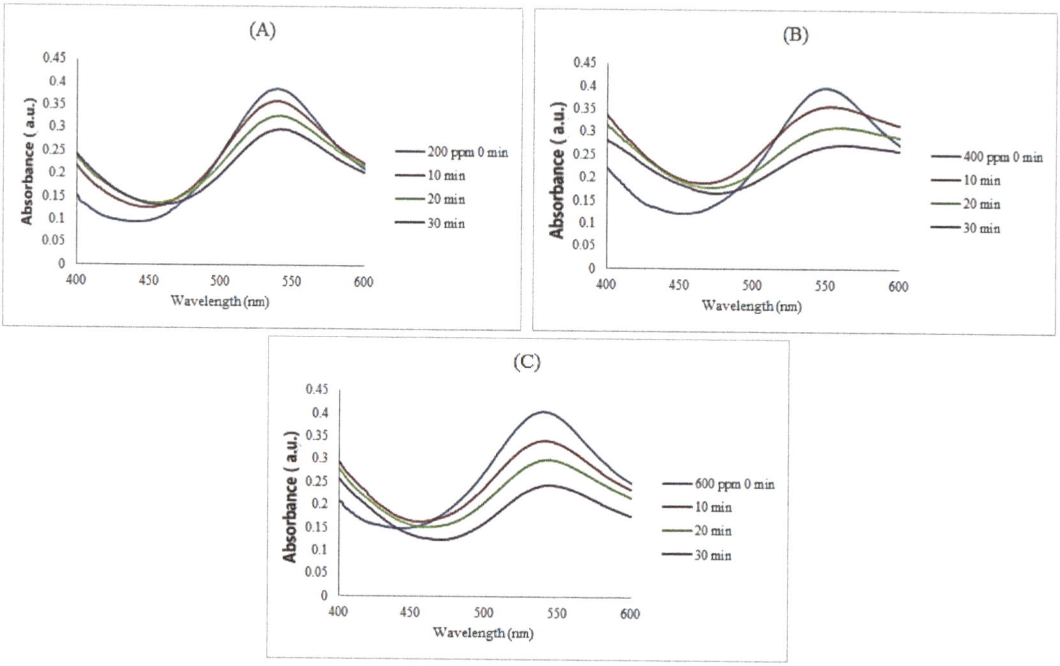

Figure 5. UV–visible spectra at different plant extract (Fe-Cu BNP) concentrations, i.e., 200 ppm (**A**), 400 ppm (**B**) and 600 ppm (**C**).

3.7. Kinetics of DPPH Radical-Scavenging Activity

Kinetics of the radical scavenging activity of BNPs were examined using a UV/Vis spectrophotometer and an increase in the percentage of scavenging was witnessed. An efficient response has been observed in the kinetics of antioxidant potential by using various catalysts [42,43]. Spectra were recorded for reactions between DPPH solution and different concentrations of BNPs (200, 400 and 600 ppm) and then percentage of scavenging was plotted against time, as shown in Figure 6a. The graph in Figure 6b shows the effective antioxidant nature of DPPH where Fe-Cu BNPs behaved as acceptors. It is evident that at each concentration of BNPs, the percentage of scavenging increased with time as DPPH captured the radicals. Additionally, Figure 6b shows that the relative inhibition of DPPH decreases by increasing plant extract concentration because the DPPH acts as a radical scavenger, i.e., antioxidant. Additionally, as the plant extract concentration increases, more DPPH is consumed.

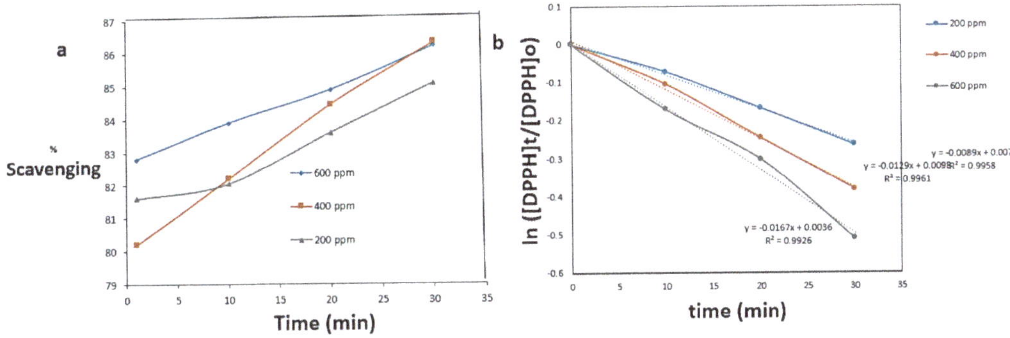

Figure 6. Comparative analysis of % of scavenging by DPPH at different concentrations (Fe-Cu BNPs) of plant extract (**a**) and antioxidant activity of DPPH as a function of time (**b**).

The kinetics of DPPH as an antioxidant were examined following a pseudo-first order reaction using the concentration of [DPPH] using the following equation [44].

$$-d[DPPH]/dt = k_{app} \cdot [DPPH] \quad (2)$$

Kinetic parameters, including apparent rate constant "k_{app} (s^{-1})" and half life "($t_{1/2}$) (s)", were determined, as shown in Table 1. It indicates that by increasing the concentration of BNPs, the rate constant increases, which refers to the increase in antioxidant activity.

Table 1. Different kinetic parameters at various concnetrations of DPPH.

Concentration (ppm)	Slope	Apparent Rate Constant k_{app} (s^{-1})	Half Life ($t_{1/2}$) (s)
200	−0.0089	0.0089	77.86516
400	−0.0129	0.0129	53.72093
600	−0.0167	0.0167	41.49700

3.8. Catalytic Activity/Dye Degradation

3.8.1. In the Absence of Fe-Cu BNP Catalysts

Different industries are continuously releasing organic dyes into water streams. Being toxic, carcinogenic and hazardous in nature, these dyes are the biggest threat to living things. The major source of methylene blue in water is the paper and cloth industry, which releases approximately 60 g/L annually. The reduction of such pollutants from waste water is a current topic of research [45,46]. Researchers have reported catalytic degradation of dyes as one of the best tools for the removal of dyes from a water medium. In the current study, an attempt was made to remove dye from a water medium using a catalytic degradation process. In an initial experimental setup, degradation of MB dye was observed in the presence of $NaBH_4$, as seen in Figure 7a. It is clear that there is no obvious decrease in the absorbance curves at 665 nm λ_{max}, showing that $NaBH_4$ were unable to reduce the cationic dye, although the reaction took place at the surface of the catalyst, which carried out electron transfer from BH_4^- to MB.

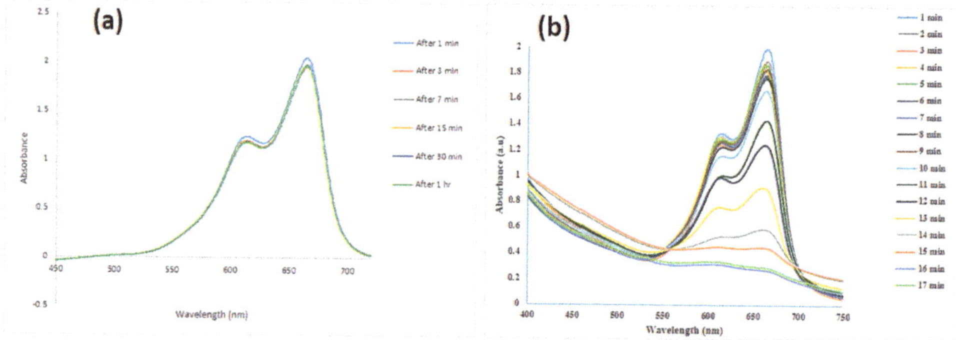

Figure 7. UV-visible spectra of reaction between MB and $NaBH_4$ in the absence of Fe-Cu BNPs (**a**) and in the presence of 200 ppm of Fe-Cu BNPs (**b**).

3.8.2. In the Presence of Fe-Cu BNP Catalysts

Degradation of MB dye in the presence of Fe-Cu BNPs was also carried out to evaluate the catalytic potential of synthesized bimetallic nanoparticles. For this purpose, optimized conditions were recorded for this reaction, i.e., 26 mM $NaBH_4$, 0.5 mL of 100 ppm Fe-Cu BNPs and 6.8 pH at 22 °C. The results in Figure 7b reveal that the reduction was effective and occurred within just 17 min, and reduction of MB was monitored every min. The reduction catalysis of MB occurred by the Langmuir–Hinshelwood (LH) mechanism that is generally followed by a reduction of organic pollutants [47,48]. According to the LH mechanism, the role of Fe-Cu BNPs in catalysis could be elucidated by electron transfer from a donor ($NaBH_4$) to an acceptor (dyes). Additionally, more active sites were available on the surface of the catalyst for the incoming substrate to adsorb, leading to the reduction reaction. The heterogeneous Fenton type reaction catalyzed by Fe-Cu BNPs occurs by surface reactions incorporating both Fe and Cu active sites [47].

The mechanism shows the –N=N– (present in MB) breakage due to an electron pair from BH_4^- on the surface of the Fe-Cu BNP, which acts as an e^- transferee. Previously, Yan Hu and coworkers reported research on the reduction of MB dye by using Fe-Cu BNPs [48], but they did not focus on the catalytic reduction of MB.

3.8.3. Kinetics of Catalytic Activity

By using the degradation spectra monitored with a UV/Vis spectrophotometer, kinetic studies were also performed. The reaction between MB and $NaBH_4$ on the surface of Fe-Cu BNPs was pseudo-first order, in which $NaBH_4$ was taken in large excess as compared to MB ($[NaBH_4] >>> [MB]$). It was found that the catalysis followed the Langmuir–Hinshelwood mechanism in which $NaBH_4$ and MB react in succession on the surface of the catalyst. The slope of the graph ($\ln A_t/A_o$ vs. time t) in Figure 8 shows the apparent rate constant from which we can calculate the rate of reaction. This k_{app} value determined from the graph is 0.2982 min^{-1} and shows the effectiveness of Fe-Cu BNPs. It is clear from the results that there is no decrease in the $\ln A_t/A_o$ value from 1 to 10 min, which is due to the fact that at the start of the reaction, the substrate and reducing agent diffuse towards the surface of the Fe-Cu BNP catalyst and this time is known as the induction time. After 10 min, a dip in the straight line gives the slope value, presented in the inset of Figure 8. This was the reaction time in which $NaBH_4$ and MB interacted with each other on the Fe-Cu BNPs and the reaction started. This reaction finally ended at 15 and 16 min when a colorless solution was obtained, which meant no reactive species were left in the reaction container.

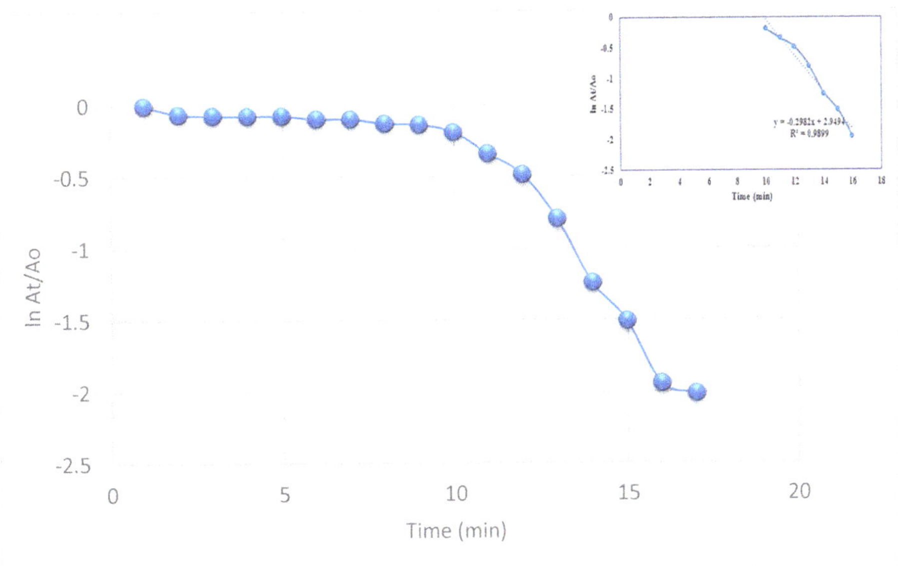

Figure 8. Dependence of absorbance (In A_t/A_o) on change in time (t) in optimized conditions.

Ixora finlaysoniana extract, when isolated from the n-hexane mixture, displayed a chemical composition comprising a hydrocarbon alcohol, nonacosanol, amyrin, 3-hydroxyhexan-5-olide, gallic acid, protocatechuic acid, β-sitostreol glucoside and sitosterol [48]. The most likely mechanism is based upon both hydrogen bond interactions and electrostatic interactions between the plant metabolites and bi-metals.

4. Conclusions

The purpose of this research work was to synthesize Fe-Cu bimetallic nanoparticles (BNPs) by a green method. *Ixora finlaysoniana*, an East Asian medicinal plant, was used as a source of biomolecules to synthesize BNPs. The confirmation for the synthesis of nanoparticles was carried out with a UV spectrophotometer. Their size range was checked by a size distribution analyzer which confirmed the presence of metal nanoparticles in the nanosized range. In addition, FTIR analysis, XRD and SEM studies confirmed the formation and structure of BNPs. The antioxidant potential in terms of radical-scavenging potential was determined by employing a DPPH assay and synthesized metal nanoparticles exhibited good antioxidant properties. The catalytic activity was determined using methylene blue dye as a substrate and sodium borohydride as a reducing agent. Results revealed that BNPs can effectively degrade the dye present in a water medium. Kinetic studies confirmed pseudo-first order reactions for both the radical-scavenging and catalytic activity. All the results of characterization and different studies proved that Fe-Cu BNPs can be successfully fabricated using *Ixora finlaysoniana* extract. Furthermore, these BNPs can be employed for radical scavenging and catalytic activities. Therefore, these Fe-Cu BNPs could attain much importance because they can meet the increasing demand for efficient and active nanoparticles in a sustainable, economical and ecofriendly way.

Author Contributions: U.Y. and S.I. designed the research scheme and performed the experimental part. S.T.H., F.A., F.H. and Z.S. analyzed the data and measurements. M.P., S.K. and S.B. wrote the manuscript. Validation, F.T.J., A.S., Z.A. and A.A.G. did the proofreading. M.O. and M.M.A.-A. performed the characterization analysis. Data Curation, S.A. All authors have read and agreed to the published version of the manuscript.

Funding: The authors are grateful to the Deanship of Scientific Research, King Saud University for funding through Vice Deanship of Scientific Research Chairs and thankful to the financial support funded by the Deanship of Scientific Research at Princess Nourah bint Abdulrahman University through the Fast-track Research Funding Program.

Institutional Review Board Statement: Not applicable.

Informed Consent Statement: Not applicable.

Data Availability Statement: Data are contained within article.

Conflicts of Interest: The authors declare no conflict of interest.

Abbreviations

Iron–copper bimetallic nanoparticles (Fe-Cu BNPs), 1,1-diphenyl-2-picrylhydrazyl (DPPH), scanning electron microscopy (SEM), X-ray diffraction (XRD), Fourier transform infrared spectroscopy (FTIR) and methylene blue (MB).

References

1. Cheng, D.; Ngo, H.H.; Guo, W.; Chang, S.W.; Nguyen, D.D.; Liu, Y.; Wei, Q.; Wei, D. A critical review on antibiotics and hormones in swine wastewater: Water pollution problems and control approaches. *J. Hazard. Mater.* **2020**, *387*, 121682. [CrossRef]
2. Bibi, S.; Ahmad, A.; Anjum, M.A.R.; Haleem, A.; Siddiq, M.; Shah, S.S.; Al Kahtani, A. Photocatalytic degradation of malachite green and methylene blue over reduced graphene oxide (rGO) based metal oxides (rGO-Fe_3O_4/TiO_2) nanocomposite under UV-visible light irradiation. *J. Environ. Chem. Eng.* **2021**, *9*, 105580. [CrossRef]
3. Mehdizadeh, P.; Orooji, Y.; Amiri, O.; Salavati-Niasari, M.; Moayedi, H. Green synthesis using cherry and orange juice and characterization of $TbFeO_3$ ceramic nanostructures and their application as photocatalysts under UV light for removal of organic dyes in water. *J. Clean. Prod.* **2020**, *252*, 119765. [CrossRef]
4. Ahmad, I.; Jamal, M.A.; Iftikhar, M.; Ahmad, A.; Hussain, S.; Asghar, H.; Saeed, M.; Yousaf, A.B.; Karri, R.R.; Al-Kadhi, N.S.; et al. Lanthanum-zinc binary oxide nanocomposite with promising heterogeneous catalysis performance for the active conversion of 4-nitrophenol into 4-aminophenol. *Coatings* **2021**, *11*, 537. [CrossRef]
5. Khatri, J.; Nidheesh, P.V.; Singh, T.A.; Kumar, M.S. Advanced oxidation processes based on zero-valent aluminium for treating textile wastewater. *Chem. Eng. J.* **2018**, *348*, 67–73. [CrossRef]
6. Robinson, T.; McMullan, G.; Marchant, R.; Nigam, P. Remediation of dyes in textile effluent: A critical review on current treatment technologies with a proposed alternative. *Bioresour. Technol.* **2001**, *77*, 247–255. [CrossRef]
7. Padhi, B. Pollution due to synthetic dyes toxicity & carcinogenicity studies and remediation. *Int. J. Environ. Sci.* **2012**, *3*, 940–955.
8. Khan, S.; Shah, S.S.; Anjum, M.A.R.; Khan, M.R.; Janjua, N.K. Electro-oxidation of ammonia over copper oxide impregnated γ-Al_2O_3 nanocatalysts. *Coatings* **2021**, *11*, 313. [CrossRef]
9. Nagajyothi, P.C.; Prabhakar Vattikuti, S.V.; Devarayapalli, K.C.; Yoo, K.; Shim, J.; Sreekanth, T.V.M. Green synthesis: Photocatalytic degradation of textile dyes using metal and metal oxide nanoparticles-latest trends and advancements. *Crit. Rev. Environ. Sci. Technol.* **2020**, *50*, 2617–2723. [CrossRef]
10. Ahmad, A.; Jini, D.; Aravind, M.; Parvathiraja, C.; Ali, R.; Kiyani, M.Z.; Alothman, A. A novel study on synthesis of egg shell based activated carbon for degradation of methylene blue via photocatalysis. *Arab. J. Chem.* **2020**, *13*, 8717–8722. [CrossRef]
11. Bassano, C.; Deiana, P.; Vilardi, G.; Verdone, N. Modeling and economic evaluation of carbon capture and storage technologies integrated into synthetic natural gas and power-to-gas plants. *Appl. Energy* **2020**, *263*, 114590. [CrossRef]
12. Vilardi, G.; Rodríguez-Rodríguez, J.; Ochando-Pulido, J.M.; Verdone, N.; Martinez-Ferez, A.; Di Palma, L. Large laboratory-plant application for the treatment of a Tannery wastewater by Fenton oxidation: Fe (II) and nZVI catalysts comparison and kinetic modelling. *Process Saf. Environ. Prot.* **2018**, *117*, 629–638. [CrossRef]
13. Astruc, D. Introduction: Nanoparticles in catalysis. *Chem. Rev.* **2020**, *120*, 461–463. [CrossRef] [PubMed]
14. Aravind, M.; Ahmad, A.; Ahmad, I.; Amalanathan, M.; Naseem, K.; Mary, S.M.M.; Parvathiraja, C.; Hussain, S.; Algarni, T.S.; Pervaiz, M.; et al. Critical green routing synthesis of silver NPs using jasmine flower extract for biological activities and photocatalytical degradation of methylene blue. *J. Environ. Chem. Eng.* **2020**, *9*, 104877. [CrossRef]
15. Sharma, G.; Kumar, A.; Sharma, S.; Naushad, M.; Dwivedi, R.P.; ALOthman, Z.A.; Mola, G.T. Novel development of nanoparticles to bimetallic nanoparticles and their composites: A review. *J. King Saud Univ. Sci.* **2019**, *31*, 257–269. [CrossRef]
16. Amen, T.W.; Eljamal, O.; Khalil, A.M.; Sugihara, Y.; Matsunaga, N. Methane yield enhancement by the addition of new novel of iron and copper-iron bimetallic nanoparticles. *Chem. Eng. Process. Process Intensif.* **2018**, *130*, 253–261. [CrossRef]
17. Viotti, P.; Tatti, F.; Rossi, A.; Luciano, A.; Marzeddu, S.; Mancini, G.; Boni, M.R. An eco-balanced and integrated approach for a more-sustainable msw management. *Waste Biomass Valoriz.* **2020**, *11*, 5139–5150. [CrossRef]
18. Nandhini, N.T.; Rajeshkumar, S.; Mythili, S. The possible mechanism of eco-friendly synthesized nanoparticles on hazardous dyes degradation. *Biocatal. Agric. Biotechnol.* **2019**, *19*, 101138. [CrossRef]

19. Tang, J.; Wang, J. Iron-copper bimetallic metal-organic frameworks for efficient Fenton-like degradation of sulfamethoxazole under mild conditions. *Chemosphere* **2020**, *241*, 125002. [CrossRef]
20. Hu, C.Y.; Lo, S.L.; Liou, Y.H.; Hsu, Y.W.; Shih, K.; Lin, C.J. Hexavalent chromium removal from near natural water by copper–iron bimetallic particles. *Water Res.* **2010**, *44*, 3101–3108. [CrossRef]
21. Yasmeen, F.; Raja, N.I.; Razzaq, A.; Komatsu, S. Proteomic and physiological analyses of wheat seeds exposed to copper and iron nanoparticles. *Biochim. et Biophys. Acta (BBA)-Proteins Proteom.* **2017**, *1865*, 28–42. [CrossRef]
22. Duan, H.; Wang, D.; Li, Y. Green chemistry for nanoparticle synthesis. *Chem. Soc. Rev.* **2015**, *44*, 5778–5792. [CrossRef]
23. Hussain, I.; Singh, N.B.; Singh, A.; Singh, H.; Singh, S.C. Green synthesis of nanoparticles and its potential application. *Biotechnol. Lett.* **2016**, *38*, 545–560. [CrossRef]
24. Devi, T.B.; Ahmaruzzaman, M. Green synthesis of silver, copper and iron nanoparticles: Synthesis, characterization and their applications in wastewater treatment. In *Green Metal Nanoparticles: Synthesis, Characterization and Their Applications*; Kanchi, S., Ahmed, S., Eds.; Wiley-Scrivener: Austin, TX, USA, 2018.
25. Ghotekar, S. A review on plant extract mediated biogenic synthesis of CdO nanoparticles and their recent applications. *Asian J. Green Chem.* **2019**, *3*, 187–200.
26. Roopan, S.M.; Surendra, T.V.; Elango, G.; Kumar, S.H.S. Biosynthetic trends and future aspects of bimetallic nanoparticles and its medicinal applications. *Appl. Microbiol. Biotechnol.* **2014**, *98*, 5289–5300. [CrossRef] [PubMed]
27. Ali, A.I.; Tohamy Ibrahim, M.; Meselhy, K.; Temraz, A.; Sleem, A. Composition and bioactivities of the essential oil of *Ixora finlaysoniana* wall. ex G. Don.(family rubiaceae). *J. Essent. Oil Bear. Plants* **2015**, *18*, 754–766. [CrossRef]
28. Abdel-Aziz, H.M.; Farag, R.S.; Abdel-Gawad, S.A. Carbamazepine removal from aqueous solution by green synthesis zero-valent iron/Cu nanoparticles with Ficus Benjamina leaves' extract. *Int. J. Environ. Res.* **2019**, *13*, 843–852. [CrossRef]
29. Shawky, M. Phytochemical and biological investigation of *Ixora finlaysoniana* wall. Ex. G. Don. growing in egypt. *Al-Azhar J. Pharm. Sci.* **2016**, *54*, 86–102.
30. Iqbal, S.; Younas, U.; Chan, K.W.; Zia-Ul-Haq, M.; Ismail, M. Chemical composition of artemisia annua L. leaves and antioxidant potential of extracts as a function of extraction solvents. *Molecules* **2012**, *17*, 6020–6032. [CrossRef]
31. Nguyen, T.B.; Dong, C.D.; Huang, C.P.; Chen, C.W.; Hsieh, S.L.; Hsieh, S. Fe-Cu bimetallic catalyst for the degradation of hazardous organic chemicals exemplified by methylene blue in Fenton-like reaction. *J. Environ. Chem. Eng.* **2020**, *8*, 104139. [CrossRef]
32. Wang, Y.; Zhao, H.; Zhao, G. Iron-copper bimetallic nanoparticles embedded within ordered mesoporous carbon as effective and stable heterogeneous Fenton catalyst for the degradation of organic contaminants. *Appl. Catal. B Environ.* **2015**, *164*, 396–406. [CrossRef]
33. Devi, H.S.; Boda, M.A.; Shah, M.A.; Parveen, S.; Wani, A.H. Green synthesis of iron oxide nanoparticles using Platanus orientalis leaf extract for antifungal activity. *Green Process. Synth.* **2019**, *8*, 38–45. [CrossRef]
34. Ramyadevi, J.; Jeyasubramanian, K.; Marikani, A.; Rajakumar, G.; Rahuman, A.A.; Santhoshkumar, T.; Kirthi, A.V.; Jayaseelan, C.; Marimuthu, S. Copper nanoparticles synthesized by polyol process used to control hematophagous parasites. *Parasitol. Res.* **2011**, *109*, 1403–1415. [CrossRef]
35. Chandra, S.; Kumar, A.; Tomar, P.K. Synthesis and characterization of copper nanoparticles by reducing agent. *J. Saudi Chem. Soc.* **2014**, *18*, 149–153. [CrossRef]
36. Zeleňáková, A.; Hrubovčák, P.; Kapusta, O.; Kučerka, N.; Kuklin, A.; Ivankov, O.; Zeleňák, V. Size and distribution of the iron oxide nanoparticles in SBA-15 nanoporous silica via SANS study. *Sci. Rep.* **2019**, *9*, 1–9. [CrossRef] [PubMed]
37. Asghar, M.A.; Zahir, E.; Asghar, M.A.; Iqbal, J.; Rehman, A.A. Facile, one-pot biosynthesis and characterization of iron, copper and silver nanoparticles using Syzygium cumini leaf extract: As an effective antimicrobial and aflatoxin B1 adsorption agents. *PLoS ONE* **2020**, *15*, e0234964. [CrossRef] [PubMed]
38. Hassanien, R.; Husein, D.Z.; Al-Hakkani, M.F. Biosynthesis of copper nanoparticles using aqueous Tilia extract: Antimicrobial and anticancer activities. *Heliyon* **2018**, *4*, e01077. [CrossRef] [PubMed]
39. Bhuiyan, M.S.H.; Miah, M.Y.; Paul, S.C.; Aka, T.D.; Saha, O.; Rahaman, M.M.; Sharif, M.J.I.; Habiba, O.; Ashaduzzaman, M. Green synthesis of iron oxide nanoparticle using Carica papaya leaf extract: Application for photocatalytic degradation of remazol yellow RR dye and antibacterial activity. *Heliyon* **2020**, *6*, e04603. [CrossRef]
40. Lien, H.-L.; Jhuo, Y.-S.; Chen, L.-H. Effect of heavy metals on dechlorination of carbon tetrachloride by iron nanoparticles. *Environ. Eng. Sci.* **2007**, *24*, 21–30. [CrossRef]
41. Herlekar, M.; Barve, S.; Kumar, R. Plant-mediated green synthesis of iron nanoparticles. *J. Nanopart.* **2014**, *2014*, 40614. [CrossRef]
42. Prasad, K.S.; Patra, A.; Shruthi, G.; Chandan, S. Aqueous extract of Saraca indica leaves in the synthesis of copper oxide nanoparticles: Finding a way towards going green. *J. Nanotechnol.* **2017**, *2017*, 7502610. [CrossRef]
43. Mou, Y.; Peng, Y.; Zhang, Y.; Cheng, H.; Chen, M. Cu-Cu bonding enhancement at low temperature by using carboxylic acid surface-modified Cu nanoparticles. *Mater. Lett.* **2018**, *227*, 179–183. [CrossRef]
44. Das, D.; Nath, B.C.; Phukon, P.; Dolui, S.K. Synthesis and evaluation of antioxidant and antibacterial behavior of CuO nanoparticles. *Colloids Surf. B Biointerfaces* **2013**, *101*, 430–433. [CrossRef]
45. Harshiny, M.; Iswarya, C.N.; Matheswaran, M. Biogenic synthesis of iron nanoparticles using Amaranthus dubius leaf extract as a reducing agent. *Powder Technol.* **2015**, *286*, 744–749. [CrossRef]

46. Momen, H.M.; Haghi, B.; Morsali, A.; Ardalan, P.; Ardalan, T. Kinetic study of DPPH scavenging in the presence of mixture of Zinc and Vitamin C as an antioxidant. *J. Chem. Health Risks* **2012**, *2*, 2.
47. Arora, B.; Sethi, S.; Joshi, A.; Sagar, V.R.; Sharma, R.R. Antioxidant degradation kinetics in apples. *J. Food Sci. Technol.* **2018**, *55*, 1306–1313. [CrossRef]
48. Hashemi, S.M.B.; Brewer, M.S.; Safari, J.; Nowroozi, M.; Abadi Sherahi, M.H.; Sadeghi, B.; Ghafoori, M. Antioxidant activity, reaction mechanisms, and kinetics of Matricaria recutita extract in commercial blended oil oxidation. *Int. J. Food Prop.* **2016**, *19*, 257–271. [CrossRef]

MDPI
St. Alban-Anlage 66
4052 Basel
Switzerland
Tel. +41 61 683 77 34
Fax +41 61 302 89 18
www.mdpi.com

Coatings Editorial Office
E-mail: coatings@mdpi.com
www.mdpi.com/journal/coatings

www.ingramcontent.com/pod-product-compliance
Lightning Source LLC
Chambersburg PA
CBHW061622210926
51850CB00009B/2401
* 9 7 8 3 0 3 6 5 1 6 6 0 8 *